Applying Qualitative Research Methods to Management Science

Badr Bentalha
National School of Business and Management, Sidi Mohammed Ben Abdellah University, Morocco

Lhoussaine Alla
Sidi Mohamed Ben Abdellah University, Morocco

A volume in the Advances in Library and Information Science (ALIS) Book Series

IGI Global
Publishing Tomorrow's Research Today

Published in the United States of America by
IGI Global
Information Science Reference (an imprint of IGI Global)
701 E. Chocolate Avenue
Hershey PA, USA 17033
Tel: 717-533-8845
Fax: 717-533-8661
E-mail: cust@igi-global.com
Web site: http://www.igi-global.com

Library of Congress Cataloging-in-Publication Data

CIP DATA PENDING

ISBN: 9798369355435
IsbnSoftcover: 9798369355442
EISBN: 9798369355459

British Cataloguing in Publication Data
A Cataloguing in Publication record for this book is available from the British Library.

Table of Contents

Detailed Table of Contents

A persuasive research proposal helps interested lecturers clearly understand the research objectives of novice researchers. This chapter focuses on research proposals in management science, particularly those employing qualitative methodologies to explore new topics. By drawing on practical examples, the authors aim to provide a guide to assist novice researchers in crafting a persuasive research proposal. In this chapter, the authors conducted a literature review from two databases to identify relevant documents on research proposals. A total of 14 documents were selected and analyzed. The findings of this research contain three key elements. The first summarizes the qualitative methods used in the selected literature. The second provides the main structure of a research proposal. The third discusses the essential elements to make the research proposal persuasive. Overall, the authors have sought to contribute to the existing body of knowledge on research proposals that utilize qualitative methods in the field of management.

The case study is one of the most important forms of qualitative research in the field of organizational management and provides a rich analysis of modern phenomena and their context. This method entails the gathering of data from different sources with a view to critically analyzing management situations. It involves skepticism of typical theories and procedures with the objectives of determining critical variables and researching organizational change and development processes. This chapter is concerned with how case studies can be defined and used in relation to research questions. The method used in the research process is systematic with the R software to cover data gathering, analysis. The chapter is supported by qualitative case studies from the published literature to stress the practical applicability and importance of this research method in expanding the body of knowledge in organizational management.

Certainly, selecting the appropriate research methodology remains a primary concern for all social science researchers. Among the less commonly used methods is participant observation (PO), considered a qualitative research approach. By using ethnography as a theoretical framework and adopting a systematic literature review approach, this chapter aims to discuss the significance and use of PO as a research method in management sciences. The main results indicate that the PO method is frequently considered in the literature as a research method to enhance analysis by providing access to the field and valuable insights into the studied objects. However, it also entails certain challenges and limitations that we have highlighted to ensure its utility, credibility, and optimize research results.

Management sciences are a complex discipline that simultaneously raises technical issues, stemming from accounting and financial approaches, and human issues, of psychosocial inspiration. Therefore, they require appropriate methodologies to capture their full complexity. It is in this perspective that we can situate the development of qualitative field research methodologies. In this chapter, the authors propose to discuss, through a case of illustration of a field research conducted within a public non-profit organization, the relevance of the qualimetical intervention research method for researchers in management sciences, as a field research, engaged, collaborative, and transformative.

This chapter aims to cover entrepreneurial motivation as a key component of the emerging field of entrepreneurship, particularly through studies developed using qualitative methods. It indicates how qualitative methods particularly contribute to the understanding of entrepreneurial motivation through examples drawn from previous publications. A synthesis of seven papers on this topic, based on qualitative research methods is presented. The chapter subsequently presents some implications and contributions of qualitative research methods on entrepreneurial motivation study.

This chapter examines the significance of qualitative methodologies in evaluating self-employment policies. The central question is how qualitative approaches enhance policy evaluation in self-employment. The chapter highlights the importance of qualitative studies, such as the biographical approach and life narratives, in offering interpretative perspectives that enrich policy assessments and inform evidence-based decision-making. Qualitative methodologies play a crucial role in providing detailed contextual insights, uncovering complexities, and understanding programmatic effects. By examining contextual influences, implementation strategies, and individual characteristics, qualitative studies help to develop comprehensive theories of change, leading to more responsive and adaptable policies. The chapter emphasizes the need to integrate qualitative and quantitative approaches for a mixed methods policy evaluation. This approach enables policymakers and researchers to develop evidence-based policies that address the diverse needs and challenges of self-employed individuals.

The chapter proposes a method for regional-scale qualitative risk assessment. To estimate critical snow heights, information was gathered from residents, documents, and back calculations of avalanche simulations based on well-known avalanche events. The elements at risk were extracted from existing digital cadastral databases, and vulnerabilities were identified. The study found that the provincial forest area had the highest annual probability of property loss. The chapter presents a model that offers an effective means of determining standardized responses and controls for common environmental issues based on the level of risk. Additionally, various potential avalanche scenarios are presented to provide an understanding of the potential threats specific to this avalanche-prone area. The resulting map provides reliable and easily understandable information, highlighting areas where avalanches pose a significant risk at the regional level, as well as areas where new avalanche paths may develop under favorable conditions within the defined region.

This chapter aims to analyze the qualitative method in management science. It examines the theoretical foundations of this approach to achieve its epistemological bases. In this perspective, the authors present an attempt to elucidate the underpinnings on which the assumptions of a knowledge production approach around a defined object in the social sciences are built. They also question why management science, which is traditionally focused on quantitative methods, adopts qualitative methods. Based on a literature review, it can be affirmed that qualitative research aims to deeply understand the context of phenomena and the mechanisms underlying them. Thus, proponents of these methodologies are interested in realities embedded in specific contexts to better grasp the complexity of human interactions within the organization and to address certain challenges, notably the management of cultural diversity and the dynamics of virtual teams. The authors illustrate this research with an exploratory study examining the effects of Big Data on the performance of companies in Morocco.

Preface

This book is a collection of qualitative research themes and methods used by researchers and practitioners in science and management. The concepts include methods and methodologies applied to qualitative research in a variety of contexts. Each concept is the subject of a separate chapter written by a researcher or practitioner with extensive experience in the field. The book aims to provide students and practitioners with an overview of the issues involved in qualitative research, based on the most recent academic research. It presents groundbreaking research by leading academics and field experts from around the world. It provides a theoretical and conceptual foundation for researchers interested in qualitative issues and for scientists wishing to gain a business, policy or industry perspective on the various themes of qualitative enquiry and analysis. Each part presents and analyses a specific topical theme, enabling readers to identify the chapters that correspond to their interests. It is an ideal reference guide for practitioners interested in the subject. The editors are specialists in qualitative research, and the contributors are also key players in the field.

The book deals with the integration of qualitative analysis into academic research and empirical projects. Balanced between application and theory, the text covers the latest technological, organizational and methodological developments in qualitative research. This book meets the need for an up-to-date reference on qualitative research, with a pedagogical approach and practical contributions directly applicable in the field.

It provides an overview of the different qualitative approaches and methods, a practical and concrete guide to implementing qualitative studies, details of the key stages of a qualitative approach (sampling, data collection and analysis, reporting) and a discussion of the advantages and limitations of qualitative studies. Its aim is to enable social science researchers to perfect their knowledge of qualitative methods, to provide students with a reference manual for their qualitative work and dissertations, to enlighten public and private decision-makers on the value of qualitative studies for understanding certain social phenomena, and to contribute to the development and enhancement of qualitative approaches in academic research. The book offers a unique synthesis of all qualitative approaches, a theoretical and practical perspective, with many concrete examples, simplified access to sometimes complex concepts and methods, an in-depth discussion of the epistemological issues of qualitative studies, and possible use in qualitative methodology training courses.

A DESCRIPTION OF THE TARGET AUDIENCE

This text is aimed at people who use qualitative studies and qualitative research. The target audience is mainly researchers and practitioners in this field, but the book may also be useful to students. These include students of the social sciences (sociology, anthropology, education sciences, social psychology, etc.), teachers and trainers in research methodology, current researchers wishing to deepen their knowledge of qualitative methods, research professionals in consultancies, polling institutes, marketing, human resources and public or private decision-makers commissioning qualitative studies.

A DESCRIPTION OF EACH CHAPTER

Chapter 1: Revealing the Subtleties the Art of Qualitative Studies in Science and Management

Qualitative research methodologies have long been undervalued and overshadowed by quantitative approaches in science and management. However, there is an inherent attractiveness and power in qualitative studies that lies in their ability to unveil the nuanced complexities and rich contextual elements often overlooked by purely numerical analyses. This paper aims to illuminate the intricate artistry of conducting impactful qualitative research within these domains. The paper also navigates the intricate process of data analysis, interpretation, and theory-building, underscoring the importance of reflexivity, triangulation, and the iterative nature of qualitative inquiry. Ultimately, this paper serves as a compelling manifesto for embracing the subtleties of qualitative research, recognizing its vital role in uncovering the rich tapestry of human existence and organizational realities within the scientific and managerial realms.

Chapter 2: Crafting a Persuasive Qualitative Research Proposal in Management Science: Practical Insights From Literature

A persuasive research proposal helps interested lecturers clearly understand the research objectives of novice researchers. This chapter focuses on research proposals in management science, particularly those employing qualitative methodologies to explore new topics. By drawing on practical examples, the authors aim to provide a guide to assist novice researchers in crafting a persuasive research proposal. In this chapter, the authors conducted a literature review from two databases to identify relevant documents on research proposals. A total of 14 documents were selected and analyzed. The findings of this research contain three key elements. The first summarizes the qualitative methods used in the selected literature. The second provides the main structure of a research proposal. The third discusses the essential elements to make the research proposal persuasive. Overall, the authors have sought to contribute to the existing body of knowledge on research proposals that utilize qualitative methods in the field of management.

Chapter 3: Case Study Methodology In Business Management: A Systematic Review

The case study is one of the most important forms of qualitative research in the field of organizational management and provides a rich analysis of modern phenomena and their context. This method entails the gathering of data from other sources with a view of critically analyzing complex management situations. It involves skepticism of typical theories and procedures with the objectives of determining critical variables and researching organizational change and development processes. This chapter is concerned with how case studies can be defined and used in relation to research questions. The method used in the research process is systematic with the R software to cover data gathering, analysis, and integration. The chapter is supported by qualitative case studies from the published literature to stress the practical applicability and importance of this research method in expanding the body of knowledge in organizational management.

Chapter 4: Participant Observation as a Qualitative Research Method in Management Sciences: Interests, Conditions, and Limitations

Certainly, selecting the appropriate research methodology remains a primary concern for all social science researchers. Among the less commonly used methods is Participant Observation (PO), considered a qualitative research approach. By using ethnography as a theoretical framework and adopting a systematic literature review approach, this article aims to discuss the significance and use of PO as a research method in management sciences. The main results indicate that the PO method is frequently considered in the literature as a research method to enhance analysis by providing access to the field and valuable insights into the studied objects. However, it also entails certain challenges and limitations that we have highlighted to ensure its utility, credibility, and optimize research results.

Chapter 5: Qualimatric Intervention Research: Interest for Management Sciences Researchers

Management sciences are a complex discipline that simultaneously raises technical issues, stemming from accounting and financial approaches, and human issues, of psychosocial inspiration. Therefore, they require appropriate methodologies to capture their full complexity. It is in this perspective that we can situate the development of qualitative field research methodologies. In this research paper, we propose to discuss, through a case of illustration of a field research conducted within a public non-profit organization, the relevance of the qualimetical intervention research method for researchers in management sciences, as a field research, engaged, collaborative and transformative.

Chapter 6: Understanding Entrepreneurial Motivation Through Qualitative Study

This chapter aims to cover entrepreneurial motivation as a key component of the emerging field of entrepreneurship, particularly through studies developed using qualitative methods. It indicates how qualitative methods particularly contribute to the understanding of entrepreneurial motivation through examples drawn from previous publications. A synthesis of seven papers on this topic, based on qualitative research methods is presented. The chapter subsequently presents some implications and contributions of qualitative research methods on entrepreneurial motivation study.

Chapter 7: Enhancing Policy Evaluation: The Role of Qualitative Method in Understanding Self-Employment Dynamics

This article examines the significance of qualitative methodologies in evaluating self-employment policies. The central question is how qualitative approaches enhance policy evaluation in self-employment. The article highlights the importance of qualitative studies, such as the biographical approach and life narratives, in offering interpretative perspectives that enrich policy assessments and inform evidence-based decision-making. Qualitative methodologies play a crucial role in providing detailed contextual insights, uncovering complexities, and understanding programmatic effects. By examining contextual influences, implementation strategies, and individual characteristics, qualitative studies help to develop comprehensive theories of change, leading to more responsive and adaptable policies. The article emphasizes the need to integrate qualitative and quantitative approaches for a mixed methods policy evaluation. This approach enables policymakers and researchers to develop evidence-based policies that address the diverse needs and challenges of self-employed individuals.

Chapter 8: Evolving Experiences in Management Education Through a Longitudinal Qualitative Study

This study employs longitudinal qualitative research methods to explore the dynamic learning experiences of students in a newly introduced management education program. Grounded in theoretical perspectives on time, lived experiences, and developmental trajectories, the paper underscores the importance of capturing the fluidity of educational journeys over an extended period. An in-depth case study was conducted at USMBA university, following a cohort of five students through their first year in the Excellence Program in Management Sciences. Data were collected through semi-structured interviews at beginning and end of the academic year. Iterative qualitative coding and analysis techniques were applied to identify evolving patterns, critical incidents, and transformations in students perspectives, skill development, and experiences. The findings reveal the complex interplay between institutional factors and personal factors shaping individual learning trajectories.

Chapter 9: Investigating the Determinants of ESG Disclosure Practices in Morocco: A Qualitative Analysis

Non-financial disclosure is publication of indicators that do not result from the financial statements. In Morocco, several companies publish their ESG reports annually, but the quality and quantity of information presented varies from one company to another. This chapter per the authors seek to understand the factors likely to impact the scale and scope of ESG reporting done by Moroccan companies. This work began with a theoretical exploration of the determinants of non-financial disclosure, which enabled the authors to draw up a theoretical model incorporating the main determinants proposed in the literature. Subsequently, an exploratory qualitative approach will be implemented in order to contextualise the theoretical model identified. In terms of results, the contextualised model contains eight explanatory variables, the size of the company, the industry sector, the governance structure, profitability, CSR performance, stakeholders pressure, the cost of producing reports and the desire to acquire a competitive advantage.

Chapter 10: Qualitative Risk Analysis Based on Avalanche Scenarios in Palandöken Ski Center, Erzurum (NE Turkey)

The paper proposes a method for regional-scale qualitative risk assessment. To estimate critical snow heights, information was gathered from residents, documents, and back calculations of avalanche simulations based on well-known avalanche events. The elements at risk were extracted from existing digital cadastral databases, and vulnerabilities were identified. The study found that the provincial forest area had the highest annual probability of property loss. The paper presents a model that offers an effective means of determining standardized responses and controls for common environmental issues based on the level of risk. Additionally, various potential avalanche scenarios are presented to provide an understanding of the potential threats specific to this avalanche-prone area. The resulting map provides reliable and easily understandable information, highlighting areas where avalanches pose a significant risk at the regional level, as well as areas where new avalanche paths may develop under favorable conditions within the defined region.

Chapter 11: The Adoption of a Qualitative Approach in Management Science: An Exploratory Survey on the Effects of Big Data on the Performance of Companies

This chapter aims to analyze the qualitative method in management science. It examines the theoretical foundations of this approach to achieve its epistemological bases. In this perspective, we present an attempt to elucidate the underpinnings on which the assumptions of a knowledge production approach around a defined object in the social sciences are built. We also question why management science, which is traditionally focused on quantitative methods, adopts qualitative methods. Based on a literature review, it can be affirmed that qualitative research aims to deeply understand the context of phenomena and the mechanisms underlying them. Thus, proponents of these methodologies are interested in realities embedded in specific contexts to better grasp the complexity of human interactions within the organization and to address certain challenges, notably the management of cultural diversity and the dynamics of virtual teams. We illustrate this research with an exploratory study examining the effects of Big Data on the performance of companies in Morocco.

A CONCLUSION OF HOW THE BOOK IMPACTS

With its broad coverage of the different qualitative approaches and its practical dimension, this book is aimed at a diverse readership, from students to professionals, who can use it both for training and as a regular reference in their research practice. It can be used to support qualitative dissertations and theses, as a methodological support for researchers designing qualitative studies, as a monitoring and self-training tool for research professionals, as a guide for decision-makers in understanding and commissioning qualitative studies, and as a source of inspiration for methodological innovation in qualitative research.

Badr Bentalha
National School of Business and Management, Sidi Mohammed Ben Abdellah University, Morocco

Lhoussaine Alla
Sidi Mohamed Ben Abdellah University, Morocco

Acknowledgment

On behalf of the editorial team, we extend our sincere appreciation to all contributors involved in the development of *Applying Qualitative Research Methods to Science and Management.*

We acknowledge the authors for their significant contributions and expertise, which form the foundation of this publication. Their dedication has been instrumental in bringing this project to fruition.

Our gratitude extends to the reviewers, whose insightful feedback has enhanced the quality, coherence, and presentation of the content. Their thorough evaluations have been crucial in refining the manuscript. We also recognize the publishing team for their professionalism and commitment throughout the production process. Their expertise has been vital in ensuring the successful completion of this endeavor.

In conclusion, we express our profound gratitude to all individuals who have played a role in the creation and publication of this book. Your collective efforts have been invaluable, and we are privileged to have collaborated with such a distinguished team of professionals. We extend our sincere thanks for your diligence and commitment to this project.

Chapter 1

Revealing the Subtleties: The Art of Qualitative Studies in Science and Management

Badr Bentalha
https://orcid.org/0000-0003-1339-542X
National School of Business and Management, Sidi Mohammed Ben Abdellah University, Morocco

Lhoussaine Alla
https://orcid.org/0000-0002-7238-1792
National School of Applied Sciences, Sidi Mohammed Ben Abdellah University, Morocco

ABSTRACT

Qualitative research methodologies have long been undervalued and overshadowed by quantitative approaches in science and management. However, there is an inherent attractiveness and power in qualitative studies that lies in their ability to unveil the nuanced complexities and rich contextual elements often overlooked by purely numerical analyses. This chapter aims to illuminate the intricate artistry of conducting impactful qualitative research within these domains. The chapter also navigates the intricate process of data analysis, interpretation, and theory-building, underscoring the importance of reflexivity, triangulation, and the iterative nature of qualitative inquiry. Ultimately, this chapter serves as a compelling manifesto for embracing the subtleties of qualitative research, recognizing its vital role in uncovering the rich tapestry of human existence and organizational realities within the scientific and managerial realms.

DOI: 10.4018/979-8-3693-5543-5.ch001

INTRODUCTION

In the vast landscape of scientific research, qualitative methods have long been considered a poor relation, overshadowed by the apparent rigor of quantitative approaches. However, as our understanding of the complexity of human and natural systems deepens, it is becoming clear that qualitative studies play a crucial role in revealing subtleties that often escape purely numerical analysis. This article aims to explore the delicate art of qualitative studies, highlighting their inestimable value in the fields of science and management.

In science, qualitative studies help to deepen our understanding of complex natural phenomena, from ecological interactions to molecular processes. Whether it's the meticulous observation of animal behavior in their natural habitat or the exploration of complex interactions within an ecosystem, qualitative methods provide a contextual richness that is essential to the interpretation of scientific results (Hmioui & Bentalha, 2021). They capture the subtle nuances that quantitative data alone cannot grasp, offering a holistic perspective that is essential for understanding natural phenomena. They offer a nuanced look at the subtle dynamics underlying these systems, catching nuances that quantitative data alone cannot capture.

In the medical sciences, the use of qualitative studies, using methods such as observation, interviews, participatory observation, and focus groups, has increased considerably in recent years. These approaches seek to gather richer, more detailed data to better understand the reasons behind medical decisions, explore patients' experiences further, and capture the complex nuances of interactions between healthcare professionals and their patients. Unlike quantitative studies, which focus on numerical data and statistical analysis, qualitative studies delve into the subjective world of human perceptions, beliefs, and behaviors. In the medical field, this translates into an in-depth exploration of various aspects. Firstly, the understanding of clinical decisions is enriched. Interviews with doctors and nurses can reveal the subtle factors that influence their treatment decisions, beyond simple protocols. These may include personal experience, clinical intuition, or even institutional constraints. Secondly, patients' experiences are brought to light. Patients' stories, gathered through interviews or focus groups, offer a unique insight into their care journey. These stories may highlight barriers to adherence to treatment, unreported side effects, or the psychosocial impact of a diagnosis. Understanding these experiences can lead to more empathetic interventions and better management of patient expectations. Thirdly, cultural and social contexts are taken into account. Qualitative studies are particularly useful for understanding how cultural beliefs, socioeconomic status, and community networks influence health behaviors (Alla et al., 2020). Fourthly, the development of new theories is stimulated. By analyzing this rich data, researchers can identify recurring themes and develop new theories about health behavior, medical decision-making, or the effectiveness of interventions. These theories, rooted in real experiences, can guide future research and the design of health policies. Participatory observation can highlight inefficiencies in workflows or communication problems between departments. These observations can lead to organizational improvements, reducing staff stress and improving the quality of care. Ethical discussions become more nuanced. Qualitative studies are essential for exploring complex ethical dilemmas. By incorporating these qualitative methods, medical science is moving away from a purely biomechanical approach towards a more holistic, patient-centered perspective. This development recognizes that health is not just a question of physiology, but also psychology, culture, and human relationships.

In the field of management, qualitative studies are essential for understanding the subtleties of human interactions, organizational dynamics, and complex decision-making processes. They enable in-depth exploration of the motivations, perceptions, experiences, and interactions of the individuals involved, revealing the nuances that shape organizational outcomes. Management research aims to describe, understand, explain, or predict phenomena related to organizations. It should not focus on a single field or discipline, but rather on different ideas and concepts that already exist or are being developed, in interaction with the research problem. Whether studying the challenges of implementing strategic change or examining the factors that influence decision-making within an organization, qualitative approaches offer valuable insights into the underlying human realities. Whether studying the challenges of implementing strategic change, examining the factors that influence decision-making within an organization, or understanding the power dynamics and cultural issues in an organizational context, qualitative approaches offer valuable insight into the underlying human realities.

In the field of engineering science, qualitative research plays a crucial role in understanding complex systems, human-technology interactions, and the societal impacts of engineering solutions. Qualitative studies in engineering science are essential for exploring the contextual factors that influence technology adoption, user experience, and the effectiveness of engineering designs. They enable researchers to delve into the perceptions, needs, and behaviors of stakeholders, revealing insights that quantitative methods alone might miss. Engineering research aims not only to develop new technologies but also to understand how these technologies integrate into and impact human systems and environments. Whether investigating the human factors in safety-critical systems, examining the user acceptance of emerging technologies, or exploring the ethical implications of artificial intelligence, qualitative approaches offer valuable perspectives on the complex interplay between technology and society. These methods allow engineers to capture the nuanced experiences of users, the unintended consequences of technological implementations, and the cultural and organizational factors that shape engineering outcomes. Qualitative research in engineering science can inform more user-centered design processes, enhance the development of sustainable technologies, and contribute to the creation of more inclusive and accessible engineering solutions. By providing rich, contextual data, qualitative studies complement quantitative analyses, leading to a more comprehensive understanding of engineering challenges and opportunities.

However, conducting high-quality qualitative research requires a mastery of rigorous techniques and a particular skill set, over and above purely technical and statistical skills. Designing in-depth interviews, participant observation, and content analysis are all essential skills that researchers need to acquire to accurately capture the nuances of human experience and natural phenomena. In addition, interpreting qualitative data requires in-depth critical reflection, and the ability to question assumptions and recognize potential biases. Beyond their contribution to the production of knowledge, qualitative studies also play an essential role in developing new theories and challenging existing models. By immersing researchers in the complexity of the phenomena studied, these approaches make it possible to identify emerging patterns, challenge established hypotheses, and generate new conceptual perspectives. This ability to challenge dominant paradigms is crucial for stimulating innovation and advancing knowledge.

Ultimately, this article sets out to explore the subtleties and richness of qualitative studies, highlighting their contribution to the advancement of knowledge in the fields of science and management. By embracing this delicate art, we will be able to deepen our understanding of the complex systems that shape our world, revealing new perspectives, opportunities for continuous improvement, and unprecedented avenues of reflection for tackling current and future challenges.

In an attempt to identify the deeper subtleties of the qualitative approach in the research process in the field of management sciences, we will begin by presenting the prerequisites of qualitative study and the profile of qualitative research, before focusing on the qualitative/quantitative choice dilemma and the prerequisites of the qualitative researcher.

1. PRELIMINARY STAGES OF THE QUALITATIVE STUDY: ROOTS AND WINGS

Research is an organized, systematic, and critical process that originates in a specific scientific question (Creswell & Poth, 2016). Its main aim is to find answers through in-depth analysis of a distinctive object of study. Methodology refers to the theoretical framework and paradigms on which the research work is based, while methods are the concrete techniques, tools, and procedures used to conduct the investigation and explain a given phenomenon (Saldaña, 2021).

The term '*qualitative research*' is used to refer both to the techniques for collecting and analyzing qualitative data and also to a broader approach to conducting research, namely an interpretative paradigm (Merriam & Grenier, 2019). In contrast to a positivist view that assumes a single objective reality, qualitative research adopts a constructivist perspective and recognizes that there are multiple socially constructed and subjective realities (Lincoln & Guba, 2013). These realities are inextricably linked to the different contexts in which they are embedded, both the immediate context of data production, such as an observation or interview, but also the wider socio-cultural, historical, political, and situational contexts influencing the phenomenon under study (Stewart, 2022). Thus, qualitative research is not limited to techniques but sees data as deeply embedded in their rich and complex contexts. Inspired by phenomenology, qualitative research considers the world not as a fixed, objective entity, but as a subjective reality constantly constructed and reconstructed by the lived experience of individuals (Vagle, 2018). According to Corbin and Strauss (2015), human behavior can only be understood and explained by the deep meanings that people attribute to things and their actions.

The qualitative approach therefore implies an inductive approach aimed at a rich explanation of meaning, constructed according to a constructivist paradigm. Knowledge thus emerges from the study of complex human processes of construction, reconstruction, and mutation of subjective reality (Charmaz, 2014). Consistent with this epistemology, qualitative research methodology conceives of the social world as being in perpetual motion. From this perspective, knowledge is not simply discovered as a pre-existing fact but emerges from the careful study of the processes by which individuals make sense of their world. The term '*construction*' emphasizes that meanings are not inherent in objects or events, but are attributed by individuals through their experiences, social interactions, and cultural frames of reference. The notion of '*reconstruction*' adds a temporal dimension to this process. It suggests that meanings are not fixed but are constantly revised in the light of new experiences or interactions (Hmioui et al., 2022). For example, the meaning a patient gives to his or her diagnosis may evolve as a result of consultations, personal research, or conversations with other patients. Similarly, the performance achieved by an organization change over time - its contours, limits, and perspectives depend on the context, the players involved, and environmental variables. This reconstruction is an active and reflexive process, in which the individual continually reassesses and adjusts his or her understanding. The term '*mutation*' evokes a more profound change. It implies that frames of reference themselves can be transformed. A public health crisis, for example, can fundamentally alter the way a society perceives vulnerability, community responsibility, or the role of experts (Alla et al., 2022). These mutations are not simple adjustments, but substantial reconfigurations of patterns of interpretation. Subjective reality should not be confused with pure individual subjectivity. It encompasses shared perspectives, social norms, and cultural discourses that influence and are influenced by individual experiences (Lamsiah & Bentalha, 2022). This epistemology has profound implications for qualitative research methodology. By conceiving of the social world as in perpetual motion, researchers recognize that their object of study is never static. This calls for flexible, iterative methods capable of capturing and interpreting these flows of meaning. This approach also implies reflexivity on the part of the researcher. Recognizing that their presence and interpretations are part of the knowledge-building process, qualitative researchers must constantly reflect on their role, their biases, and how they influence and are predisposed by the field. Furthermore, this vision of the social world in constant flux calls into question the notions of generalization or replicability often advocated in quantitative research. Instead of looking for universal laws, qualitative research aims to understand how meanings emerge and are transformed in specific contexts. This does not diminish its rigor or value; on the contrary, it provides a nuanced understanding of social processes that can inform theory and practice (Bentalha, 2024a). The qualitative researcher is then seen as a fully-fledged player in this interpretative process, hence the need to replace quantitative systematic observation with committed participant observation (Tedlock, 2020). The fundamental epistemological dimension of this approach therefore lies in these two key words: constant evolution and mutation of subjective realities. Qualitative research is therefore based on subjectivist epistemology, in which the researcher is an involved actor, and pluralist epistemology, in which the diversity of realities depends on the actors involved.

Within the qualitative perspective, there is an important distinction between, on the one hand, empirical, objective, and material reality, and on the other hand, the knowledge we can have of this reality, i.e. the epistemic reality through which empirical reality becomes accessible (Schwandt, 2000). While empirical reality may have an existence independent of the knowing subject, epistemic reality exists only through the presence of the subject (Lincoln et al., 2018). Thus, the subject's localized and historical context in some way conditions an epistemic reality. Indeed, the subject's definition, understanding, and analysis of reality depend on the situated ways of perceiving, thinking, feeling, and acting that are specific to him or her at a given time and place (Cunliffe, 2003). From this point of view, knowledge is a shared construction based on researcher/participant interaction, an interaction traversed by values that influence the knowledge produced and its production process (Guba & Lincoln, 2005). In management research, this viewpoint aligns with interpretive and constructivist approaches, which recognize that organizational realities are socially constructed and that knowledge about management practices and organizational phenomena emerges through interaction and interpretation. This approach emphasizes that management knowledge is not simply objective facts waiting to be discovered, but rather is created through the research process itself. When researchers engage with managers, employees, or other organizational stakeholders, the resulting insights are co-created through their dialogue and shared experiences. For example, when studying leadership styles or organizational culture, the researcher's questions and the participants' responses mutually shape the understanding that emerges. The values and experiences of both researchers and participants significantly influence this knowledge construction process. A researcher's theoretical background, personal experiences with management, and cultural context can affect how they frame questions, interpret responses, and draw conclusions. Similarly, participants' positions within the organization, past experiences, and personal values color their perceptions and the information they choose to share. This perspective encourages management researchers to practice reflexivity – critically examining their own role in the research process and how their presence and actions might be influencing the data collected and the conclusions drawn. It calls for transparency about the researcher's position and potential biases, acknowledging that complete objectivity is neither possible nor necessarily desirable in understanding complex organizational phenomena. In practical terms, this approach might lead management researchers to employ methods that facilitate deeper engagement with participants, such as in-depth interviews, participant observation, or action research. These methods allow for a richer exploration of context and meaning, recognizing that management practices and their effects are deeply embedded in specific organizational and cultural contexts. This view also has implications for how management knowledge is applied. Instead of seeking universal "best practices," it suggests that management solutions should be adapted to specific contexts, taking into account the unique values, cultures, and dynamics of each organization. It emphasizes the importance of engaging with various stakeholders in the process of implementing management changes, recognizing that their perceptions and values will shape the success of any intervention. By acknowledging the co-constructed nature of management knowledge, this perspective can lead to more nuanced and context-sensitive understandings of organizational phenomena, potentially resulting in more effective and ethically grounded management practices. Subjectivity and intersubjectivity are seen not as obstacles, but as essential 'means' for constructing knowledge. This type of research adopts an '*emergent design*' rather than a fixed, pre-established design (Creswell & Poth, 2018). It is structured based on successive discoveries made during the process, through constant dialogue, negotiation of meanings, interpersonal critical reflection, and interaction between the researcher's experience and that of the participants, to co-construct a consensual reading of the reality being studied (Tracy, 2020).

The field of qualitative research has evolved significantly over the years, with various scholars proposing different frameworks to categorize and understand its diverse approaches. In the late 1980s, one researcher identified six primary qualitative traditions, including areas such as human ethology and cognitive anthropology (Jacob, 1987; Jacob, 1988). As the field progressed, other academics expanded on this work, with one notable scholar in the early 2000s presenting a more comprehensive list of theoretical orientations, encompassing a wide range of approaches from ethnography to chaos theory (Patton, 2002). More recent classifications have streamlined these categorizations. For instance, some contemporary researchers have proposed four main qualitative research designs, while others have recognized a broader array of emerging inquiry strategies (Creswell & Poth, 2018). These newer typologies include methods such as narrative research, phenomenology, and participatory action research, among others. The variety of classification systems proposed over time reflects the complex historical development of qualitative research. It highlights the field's diverse theoretical foundations and the range of philosophical perspectives that inform different approaches (Denzin & Lincoln, 2018). Moreover, this ongoing evolution in categorization demonstrates the dynamic nature of qualitative research and its continuous push to expand our understanding of social phenomena. This diversity in typologies not only showcases the richness of qualitative research traditions but also underscores the field's vibrant nature and its persistent efforts to advance knowledge in various disciplines (Ravitch & Carl, 2021). As the landscape of qualitative inquiry continues to evolve, it's likely that new frameworks for understanding and organizing these approaches will emerge, further enriching our comprehension of this multifaceted field of study.

Qualitative methods have their origins in various disciplines in the human and social sciences, such as anthropology, sociology, and psychology. These disciplines have helped to shape the theoretical and methodological foundations of qualitative research. Anthropology pioneered the key method of participant observation in the early 20th century. This immersive approach, in which researchers integrate themselves into the cultural groups studied to understand their codes and practices from the inside, laid the foundations for the qualitative study of social phenomena in their natural context (Tedlock, 2003). Sociology, for its part, made a major contribution by conceiving of social reality as an intersubjective construct based on interactions and shared meanings (Berger & Luckmann, 1966). This perspective paved the way for the use of qualitative methods to capture the lived experiences and meanings attributed by social actors. At the same time, certain currents in psychology, such as phenomenological psychology, emphasized a deeper understanding of subjective experiences, encouraging the use of in-depth interviews and other qualitative techniques (Moustakas, 1994). The Chicago School also played a key role in developing case studies and direct observation techniques in the urban field in the early 20th century, contributing to the rise of qualitative methods in the study of social phenomena (Park et al., 1967). Qualitative methods thus emerged from the desire of various disciplines in the human and social sciences to grasp the richness, complexity, and significance of human experience in its real context, by combining complementary theoretical and methodological approaches.

2. CHARACTERISTICS AND SPECIFIC FEATURES OF QUALITATIVE RESEARCH

The qualitative approach is characterized by its deep rootedness in discursive data, the words, stories, and testimonies expressed by participants. The raw material of qualitative research lies in the lived experiences of individuals in their everyday reality, as well as in the direct observation of social, cultural, or organizational phenomena taking place in their natural context (Patton, 2015). The aim is to gain an in-depth understanding of the richness and complexity of human and social processes from discursive data collected in the field (Denzin & Lincoln, 2011). The qualitative researcher attaches primary importance to language and the meanings conveyed by the discourse of individuals (Lamsiah et al., 2021). The purpose of hypothesizing, in other words, is to construct conceptual models and interpretations rooted in the experiences reported by the participants (Strauss & Corbin, 1998). These elements are intimately linked and reveal a profoundly humanistic and interpretive approach to research. Firstly, the emphasis on *'language and the meanings conveyed by people's discourse'* goes far beyond a simple transcription of words. It recognizes that language is not a neutral medium, but a complex system of symbols charged with cultural, historical, and personal meaning. Every word chosen, every turn of phrase, even every silence, can be revealing. The qualitative researcher listens not only to what is said but how it is said. They are attentive to the narratives that individuals construct to give meaning to their experiences. These narratives can be analyzed in a variety of ways: narrative analysis, content analysis, discourse analysis, etc. Each approach offers tools for deconstructing language and identifying recurring themes, dominant metaphors, or narrative structures. The purpose is not to reduce these rich expressions to simplistic categories but to grasp their full complexity and ambiguity. Secondly, the aim of *'theorizing'* distinguishes qualitative research from mere description. The intention is not to catalogue experiences, but to understand them in depth to build conceptual models that can inform theory and practice. The notion of *'conceptual models'* implies a systematic organization of the insights gained through discourse analysis. These models can take various forms: a typology of patients' experiences of a diagnosis, an outline of the stages in a family's adaptation to a chronic illness, or a theoretical framework for understanding the power dynamics involved in end-of-life decisions (Bentalha, 2023c).

Qualitative research, therefore, involves direct personal experience by the researcher, to deepen understanding of 'externally observable behaviors and internal states' in context (Patton, 2015). It often ventures into uncharted territory, recognizing the emergent and malleable nature of the knowledge produced (Norum, 2008). The objective is to gain an in-depth understanding of the richness, complexity, and meanings of human and social processes from discursive data collected in the field. Unlike the positivist approach, which aims for '*objective*' and '*uncontaminated*' knowledge, qualitative research recognizes the existence of the researcher's prejudices, values, and subjectivities, and incorporates them into the analysis (Ackerly & True, 2010). It values the subjectivity of data and interpretations, considering that meanings are co-constructed by social interactions, and political, historical, and cultural contexts (Creswell & Poth, 2018).

The central logic is not to test existing theories, but rather to research or construct new theories rooted in the data. Qualitative researchers often start with intuitions and adopt an open, exploratory attitude, seeking to understand participants' perceptions, experiences, and meanings, which can lead to unexpected results and the development of new theories (Strauss & Corbin, 1998). Qualitative research, in seeking to construct new theories rooted in the data, adopts an open and exploratory stance. Researchers do not enter the field with rigid hypotheses to test but with broad questions and genuine curiosity. The importance attached to researchers' "*intuitions*" deserves particular attention. Contrary to the notion of detached objectivity often advocated in the hard sciences, qualitative research recognizes that researchers bring their experiences, theoretical sensibilities, and intuitions to the research process. These intuitions are not biases to be eliminated, but valuable resources that guide the investigation, while remaining open to being challenged by the data. This inductive, open, and flexible approach can lead to unexpected results and the emergence of new theoretical perspectives. Qualitative research recognizes the emergent, iterative, and malleable nature of the research process, where questions, methods, and analyses can evolve as the project progresses (Norum, 2008).

3. THE QUALITATIVE/QUANTITATIVE CHOICE DILEMMA: REASONS AND PROSPECTS

The choice between a qualitative or quantitative approach to research depends largely on the nature of the research question, the objectives pursued, and the epistemological assumptions of the researcher. Each approach has strengths and limitations that need to be carefully considered.

The distinction between quantitative and qualitative methods is fundamentally the result of the influence of interpretation and subjective perception, which in turn is reflected in the research methods employed in carrying out the study. The methodologies used in management science are varied. A major trend is the increasing frequency of use of quantitative methodologies. There is currently a craze among researchers for quantitative rather than qualitative models for several reasons (Maurand-Valet, 2011). One argument to explain the importance of quantitative methods in management science is the fact that for some people, what cannot be measured using known and current techniques does not exist. So, measurement using a tool alone would make observation scientific and therefore valid, this measurement being expressed in figures. The human mind and senses would therefore not be sufficient to validly measure a phenomenon, or even to observe it. The possibility of expressing observations using figures could therefore contribute to the validity of management science research, if not consciously, then at least unconsciously among researchers. Another argument to explain the importance of quantitative methods in management science is that quantification benefits from the appeal of the symbolic nature of numbers. Symbols have the characteristic, or at least the reputation, of being able to concentrate a maximum of meaning in a few strokes. Quantitative methods are therefore rich in tools. At the same time, developments in data analysis, including correspondence factor analysis, have made it possible to process qualitative data. All these developments are helping to increase the importance of quantification in management science, especially as interpreting a probabilistic test is much easier than verifying the plausibility of a qualitative hypothesis. Quantitative methods make it possible to generalize results to a wider population using representative samples. However, this approach may lack nuance and depth in understanding complex phenomena and the meanings attributed by individuals (Bentalha, 2024b).

Conversely, qualitative methods, engrained in the interpretive and constructivist paradigms, are preferred when the aim is to explore in depth the experiences, meanings, and processes as lived by the participants (Merriam & Tisdell, 2016). They provide a rich and nuanced understanding of social phenomena in their natural context. In reality, the debate between quantitative and qualitative methods reflects deeper differences in the conception of knowledge and the nature of the phenomena studied. Quantitative methods, rooted in positivism, seek to identify universal laws and causal relationships through statistical analysis of numerical data. They make it possible to test hypotheses, measure the extent of effects, and generalize results to larger populations. However, they can lack depth and overlook individual and contextual particularities. Nevertheless, the results are difficult to generalize, and the researcher's subjectivity can influence the interpretation of the data (Patton, 2015). The choice between the two approaches therefore depends on the research question and the nature of the phenomenon being studied. Qualitative methods have developed strategies to reinforce the rigor and credibility of their results. Data triangulation, peer review, and researcher reflexivity are techniques designed to reduce bias and deepen analysis (Lincoln & Guba, 1985). At the same time, certain qualitative approaches such as grounded theory aim to develop substantive theories which, although specific to one context, can inform similar situations (Charmaz, 2006).

One of the major advantages of qualitative research is its ability to capture the richness and complexity of human experience. Unlike quantitative data, which often reduces phenomena to numbers, qualitative data captures the nuances, contradictions, and ambiguities inherent in social life. This depth is particularly valuable when studying complex issues such as identity, power, or the meaning given to experiences. Qualitative methods, such as semi-structured interviews or participant observation, offer a flexibility that rigid quantitative protocols do not. Researchers can adjust their questions or focus according to participants' responses, exploring unexpected but revealing avenues (Denzin & Lincoln, 2011). This adaptability is crucial in uncovering aspects of the phenomenon that the researcher had not anticipated, thus contributing to a more holistic understanding. Qualitative methods are particularly well suited to studying dynamic processes and social change. Approaches such as longitudinal case studies or life histories capture how individuals and communities evolve, negotiate transitions, and construct meaning in the face of change (Bentalha & Hmioui, 2021). This processual understanding is essential in fields such as organizational development, education, and the study of social movements.

In almost all fields of research, and particularly in management sciences, we are gradually seeing a kind of integration of the two approaches, with more '*qualitization*' of quantitative studies and quantification of qualitative studies. Some researchers recommend a mixed approach, combining qualitative and quantitative methods, to take advantage of the respective strengths of each approach and obtain a more complete understanding of the phenomenon under study (Creswell & Plano Clark, 2018). However, this approach can be more complex and demanding in terms of time and resources. Thus, the choice depends on the specific research questions, the objectives pursued, and the epistemological orientation of the researcher. A qualitative approach will be preferred for exploring complex human experiences in depth, while a quantitative approach will be more appropriate for measuring variables and establishing causal relationships.

On the one hand, the first trend tends to give greater meaning and significance to numerical data, using complementary investigative tools such as interviews, focus groups, or observations (Miles & Huberman, 1994). Admittedly, this research strategy makes it possible to better understand the context and motivations behind the quantitative data, to give more meaning to the numerical data to interpret them more richly, and to discover information hidden in the quantitative data (Bentalha, 2023d). However, the researcher runs multiple risks in terms of subjectivity in interpretation, bias in the selection of participants, difficulty in generalizing results, etc. On the other hand, the second trend seeks to transform textual, narrative, and descriptive data into numerical data, using several techniques, such as coding, content analysis, lexical analysis, etc. (Krippendorff, 2018; Huberman, 2014). This has the advantage of making it easier to compare data from different groups or contexts, facilitate statistical analysis and the search for trends, and be able to generalize the results to a wider population. However, there are major risks inherent in this, such as the loss of the richness of qualitative data, over-simplification of data and loss of information, biases in data interpretation, etc. Quantification and qualitative research are not exclusive methods (Bryman, 2016), they remain complementary tools that can be used to mutually enrich the analysis of qualitative and quantitative data in the hope of obtaining a more complete understanding of data relating to more complex problems as in the case of the social sciences (Creswell & Clark, 2017; Tashakkori & Teddlie, 2010).

4. THE PREREQUISITES OF A QUALITATIVE RESEARCHER

The fundamentals of a qualitative researcher form a complex constellation of skills, attitudes, and personal qualities, reflecting the deeply human and interpretive nature of this research approach. Far from being a simple list of technical skills, these prerequisites paint a picture of an individual engaged in a quest for understanding that requires as much introspection as external observation.

At the heart of these prerequisites lies a productive tension between openness and rigor. Open-mindedness and curiosity are essential. Researchers must approach the field with a willingness to be surprised, to question their hypotheses, and to follow the unexpected leads revealed by the data. This openness is not naivety, but a disciplined receptiveness, a willingness to suspend judgment to allow new perspectives to emerge. At the same time, this openness is counterbalanced by methodological and analytical rigor. The qualitative researcher is not simply a collector of stories, but an attentive and systematic interpreter (Lamsiah & Bentalha, 2023b). They must have refined linguistic skills to detect the nuances of language, eloquent silences, and revealing metaphors. When analyzing data, they must think both analytically and creatively, identifying recurring patterns while remaining open to contradictions and atypical cases. This rigor extends to the validation of interpretations, often involving feedback to participants and confrontation with peers.

Another central tension lies between empathic involvement and reflective distance. Empathy and active listening are crucial to creating a space where participants feel confident enough to share experiences that are sometimes intimate or painful. However, the researcher must also maintain a reflexive distance, recognizing that his own experiences, biases, and social positions influence his perception (Bentalha, 2023b). This reflexivity is not an obstacle to empathy, but a safeguard against projection or hasty interpretation. This tension also manifests itself in managing the emotional impact of research. Working on subjects such as chronic illness, bereavement or ethical dilemmas at the end of life can be emotionally demanding.

The temporality of qualitative research also requires specific qualities. Patience and persistence are essential in the face of often lengthy and iterative processes. Unlike quantitative research, where hypotheses and methods are often fixed in advance, qualitative research requires constant flexibility. The researcher must be comfortable with ambiguity, resist the temptation to jump to premature conclusions, and accept that understanding is built up gradually, sometimes through unexpected detours. This patience goes hand in hand with a particular theoretical sensitivity. Qualitative researchers do not approach the field as a tabula rasa; they are informed by existing theories. However, their relationship with these theories is dialectical. They must be sufficiently familiar with them to recognize when the data challenges or enriches them, without allowing themselves to be constrained by them.

An interest in process and meaning, beyond cause and effect, is essential in qualitative research. The aim is to gain an in-depth understanding of the meanings, experiences, and processes experienced by participants (Merriam & Tisdell, 2016). A critical and questioning approach is crucial (Bentalha, 2023a). The qualitative researcher must not take things for granted but question assumptions, and underlying interests and consider alternative perspectives (Kincheloe et al., 2018). The ability to step back and adopt a cultural commentator posture is paramount. This involves identifying and bracketing one's assumptions and values so as not to be influenced by them when analyzing data (Guba & Lincoln, 1994). Developing a 'double consciousness' or 'analytical eye' to listen carefully while thinking critically and analytically is a key skill for producing rich and complex data (Patton, 2015). Reflexivity, i.e. critical reflection on the research process and one's role as a researcher, including insider and outsider positions, is essential (Gallais, 2008). Good interpersonal skills, and a warm and caring attitude that enables a relationship of trust to be established with participants, are crucial, without necessarily being extroverted (Rubin & Rubin, 2012).

Relational and ethical skills are also crucial. Researchers navigate complex terrain - hospital hierarchies, family dynamics, and diverse communities. They must be sensitive to power dynamics and capable of building relationships of trust while respecting ethical limits (Lamsiah & Bentalha, 2023a). Finally, the qualitative researcher must possess intellectual humility and robust integrity. Humility manifests itself in the recognition that participants are the experts of their own experience. It implies an openness to learning and to being transformed by the research process. Integrity, for its part, translates into a willingness to report results faithfully, even if they are uncomfortable or call into question established practices.

So the prerequisites of a qualitative researcher paint a picture of an individual committed to a practice that is both rigorous and deeply humanistic. These different qualities of the qualitative researcher are not isolated skills but form a dynamic network. They reinforce and temper each other, enabling the researcher to navigate the complexities of human experience. This constellation of prerequisites reflects the unique nature of qualitative research. It is not simply an alternative or complement to quantitative research, but a distinct approach that recognizes that understanding health and illness requires more than measurement and statistics. It requires a dive into the lived world of individuals, and an exploration of the meanings, emotions, and social contexts that shape the experience of health. By requiring this unique combination of qualities, qualitative research is not just a methodology, but a vocation. It invites researchers to embark on a journey of discovery that can be demanding, sometimes trying, but profoundly rewarding. It is a path that demands ongoing personal and professional growth, a refined sensitivity, and a commitment to human understanding. In the landscape of critical research, the qualitative researcher is not only a scientist but also an interpreter, a witness, and a defender of the human experience in all its richness and diversity.

CONCLUSION

Qualitative studies represent a subtle but essential art in the quest to understand the complex phenomena that shape our world. Although long overshadowed by the apparent rigor of quantitative approaches, these methods have demonstrated their inestimable value in the fields of science and management.

In the natural sciences, qualitative studies have deepened our understanding of ecological systems, biological processes, and animal behavior, capturing the subtle nuances that numerical data alone cannot capture. They offer a holistic perspective that is essential for interpreting scientific results, revealing the complex dynamics underlying the phenomena studied (Bentalha et al., 2023). In the field of management, qualitative approaches have proved crucial for exploring the subtleties of human interactions, organizational dynamics, and decision-making processes. By giving a voice to the actors involved, these studies have shed light on motivations, perceptions, and lived experiences, providing valuable insight into the human realities that shape organizational outcomes (Hmioui et al., 2019).

However, conducting high-quality qualitative research requires rigorous mastery of survey, observation, and analysis techniques. Researchers must demonstrate particular sensitivity to accurately capture the nuances of human experiences and natural phenomena while remaining vigilant for potential biases and contextualizing their findings within an appropriate theoretical and cultural framework. Beyond their contribution to the production of knowledge, qualitative studies have played an essential role in developing new theories and challenging existing models (Bouhtati et al., 2024). By immersing researchers in the complexity of the phenomena studied, these approaches have made it possible to identify emerging patterns, challenge established hypotheses, and generate new conceptual perspectives, thereby stimulating innovation and the advancement of knowledge.

Through our reflections, we have sought to explore the crucial role of qualitative research methods in the fields of science and management, while highlighting their ability to bring to light the nuances and rich contextual elements often overlooked by purely quantitative approaches. By embracing the delicate art of qualitative research, we can deepen our understanding of the complex systems that shape our world, revealing new perspectives, opportunities for continuous improvement, and fresh ways of thinking about current and future challenges. Far from being a simple complement to quantitative approaches, qualitative studies are an essential component of scientific research and management, offering a unique insight into the subtleties that often escape purely numerical analysis.

To deepen the understanding and application of qualitative methods and highlight their crucial subtle and contextual aspects in science and management, we consider theoretical, methodological, and empirical perspectives to this research. First, by deepening our understanding of the philosophical underpinnings of qualitative research by exploring its epistemological and ontological foundations. Secondly, by further discussing the frameworks and protocols for collecting and processing qualitative data to gain a better understanding of the nuances of the context in each of the management science disciplines studied. Finally, by conducting empirical research comparing existing qualitative and quantitative results in a field of management sciences to identify the complementarities and convergences in the conclusions obtained.

REFERENCES

Ackerly, B., & True, J. (2010). *Doing feminist research in political and social science*. Palgrave Macmillan. 10.1007/978-1-137-05442-5

Alla, L., Bentalha, B., Megzari, G., & Hmioui, A. (2022). The effects of COVID-19 on supply chains: A qualitative analysis. Reconfigurations du tourisme en contexte de la crise du Covid-19: Quelles démarches pour quelles résiliences au Maroc et ailleurs, 1(1), 130-153.

Alla, L., Hmioui, A., & Bentalha, B. (2020). La netnographie dans les recherches marketing: La communauté virtuelle comme consom'acteur vecteur d'efficacité marketing. *Alternatives Managériales Economiques*, 2(4), 631–652.

Bentalha, B. (2023a). Les fondements théoriques du supply chain management: Une analyse critique. *Alternatives Managériales Economiques*, 5(2), 1–22.

Bentalha, B. (2023b). Consumer Perception of Robotic Mobile Fulfillment Systems: A Netnographic Case Study of Amazon. In Innovation, Strategy, and Transformation Frameworks for the Modern Enterprise (pp. 243-264). IGI Global. 10.4018/979-8-3693-0458-7.ch010

Bentalha, B. (2023c). A Framework for Exploring the Global Performance of Service Supply Chain Management: A Qualitative Study. Government Impact on Sustainable and Responsible Supply Chain Management, 106-129.

Bentalha, B. (2023d). Influence des actions sociales de la RSE sur les valeurs perçues des employés: Étude qualitative des œuvres sociales de l'entreprise Taajmouati. Repères et Perspectives Economiques, 7(2).

Bentalha, B. (2024a). Contribution des randonnées au développement d'un tourisme durable dans la région Fès Meknès. Numérisation et durabilité, Vers une économie responsable et innovante, 226-245.

Bentalha, B. (2024b). Employees and Robots in Amazon's Robotic Mobile Fulfillment Systems: A Netnographic Analysis of a Supply Chain Transformation. In Blockchain Applications for Smart Contract Technologies (pp. 188-207). IGI Global. 10.4018/979-8-3693-1511-8.ch009

Bentalha, B., & Hmioui, A. (2021). Smart service supply chain and Just Walk Out technology: a netnographic approach. In *The Proceedings of the International Conference on Smart City Applications* (pp. 223-236). Cham: Springer International Publishing.

Bentalha, B., Hmioui, A., & Alla, L. (Eds.). (2023). *Integrating Intelligence and Sustainability in Supply Chains*. IGI Global. 10.4018/979-8-3693-0225-5

Berger, P. L., & Luckmann, T. (1966). *The social construction of reality: A treatise in the sociology of knowledge*. Anchor.

Bouhtati, N., El Yaacoubi, Y., Bentalha, B., & Lhoussaine, A. L. L. A. (2024). Leviers logistiques de promotion du commerce équitable des produits de terroir: Une esquisse de modèle d'analyse. Alternatives Managériales Economiques, 6(Spécial 1), 17-35.

Bryman, A. (2016). *Social research methods*. Oxford University Press.

Charmaz, K. (2014). *Constructing grounded theory* (2nd ed.). Sage Publications.

Corbin, J., & Strauss, A. (2015). Basics of qualitative research (4e éd.). Sage.

Creswell, J. W., & Clark, V. L. P. (2017). *Designing and conducting mixed methods research*. Sage publications.

Creswell, J. W., & Creswell, J. D. (2018). *Research design: Qualitative, quantitative, and mixed methods approaches* (5th ed.). Sage Publications.

Creswell, J. W., & Plano Clark, V. L. (2018). *Designing and conducting mixed methods research* (3rd ed.). Sage Publications.

Creswell, J. W., & Poth, C. N. (2016). Qualitative inquiry and research design: Choosing among five approaches. *Sage (Atlanta, Ga.)*.

Creswell, J. W., & Poth, C. N. (2018). Qualitative inquiry and research design (4e éd.). Sage.

Cunliffe, A. L. (2003). Intersubjective voices: The role of the "theorist". *Administration & Society*, 46(4), 504–533.

Denzin, N. K., & Lincoln, Y. S. (2011). *The SAGE handbook of qualitative research*. Sage Publications.

Denzin, N. K., & Lincoln, Y. S. (2018). The Sage handbook of qualitative research (5e éd.). Sage.

Flick, U. (2018). *An introduction to qualitative research* (6th ed.). Sage Publications.

Gallais, T. L. (2008). Wherever I go there I am: Reflections on reflexivity and the research stance. *Reflective Practice*, 9(2), 145–155. 10.1080/14623940802005475

Guba, E. G., & Lincoln, Y. S. (1994). Competing paradigms in qualitative research. In Denzin, N. K., & Lincoln, Y. S. (Eds.), *Handbook of qualitative research* (pp. 105–117). Sage Publications.

Guba, E. G., & Lincoln, Y. S. (2005). Paradigmatic controversies, contradictions, and emerging confluences. In The Sage handbook of qualitative research (3e éd., pp.191-215). Sage.

Hmioui, A., Alla, L., & Bentalha, B. (2019). Évaluation des formations en ligne: Étude qualitative des perceptions des utilisateurs des MOOC au Maroc. In ICT-Days' 19.

Hmioui, A., Alla, L., & Bentalha, B. (2022). Perception du tourisme durable par les clients étrangers: Cas du séjour dans un établissement hôtelier. Reconfigurations du tourisme en contexte de la crise du Covid-19: Quelles démarches pour quelles résiliences au Maroc et ailleurs?, 1(1), 5-38.

Hmioui, A., & Bentalha, B. (2021). Service supply chain management: A literature review. *International Journal of Logistics Systems and Management*, 40(3), 332–353. 10.1504/IJLSM.2021.120253

Huberman, A. (2014). Qualitative data analysis a methods sourcebook.

Jacob, E. (1987). Qualitative research traditions: A review. *Review of Educational Research*, 57(1), 1–50. 10.3102/00346543057001001

Jacob, E. (1988). Clarifying qualitative research: A focus on traditions. *Educational Researcher*, 17(1), 16–24. 10.2307/1175461

Kincheloe, J. L., McLaren, P., & Steinberg, S. R. (2018). Critical pedagogy and qualitative research: Advancing the bricolage. In Denzin, N. K., & Lincoln, Y. S. (Eds.), *The landscape of qualitative research* (5th ed., pp. 235–260). Sage Publications.

Krippendorff, K. (2018). *Content analysis: An introduction to its methodology*. Sage publications.

Lamsiah, A., Benabderrazik, Y., & Bentalha, B. (2021). The Economic and political components of diplomatic discourse: State actors in Morocco-UK Free Trade Agreement as a case study. *Alternatives Managériales Economiques*, 3(4), 496–517.

Lamsiah, A., & Bentalha, B. (2022). Economic Diplomacy and Economic Intelligence Amalgamation: The Moroccan-Nigerian Economic Exchange as a Case Study. *Alternatives Managériales Economiques*, 4(4), 582–603.

Lamsiah, A., & Bentalha, B. (2023a). 2022 Qatar World Cup: A Netnographic Analysis of the Relationship Between Sport, Media and Politic . Réflexions sportives, 1(3), 69-95.

Lamsiah, A., & Bentalha, B. (2023b). Moroccan Jews And Economic Intelligence: The Implicit Power Of Belonging. Le concept de l'intéligence en sciences juridiques, économiques et sociales, 71-90.

Lincoln, Y. S., & Guba, E. G. (2013). *The constructivist credo*. Left Coast Press.

Lincoln, Y. S., Lynham, S. A., & Guba, E. G. (2018). Paradigmatic controversies, contradictions, and emerging confluences, revisited. Dans N.K. Denzin & Y.S. Lincoln (Eds.), The Sage handbook of qualitative research (5e éd., pp.108-150). Sage.

Maurand-Valet, A. (2011). Choix méthodologiques en Sciences de Gestion: pourquoi tant de chiffres?. Revue management et avenir, (3), 289-302.

Merriam, S. B., & Grenier, R. S. (2019). *Qualitative research in practice: Examples for discussion and analysis*. Jossey-Bass.

Merriam, S. B., & Tisdell, E. J. (2016). *Qualitative research: A guide to design and implementation* (4th ed.). Jossey-Bass.

Miles, M. B., & Huberman, A. M. (1994). Qualitative data analysis: An expanded sourcebook. Sage.

Moustakas, C. (1994). *Phenomenological research methods*. Sage Publications. 10.4135/9781412995658

Norum, K. E. (2008). Artifact analysis. In Given, L. M. (Ed.), *The SAGE encyclopedia of qualitative research methods* (pp. 25–26). Sage Publications.

Park, R. E., Burgess, E. W., & McKenzie, R. D. (1967). *The city*. University of Chicago Press.

Patton, M. Q. (2002). Qualitative research and evaluation methods (3e éd.). Sage.

Patton, M. Q. (2015). *Qualitative research & evaluation methods: Integrating theory and practice* (4th ed.). Sage Publications.

Ravitch, S. M., & Carl, N. M. (2021). Qualitative research: Bridging the conceptual, theoretical, and methodological (2e éd.). Sage.

Rubin, H. J., & Rubin, I. S. (2012). *Qualitative interviewing: The art of hearing data* (3rd ed.). Sage Publications.

Saldaña, J. (2021). The coding manual for qualitative researchers. *Sage (Atlanta, Ga.)*.

Schwandt, T. A. (2000). Three epistemological stances for qualitative inquiry. In Handbook of qualitative research (2e éd., pp.189-213). Sage.

Stewart, D. W. (2022). *Qualitative research methods*. University Press.

Strauss, A., & Corbin, J. (1998). *Basics of qualitative research: Techniques and procedures for developing grounded theory* (2nd ed.). Sage Publications.

Tashakkori, A., & Teddlie, C. (2010). Putting the human back in "human research methodology": The researcher in mixed methods research. *Journal of Mixed Methods Research*, 4(4), 271–277. 10.1177/1558689810382532

Tedlock, B. (2003). Ethnography and ethnographic representation. In Denzin, N. K., & Lincoln, Y. S. (Eds.), *Strategies of qualitative inquiry* (pp. 165–213). Sage Publications.

Tedlock, B. (2020). Avances en etnografía de la observación participante. *Revista Colombiana de Antropología*, 56(2), 91–115.

Tracy, S. J. (2020). *Qualitative research methods: Collecting evidence, crafting analysis, communicating impact*. Wiley-Blackwell.

Vagle, M. D. (2018). Crafting phenomenological research (2e éd.). Routledge.

Chapter 2
Crafting a Persuasive Qualitative Research Proposal in Management Science:
Practical Insights From Literature

Mohammed Amine Benarbi
https://orcid.org/0000-0002-5659-9695
Sidi Mohammed Ben Abdellah University, Morocco

Mouhcine Rhouiri
https://orcid.org/0009-0007-7115-871X
Sidi Mohammed Ben Abdellah University, Morocco

Mohamed Habiboullah Meyabe
Sidi Mohammed Ben Abdellah University, Morocco

ABSTRACT

A persuasive research proposal helps interested lecturers clearly understand the research objectives of novice researchers. This chapter focuses on research proposals in management science, particularly those employing qualitative methodologies to explore new topics. By drawing on practical examples, the authors aim to provide a guide to assist novice researchers in crafting a persuasive research proposal. In this chapter, the authors conducted a literature review from two databases to identify relevant documents on research proposals. A total of 14 documents were selected and analyzed. The findings of this research contain three key elements. The first summarizes the qualitative methods used in the selected literature. The second provides the main structure of a research proposal. The third discusses the

DOI: 10.4018/979-8-3693-5543-5.ch002

essential elements to make the research proposal persuasive. Overall, the authors have sought to contribute to the existing body of knowledge on research proposals that utilize qualitative methods in the field of management.

INTRODUCTION

Management science is a wide and interdisciplinary field that focuses on solving problems and making strategic decisions within institutions, corporations, governments, and other organizational entities. Management science applies scientific methods to address management issues and improve organizational performance. It enables managers to make informed decisions based on data and analysis (Ma, 2023).

Research questions in management science can be addressed using either quantitative or qualitative approaches. To begin a study with the goal of addressing a problem or filling a research gap in the literature, researchers may start with a research proposal that they submit to the party interested in working on a specific topic. This party could be a university or even an organization (such as enterprise).

Thus, research proposals are seen as an important tool in management science due to their ability to outline the research project, its objectives, methodology, and even contributions or outcomes. Well-developed proposals particularly show how they can advance academic knowledge and solve actual problems (McNabb, 2017). This is achieved by blending relevant theoretical points of view in management science and practice, undertaking research projects by seeking funds or collaborative partners, promoting innovations, encouraging rigorous research designs, expanding the analyze of phenomenological realities by exploratory thinking and enhancing stakeholder involvement. Moreover, proposals guide research projects along the way and help achieve the expected results.

This book chapter is intended for novice researchers seeking to apply for a PhD program or carry out a project to address a practical problem in a specific entity such as a commercial company or enterprise, particularly those who wish to employ qualitative methods to explore their research subject.

This category of researchers aims to craft a persuasive research proposal to secure approval for their project. To achieve this goal, the authors believe that these novice researchers must understand three key elements: first, the qualitative methods, especially the most commonly used ones in the literature; second, the structure of a research proposal; and third, how to make their project proposal persuasive.

Despite the relevance of resources on how to write research proposals, one still feels inadequately served in terms of fully guiding both the new researcher and, more so, one wanting to apply qualitative approaches in management science. Thus, this chapter is intended to fill this gap by pointing out a step-by-step framework and best practices that would link the aspiring researcher to develop a well-structured, persuasive, and impactful qualitative research proposal.

This book chapter is structured in a simple way. First, the authors have defined the three concepts of this study which are *"management science"*, *"research proposal"* and *"qualitative methodologies"*. Next, they outline the methodology used to answer the problematic. Finally, they present the findings and discussion before they conclude.

The methodology used in this book chapter is a literature review using two databases (Web of Science and Scopus) to select relevant documents for this research. By reviewing the literature about this subject, the authors consider that they could extract valuable and practical information about research proposals and qualitative methodologies in the field of management. In other terms, the authors could address the research problem of how to craft a persuasive research proposal in the field of management science that uses qualitative methodologies.

A total of 14 documents were selected, divided into two categories. The first category provides 4 documents considered as guides for research proposals, the second category contains 10 documents as examples of research proposals using qualitative methods.

Three main elements make up the findings of this study. The first presents insights from literature about qualitative methods used in research proposals selected from the two databases. The second presents the different structures identified from the literature and offers the main elements that must appear in a research proposal. The third point provides the main elements that make the research proposal persuasive.

1. CONCEPTUAL FRAMEWORK

1.1. Management Science

Management science is a multifaceted field that encompasses a diverse range of research focused on organizational leadership, employees, consumers, and shareholders. It examines complex issues related to decision-making, resource allocation, and performance evaluation within organizations (Kedia et al., 2002).

Management science research often includes a comprehensive analysis of a wide range of data sources, including surveys, financial data, operational systems, business literature, and theoretical frameworks (Zerbino et al., 2021). The ultimate goal of management science research is to develop a more comprehensive and nuanced understanding of organizational administration and to address real-world challenges facing businesses and industries through the application of evidence-based insights and practical solutions (Buffington et al., 2016).

1.2. Research Proposal

Research proposal can be used in both quantitative and qualitative studies. The objective from research proposal in quantitative studies is to design a study that will collect and analyze numerical data to test hypotheses and make generalizations as noticed by Silyn-Roberts (2013). On the other hand, in qualitative research, the research proposal aims to design an exploratory study to understand the underlying reasons, motivations, and processes behind organizational behaviors and decision-making (Vasanthakumari, 2021). In this chapter, we will explore research proposals that uses qualitative methodologies in the field of management science.

In this context, we can say research proposals play a crucial role in advancing management science, especially qualitive one, because the help to provide in-depth insights to complex organizational phenomena. This research proposals offers a deeper understanding of the subjective experiences, perceptions, and motivations of individuals within organizations. So, using elaborating such as proposals in qualitative studies allows researchers to design and implement studies that capture rich, detailed data and explore the nuances of organizational dynamics (Vasanthakumari, 2021).

A research proposal outlines the purpose, methodology, and significance of a study base on qualitative research (Vasanthakumari, 2021). This research contributes to examine subjects in their natural environments and interpret phenomena based on people's interpretations (Javadian et al., 2020). The proposal requires investigation, it help to describe methods for extracting valuable information, summaries methods, present tools to do the research and the procedure to analysis data. It gives also an explanation of the significance of the study and considers ethical aspects concerning respectful treatment of participants and data (Oddershede et al., 2018). Based on this research proposals, researchers could benefit from stakeholder support (McNabb, 2017).

1.3. Qualitative Methodologies

Qualitative research methods are commonly used in management science to investigate complex phenomena, explore new domains, and develop theories. Some common qualitative methodologies in management science include case studies, ethnography, interviews, focus groups and grounded theory (Jabar et al., 2009) as described in the table below.

Table 1. Common Qualitative methodologies in Management Science

Type	Description
Case studies	In-depth analysis of a single or multiple organizations, to understand complex processes and dynamics.
Ethnography	Immersive study of organizational cultures, behaviors, and interactions over an extended period.
Interviews	In-depth, semi-structured conversations to understand perspectives, experiences, and decision-making.
Focus Groups	Facilitated group discussions to explore attitudes, opinions, and beliefs around a specific topic.
Grounded Theory	Iterative data collection and analysis to inductively develop new theories.

Source: (Jabar et al., 2009).

These qualitative methods are commonly used to explore new subjects, while quantitative methods are typically employed to generalize about known subjects. In other terms, qualitative methodologies allow researchers to gain rich, contextualized insights that are difficult to capture through quantitative methods alone (Patrik Aspers & Ugo Corte, 2019).

2. METHODOLOGY

A comprehensive review of the literature was performed to enhance comprehension of research proposals in management science and to delineate the fundamental framework and persuasive elements of this document. The methodology for this study drew partial inspiration from (Tijan et al., 2021).

Figure 1 determines the step-by-step methodological approach utilized in the study. The search was conducted by using two research databases: Web of Science and Scopus. In this study, we concentrated on the following key terms:

- **"Research proposal"**
- **"Qualitative research proposal"**

A search was conducted in the mentioned databases to find out the existence of publications including any of the search terms just in their *"titles"*. The criteria for inclusion and exclusion, the sources we used in this research and an analysis of the identified literature, and other sources are elaborated in the following sub-sections.

Figure 1. Methodological steps

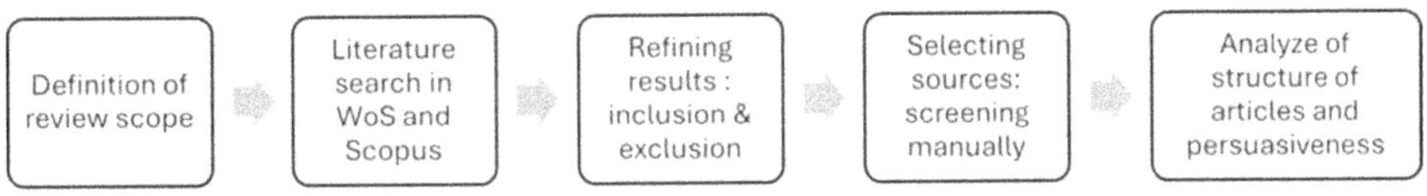

Source: *designed by the authors*

2.1. Inclusion and Exclusion Criteria

In order to select the most appropriate documents for this topic, specific criteria for inclusion and exclusion were established. Considering the limited number of scientific papers focusing on qualitative research proposals in management science, the authors also examined papers related to qualitative research proposals in social science.

Two databases were searched for articles published within the 2019-2023 time frame, in order to capture the latest trends and practices in this topic. We excluded documents of 2024 to facilitate the reproducibility of this study by future researchers.

In the Web of Science database, searches were conducted for publications with *"research proposal"* as a keyword in the Title, limited to the Management category and English language, hence, the results were 4 documents.

The search was conducted in the Scopus database to identify publications containing the keyword *"research proposal"* specifically in the article titles. The results were filtered for documents published between 2019 and 2023, with a focus on the subject areas of Business and Management as well as accounting, resulting in a total of 23 English-language documents.

From the selected sources, one document was identified as indexed in both databases (Scopus and WoS).

Additionally, a search was performed on *"qualitative research proposal",in article titles only,* without specifying years, yielding just one article from the WoS database, published in 2017, in the field of management.

The same inquiry has yielded 5 articles from Scopus database, within the field of social science. These documents are considered as additional sources despite not being directly related to management and not meeting the specified publication time frame requirements.

Excluding articles written in languages other than English helps to prevent a potential bias towards certain regions in the formal analysis (Tijan et al., 2021).

Figure 2 shows the literature search process, partially adapted from (Tijan et al., 2021). After screening manually, the authors have selected 14 sources as pertinent for this study, including 4 sources considered as guides for preparing research proposals.

There are two principal reasons behind excluding the 17 other sources. First, some sources are not employing qualitative methods. Second, there are some ones that are written as essay and don't present a structure of titles.

Figure 2. Literature search process and search results

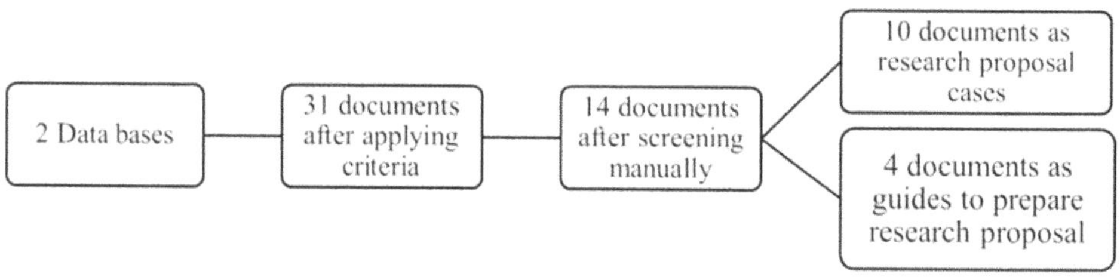

Source: *designed by the authors*

2.2. Sources Used in the Research

In order to address the research gap more comprehensively, the authors have not only examined articles and conference papers but also considered proceeding papers and review sources. All sources are categorized in Table 2.

Table 2. Sources used in the research

Sources considered as cases of research proposal using qualitative methods in management science		Sources as guide for preparing research proposal	
Articles	7	Articles	3
Conference papers	1	Notes	1
Proceeding papers	1		
Review	1		
Total	**10**		**4**

Source: designed by the authors

The selected sources from these two databases were used to know about the most commonly employed qualitative methods in the literature on research proposals. They were also used to identify the usual structure of research proposals and the elements that contribute to their persuasiveness. Additionally, the authors consulted supplementary sources to explain and justify certain elements of the research proposal structure.

2.3. Analysis of the Identified Literature and Other Sources

Priority was given to papers whose title included the term *"research proposal"*. To identify the framework of a research proposal and explore its persuasive elements, a comprehensive analysis of all available papers was essential.

The authors initially found it necessary to identify the main qualitative methods used in the selected research proposal literature. They then focused on understanding the typical structure of qualitative research proposals, and finally, they explored the elements that contribute to making research proposals formulated using qualitative methodologies more persuasive. All these steps were aimed at answering the central question of this book chapter: *"What are the elements that enhance the persuasiveness of proposals formulated through a qualitative methodology?"*

3. FINDINGS AND DISCUSSION

3.1. Qualitative Methods Used in Literature

From the selected literature, the authors found that the most widely used qualitative methods are *"literature reviews"*, with their different types (*systematic, comprehensive, in-depth literature reviews, ... etc*) (da Silva & Silva Cirani, 2020; González-Sánchez et al., 2023; Tortora et al., 2021). The authors found also that *"cases studies"* or *"multiple case studies"* are used in this literature to explore the studied topics (Leo, 2019; Marican et al., 2023; Suyadi et al., 2020; Vollero et al., 2019) by using *"face-to-face interviews"* or "*focus group discussions*" or *"in-depth interviews*" or even *"on-site observation"* (Marican et al., 2023; Nanuru & Utama, 2020; Suyadi et al., 2020; Tran & Yeomans, 2009).

There are also some articles that uses *"documents searches"* and *"content analysis"* to well highlight their topics (Beck et al., 2021; Nanuru & Utama, 2020). In other side, there is a study that have used "*Survey*" that uses research and development analytical methods by applying Rapid Application Development (RAD), we talk here about the study of Nasution et al. (2019).

Finally, the authors found a chapter using *"longitudinal analysis"* from 1992 to 1995. It the study of Jacobs et al. (1999) that studied the impact of family formation on the division of labor in couple households. We notice here that the longitudinal analysis could be used also as quantitative method as mentioned by (Miller & Friesen, 1984).

3.2. Structure of Research Proposal in Qualitative Study

The analysis of structures contributed by the chosen articles in the literature yields different structures from various documents. The authors condensed it all in one table for analysis. Table 3 presents the overview of the major titles and elements that are commonly found in research proposals using qualitative methods of the recent literature. To make explanation easier, the authors have selected 5 documents from among the many "research proposal cases," which best bring out the differences in how these research proposals have been structured.

Table 3. Main elements in the structure of documents selected

Main elements	González-Sánchez et al. (2023)	Tortora et al. (2021)	da Silva & Silva Cirani (2020)	Nanuru & Utama (2020)	Nasution et al. (2019)	Total
Title	X	X	X	X	X	5
Abstract	X	X	X	X	X	5
Introduction	X	X	X	X	X	5
Context	X					1
Literature review / definitions		X	X		X	3
Methodology/Methods	X	X	X	X	X	5
Results /Expected results	X	X	X	X	X	5
Research proposal		X				1
Research trend topics	X					1
Review analysis	X					1
Discussion	X			X	X	3
Research agenda	X		X			2
Conclusions	X	X	X	X	X	5
Limitation and future research		X				1
Acknowledgments				X	X	2
Declarations/Fundings	X					1
References	X	X	X	X	X	5

Source: designed by the authors

By analyzing table 3, we can observe that the main elements cited in all documents are title, abstract, introduction, methodology, results and references. The other elements cited in the majority of documents are results and conclusions. In the half of documents selected we found literature review / definition highlighted and discussion in a title.

We can say also that there are elements like context, research proposal, research trend topics, review analysis that are highlighted as title. This minority of articles using those titles are perhaps responding to exigence of journals where they have published their articles, or it depend on the topic they studied, or maybe they had the interest to be clearer or for other reasons that we didn't think about.

At all, the structure or plan of article could be different from document to another one. But there are some specific elements that should necessarily be cited in a research proposal, but not necessarily separated as title. Novice researchers could join many elements in one title, it depends on how they want to present their research proposal like making in the introduction the context, the objectives of research, the literature review and theorical framework. Another example we could give here is about the conclusion that may include even principles contributions (theorical and practical), limitations of study and future research.

Novice researchers have to pay attention to those elements except if they do not need one of them. We talk here about suggested time and budget (research agenda), acknowledgments, declarations/fundings. Those elements depend on what they want to do and what journal or organization you want to apply for.

Now, we will explain every element that should be in the research proposal, as separated title or not.

3.2.1. Title

A good title must be both informative and striking. One needs to be very careful when constructing the title, for literally, it is the first interface and entry to a research proposal. An effective title is one that briefly suggests the major components of the research to be proposed, explicitly indicating the inquiry, approach, and environment (Galdas, 2017; Piperopoulos, 2010).

In the process of forming a relevant title, the novice researcher will be able to capture the interest of potential funders, collaborators, and other academic peers, and increase the likelihood by very significant margins that the proposal will actually be looked at and seriously considered. Apart from increasing the chances of a research proposal to get all necessary help and resources for conducting a study, a good title may make all the difference to its whole effect.

3.2.2. Abstract/Summary

The abstract acts as a sort of gateway into the research proposal, giving a brief overview of the various important elements (Derntl, 2014). This part, therefore, needs to be written as clearly and concisely as possible and should have key components captured within usually a short word count, as specified almost by journals and editors.

The abstract summarizes the general background and context of the research, specific research questions or objectives, methodology to be used, expected findings or results, and what the proposed study is likely to produce in general (Pearce & Ferguson, 2017). These principal aspects will be interestingly communicated by simply putting them across brightly to capture the attention of readers and the importance associated with the qualitative research pursuit.

3.2.3. Introduction

It is incumbent upon novice researchers to elucidate the context of their study, conduct a thorough a review of the existing literature, and identify a gap in the current body of knowledge. This identified gap will then serve as a basis for justifying the research question and formulating the research objectives (Ajemba et al., 2022).

3.2.3.1. Literature Review

We need in this part of research proposal to review the main topics from the literature. In other words, the objective is to select relevant works for the study. In this literature review, we have to explain the interesting points from general to specific (Chigbu et al., 2023) . The aim of a literature review is to find the research gap. So, the elements that we select for our literature review must have a relation with the research question and the aim of study (Deb et al., 2019).

There are the steps that enable one to conduct an effective literature review:

- Identify key concepts and themes relevant to your topic. This step requires searching at least one academic database to find various documents, including journal articles, books, and conference papers, relating to the topic of study.
- After this, the researcher should go through the titles and abstracts to screen the relevance and usefulness of the retrieved documents for the study. Such documents that may have valuable insights or findings should be kept for further review.
- Through a critical reading of the chosen literature, key ideas, theories, methodologies, results, …etc. can be identified that aid in justifying the relevance and/or significance of the research to be put forward. It will aid them in building up the clear foundation of the study and pinpoint areas where efficiency in knowledge is absent.

Researchers must analyze the relationship between different studies and compose paragraphs that synthesize insights from multiple sources, rather than focusing solely on individual studies.

3.2.3.2. Justification of Research Questions

In this part of research proposal, we have to justify the research question. To do that, it important to write two or three paragraphs about the research gap. This research gap needs to be connected to each research question/aim.

There are seven principal types of research gap in general. Table 4 highlights these 7 types of search gap with a concise definition of each type as cited by Miles (2017).

Table 4. The Seven Research Gaps

Research Gap Type	Definition
Evidence Gap (Contradictory Evidence Gap)	Results from studies allow for conclusions in their own right but are contradictory when examined from a more abstract point of view (Jacobs, 2011; Müller-Bloch & Kranz, 2014; Miles, 2017).
Knowledge Gap (Knowledge Void Gap)	Desired research findings do not exist (Jacobs, 2011; Müller-Bloch & Kranz, 2014; Miles, 2017).
Practical-Knowledge Gap (Action-Knowledge Conflict Gap)	Professional behavior or practices deviate from research findings or are not covered by research (Jacobs, 2011; Müller-Bloch & Kranz, 2014; Miles, 2017).
Methodological Gap (Method and Research Design Gap)	A variation of research methods is necessary to generate new insights or to avoid distorted findings (Jacobs, 2011; Müller-Bloch & Kranz, 2014; Miles, 2017).
Empirical Gap (Evaluation Void Gap)	Research findings or propositions need to be evaluated or empirically verified (Jacobs, 2011; Müller-Bloch & Kranz, 2014; Miles, 2017).
Theoretical Gap (Theory Application Void Gap)	Theory should be applied to certain research issues to generate new insights. There is lack of theory thus a gap exists (Müller-Bloch & Kranz, 2014]) (Jacobs, 2011; Müller-Bloch & Kranz, 2014; Miles, 2017).
Population Gap	Research regarding the population that is not adequately represented or under-researched in the evidence base or prior research (e.g., gender, race/ethnicity, age and etic). (Robinson, et al, 2011).

Source: Miles (2017)

3.2.3.3. Research Objectives / The Aim of the Study

Research objectives or *"research questions"*, *"research aims"* (or even *"research hypothesis"* if the research proposal is based on quantitative method) mean the same thing. We have drawn attention to this point because there are those who put each of these elements as a separate title, even though all these forms mean the same thing.

The aim here is to number the research questions in relation to the research gaps identified previously. The novice researcher must write a sentence at the beginning to introduce these questions and finally put a small transition sentence about the next point.

As noticed by (Ratan et al., 2019), a good research question is represented by the acronym FINERMAPS. Figure 3 highlights the meaning of these letters.

Figure 3. Good research question represented by acronym FINERMAPS

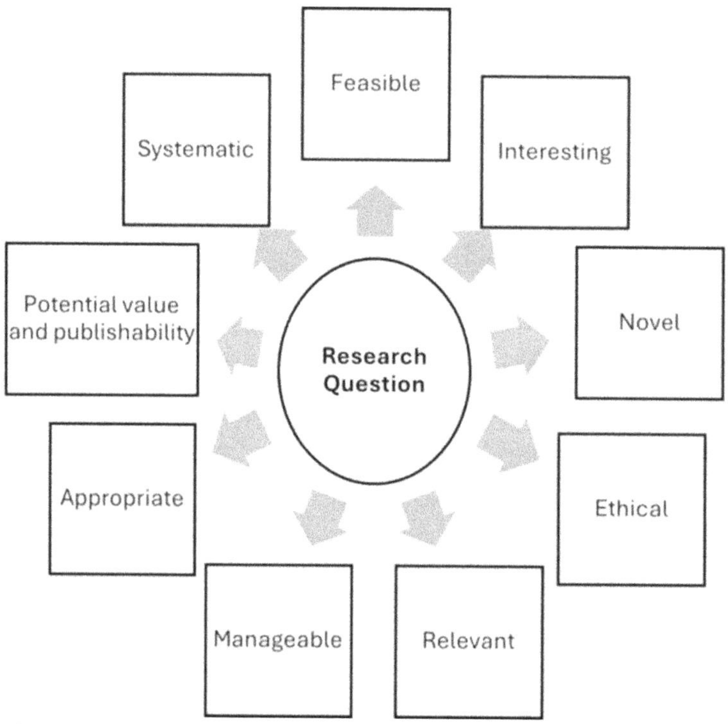

Source: *designed by the authors from (Ratan et al., 2019)*

These research questions will guide the investigation and help to ensure the study makes a meaningful contribution to the field. The next section will outline the methodological approach and data collection strategies that will be employed to answer these inquiries and achieve the overarching research objectives.

3.2.4. Methodology

The methodology is the approach and rationale for how the research will be conducted (Yin, 2018). Generally, in this part of research proposal, the researcher highlights those elements:

3.2.4.1. Participants

Remaining to this fact, the novice researcher must clearly justify the sample when choosing participants for the qualitative research study. It should give reasons concerning the selected criteria that identify and enlist participants such as their expertise, background expectations, or role in the research context (Boddy, 2016).

The novice researcher should also include a description of participants, which includes demographic information, professional experience, or any other relevant features that define the choice of participants for a study (Rowley, 2012). Giving the reasons for the choice of participants and delineating their characteristics helps a researcher to communicate the appropriateness and representativeness of the sample in a way that is important to ensure maximum validity and credibility for the qualitative findings (McNabb, 2017).

3.2.4.2. Methods

In this sub section, novice researcher will write about data collection and how it will be analyzed, he must also explain and justify why he choose this method specifically. We have to notice also that in this sub-section, the research must refer to the literature to define and explain the method, and by the way justify the use of this specific method /design.

Our book chapter focuses on qualitative research in management science, the chapter of Bentalha (2023) is an example of qualitative study in this field. The authors have identified that's the most used methods in these proposals using qualitative methods are literature review with its different types (narrative, systematic, comprehensive, ...etc) ; case studies (single or multiple case studies) ; interviews (structured, semi-structured or even open interviews) ; focus groups ; documents searches ; content analysis ; observations ; and longitudinal analysis. The researcher could use one (or other ones that don't exist in our documents requested) or combine multiple methods to make the study as interesting as possible.

For example, the novice researcher may decide to adopt a case study methodology, in which a single or multiple organizational or social phenomena will be studied in detail. Drawing on the ground-breaking works, for instance, of Yin (2018) and Stake (1995), the researcher may offer a definition of case study approach and detail how it is quite relevant and applicable in answering the set research questions and objectives. Furthermore, the researcher will be in a better position to justify this method on the basis that it has an added advantage of producing rich contextual insight to inform theory development and practical decision-making within management science. There are some useful examples of case study (Bentalha, 2023b; Lamsiah & Bentalha, 2022).

Another case would be the methodology of grounded theory, in which data are collected in a systematic way but analyzed subsequently to induce the generation of new theoretical frameworks. In this case, the researcher can refer to and make use of a landed methodological basis, which was created by Glaser & Strauss (2017) and Charmaz (2014), to explain the procedures that will be used in data collection and analysis in the study, among others, theoretical sampling, constant comparative analysis, and generation of conceptual categories.

Only through the proper understanding of the chosen method and its correct application in management science research is a researcher able to do so using any of the many qualitative approaches available. A researcher must carefully explain the procedure for data collection and the analysis methods applied, anchoring them in the relevant academic literature, to communicate the rigor and validity of the proposed qualitative study to the targeted audience (Sousa, 2014).

3.2.4.3. Research Tools and Procedures

In this subsection, the novice researcher needs to outline in detail the procedures of data analysis and research tools that will be used in this study. This may include developing an interview guide, checklists, coding frameworks, qualitative data analysis software like NVivo, Atlas.ti, or MAX QDA in organizing the information in order to group, label, and make sense of data systematically (Robertson, 2008).

This novice researcher will have to explain in detail how such diverse methods and tools will help in the effective application of qualitative data, its analysis in a strict way, and in drawing significant insights for addressing the core research questions. The manner in which data will be analyzed and reasons for opting for that method help the researcher to validate his or her methodological choice for the proposed qualitative study (Gaus, 2017).

3.2.4.4. Data Analysis

At the data analysis stage, novice researchers must provide a clear description of the analytical methods and techniques that will be applied to the qualitative data collected. This will involve a general description of which data analysis approaches, thematic analysis, content analysis, or discourse analysis the researchers intend to use (Nowell et al., 2017).

Novice researchers will further explain how they would identify themes, patterns, and insights from the data in a systematic way. In addition, the research has to state how they would ensure strength, reliability, and credibility of the qualitative data analysis (Nowell et al., 2017). They could achieve these through multiple coders, member checking, or other quality control measures. This will establish, through a comprehensive elaboration of the strategy of data analysis, that there is considerable methodological expertise and analytical rigor that will be brought to bear on the proposed qualitative study by the researchers (Grossoehme, 2014).

3.2.5. Expected Results

For novice researchers, it becomes very essential to provide a detailed overview of the imperative findings and an anticipated impact for the proposed qualitative research study. They should explain in detail how the research would meaningfully add to and substantially advance a field of study (McNabb, 2017). This explains some theoretical and practical implications of the research and, furthermore, shows how the findings might give rise to novel insights, inform policy decisions, or drive the development of innovative practices (McNabb, 2017).

By elaborating on the importance and impact to society in this study, young researchers are better placed to communicate such value and purpose of the proposed study to their intended audience in the best way, thereby maximizing the opportunity for the qualitative research with significant impact to successfully secure all the funding and resources required for its implementation.

3.2.6. Discussion

The discussion forms a summary of all the major findings and results that one would expect to get from the study (Busse & August, 2021). This section gives the reader an analysis of the implications and importance of research, relating the results to the wider literature in the area, and identifying the limitations and new contributions brought forward by the study.

The discussion is where the findings are related to the original research questions, and further consideration of limitations or any unexpected outcomes that could have been found out during the investigation could find a place. Further, it is the researcher who presents recommendations for research directions which could be conducted based on current work and advance further in the field of management science (Busse & August, 2021).

3.2.7. Conclusion

The conclusion forms an integral part of any research proposal. It is an element that summarizes all critical elements of significance in the study. It therefore serves the target audience, usually researchers and academics, by revealing what the qualitative inquiry it proposes may do, why it is necessary, and what effect it will have in general (McNabb, 2017).

In conclusion, it is necessary for the researcher to state categorically what problem or gap in research the study is to address and what theorical and practical contributions the work is to make. The section should also be used to stress the significance of the research topic, showing that it is relevant to the broader field, particularly when an external fund for the study is called for (Busse & August, 2021).

Novice researchers could integrate in this conclusion can also the possible implications of the study (sub-section 3.2.11), for example, including how the findings might inform policy or best practices by therefore probably extending the underpinning knowledge base. They could also want to make some statements of limitations or challenges (sub-section 3.2.10) arising out of their envisaged research and, through that, show their awareness of such obstacles and how they will mitigate them.

A good conclusion will wrap up the best final impression on the audience and maximize chances of receiving the funding needed for the study.

3.2.8. Ethical Considerations

It's important that qualitative researchers have to fully address some ethical issues. Every researcher has to pledge to care and safeguard the personal information and rights of all participants involved in the study. These include informed consent, confidentiality, respect, and protection of the autonomy and privacy of people participating in the research (Zhang, 2020).

Ethically sensitive handling of all this will enhance trust among the subjects and yield results that are scientifically and morally justifiable. It is in maintaining the highest ethical standards that one can ensure qualitative research does have an impact, producing a kind of trustworthy knowledge of societal benefit (Taquette & Borges da Matta Souza, 2022).

3.2.9. Suggested Time Frame

It could be expressed as a table, wherein the process will be broken down into discrete phases of literature review, research design, data collection, data analysis, and reporting. Then, in all these phases, one could indicate the duration that will be taken up by the work, tasks involved, and the personnel to do them. Basically, this rather structured approach enables the researcher to show he or she has planned everything properly and this is not lacking in project management skills elements that will instill confidence in the organization interested by the project, thereby enhancing the persuasiveness and fundability of the research proposal (McNabb, 2017).

The researcher should be able to include diagrams or visualizations, such as Gantt charts or project management timelines, to draw a vivid outline of the study's timeline and workflow. This could mean that with its inclusion, the target audience could get the whole structure and timeline in no time, thus further enhancing the appeal of the research proposal. In this respect, by coming up with a developed, realistic, and comprehensive work plan, the feasibility or viability of the proposed qualitative research study can be effectively communicated by the researcher. The possibility will, therefore, be enhanced for the proposed qualitative research study to secure the needed funding and support (McNabb, 2017).

3.2.10. Research Limitations

This section provides researchers with the opportunity to, through anticipation, deal in advance with potential drawbacks or expected challenges that might arise during the proposed qualitative research study. This ability to foresee and make proactive plans for such issues helps researchers demonstrate their level of preparedness and premeditation, which consequently improves the overall credibility and viability of the research proposal (Klopper, 2008).

Limitations that can be taken into consideration at this stage include access to targeted research participants, confidentiality and ethical issues of data, avoiding bias in data collection and analysis. The researchers can thus mention specific strategies that will be implemented at each limitation to solve it. These can be the requirement of strong partnerships, robust data management procedures, triangulation techniques, a good discussion of the limitations and boundaries of the study along with future research opportunities (Busse & August, 2021).

The full address of possible limitations of the present work and their mitigation indicates a good grasp of the research process, commitment to rigor and integrity, and capacity to yield quality and high-impact findings that will help further the advancement of management science. This essentially affords a greater degree of persuasiveness and fundability of the research proposal (Lindgreen et al., 2019).

3.2.11. Possible Implications

This is a very important section in trying to convince targeted readership of a qualitative research study to be undertaken through clear expression of the value, significance, and expected contributions that will be made. In the same breath, the researcher can defend the significance of the study by explaining the potential practical insights that should inform organizational decision-making and theoretical improvements that can help grow scholarly understanding of the topic material at hand, hence helping in defending why this study needs to be supported and prioritized (McNabb, 2017).

It can position the qualitative inquiry to lead to impactful results in a way that meaningfully advances the field of study by emphasizing both the value and impact of the expected findings (Busse & August, 2021). This would also increase focus on some of the broader applicability of the research methodologies and concepts, therefore being able to serve as a template for conducting impactful qualitative research across various management science contexts, hence enhancing the overall significance and reach of the given study (Cassell & Bishop, 2019).

3.2.12. References

Every scientific work must have its own references. But the way we structure these references can be different from document to another one. It depends on exigences of publishers.

In general, there are several ways to give references such as APA, Harvard or Vancouver style. For example, for this chapter, our editor "IGI Global" requires writing with the APA style version 7. In this step, we have presented the main elements that must exist in the structure of research proposal. By now, we will explore elements that help researchers to craft a persuasive research proposal.

3.3. Crafting Persuasive Elements for Research Proposal

Crafting a well-structured research proposal was the first step to guarantee the attractiveness of the project. Now, after identifying and analyzing this structure, it is important to discuss the elements that can make the research proposal more persuasive.

By reading documents, categorized as *"guide for research proposals"*, from selected literature, we have extracted these persuasive elements, explicitly cited in these documents, as summarized in table 5.

Table 5. persuasive elements in crafting research proposals from literature

Persuasive elements from literature	Authors
Coherence	(Cavalcanti, 2017)
Transparency	(Cavalcanti, 2017)

continued on following page

Table 5. Continued

Persuasive elements from literature	**Authors**
Reflexivity	(Cavalcanti, 2017)
Contributions	(Beck et al., 2021; Cavalcanti, 2017)
Transferability of findings	(Cavalcanti, 2017)
Simplicity	(Cristina García Vivar et al., 2011)
Risking	(Franzoni & Stephan, 2023)
Scientific collaboration	(Beck et al., 2021)

Source: designed by the authors.

As we can see in table 5, the elements that could make the research proposal persuasive are:

3.3.1. Coherence

Coherence incorporates logical flow and organizational structure of the research proposal. A well-organized proposal gives assurance regarding the rigorousness and suitability of the methods selected for examining the study issue. For instance, if it aims to delve deeper into an investigation of a new or complex issue, a qualitative method is normally more fitting than purely quantitative (McNabb, 2017).

While quantitative methods are useful in the testing of hypotheses and generalizations, qualitative ones let one get closer to subtleties, meanings, and contextual aspects associated with phenomena. In doing so, a coherent and persuasive proposal must choose one of these research methodologies in terms of the nature of the research question underlying the study for the presentation of arguments to prove the validity and reliability of the conclusion (McNabb, 2017).

3.3.2. Transparency

It is in elaborating on the methodology and approaches to data analysis that such transparency can be attained in a research proposal. This will call for explanation of how the selected qualitative methods, sampling strategy, data collection techniques, and analytical approaches all fit into their place in answering the study objective and questions. An overt explanation of the rationality behind these methodological decisions will enable the reviewer to evaluate the coherence and feasibility of the research design as a whole, which would also partly demonstrate the rigor and validity of the proposed study (Suryani & Utami, 2020).

In this context, transparency denotes the act of providing a description that is detailed and logically explained of how the qualitative investigation is to be undertaken with meaningfully valid and dependable results contributing to knowledge accumulation in the respective field (Kapiszewski & Karcher, 2021).

3.3.3. Reflexivity

A reflexive researcher should at least demonstrate an awareness of the presence of one's biases, assumptions, and limitations in interpretation and of the investigation in general. It constitutes the act of self-awareness and commitment to the maintenance of objectivity and credibility of the study by reflections on and transparent communication of all personal and methodological considerations (Dodgson, 2019).

The reflexive approach, therefore, helps the reader understand a researcher's positionality and the context that is, factors that might have shaped the findings so proper judgment is made through a more nuanced critical view of the research. This will enhance the trustworthiness of the qualitative inquiry and, importantly, the persuasiveness of the whole research proposal (Joseph et al., 2021).

3.3.4. Contributions

One of the integral keys to a persuasive qualitative research proposal is to make articulations of potential contributions of the proposed research, which has been well-voiced. The contribution here forms a basis for demonstrating the capability of the researcher to add on what exists in terms of knowledge and argues salient gaps in literature (Beck et al., 2021). Contributions stand to be either practical, with insights allowing organizational decision-making processes, or theoretical in nature to augment scholarly understanding of the subject matter (McNabb, 2017).

The broad aim will be to set the qualitative inquiry as one of the means targeted at developing high-impact findings that make meaningful advances in the field of study. In emphasizing the value and significance of the expected contributions, the researcher can strengthen the resulting case for why their study deserves support and attention from reviewers and funding bodies (Busse & August, 2021).

3.3.5. Transferability of Findings

One of the central strengths of the qualitative research proposal projects is that the conceptual potentials have wider applications and greater application beyond the immediate context (Yadav, 2022). Also, considering the use of an epistemically rigid and totally transparent methodological approach, the findings from the inquiry will be a great value and transferable to similar management science and organizational settings.

Such contextual, deep understanding elicited through the qualitative investigation could unravel nuanced dynamics and themes that would ring a bell with more practitioners and researchers. It is the capacity of concepts and methodologies of the study to be adapted to and applied across diverse contexts that further enhances its significance and impact as a valuable resource in pushing forward qualitative research in the area of study (Patrik Aspers & Ugo Corte, 2019).

3.3.6. Simplicity

The problem or research gap that the researcher intends to address should be clearly stated. In order to give a better overview, the researcher may use graphs or charts or some other type of illustration. These will help the reader understand the scale and importance of the research problem and what contribution the qualitative inquiry being proposed can make (Maxwell, 2012).

It is also a good way that may introduce enthusiasm into the proposal, making very sharp the logic and sound reasoning for conducting the study. Such visual aids will, therefore, help a scientist to elaborate clearly, coherently, and penultimately with persuasiveness about his research proposal (McNabb, 2017).

3.3.7. Risking

This concept is cited by Franzoni & Stephan (2023). In their relevant article entitled *"Uncertainty and risk-taking in science: Meaning, measurement and management in peer review of research proposals"*, the authors propose a framework designed to understand and assess the role that risk and uncertainty can play in scientific research. This framework can also make a case for funding bodies in evaluation and choice of research proposals, encouraging them to fund such studies which may offer significant theoretical or practical contributions in cases that bear a higher degree of risk.

The vision is to help shape a landscape of funding that is more open toward such bold, potentially transformative research, while attaining the purpose of advancing knowledge and practice in management science. Ultimately, funders could help strike a balance between supporting low-risk incremental research and funding high-risk, high-impact studies that really move the boundaries of knowledge.

3.3.8. Scientific Collaboration

Every research proposal immediately takes on more persuasiveness and impact when coupled with scientific collaboration (Beck et al., 2021). These collaborative efforts in research bring together several different investigators with varied areas of expertise, perspectives, and resources to strengthen the overall design of the research project and increase credibility for the proposed inquiry. Such a research proposal, by way of its nurture of a multidisciplinary team composed of scholars and practitioners, reflects a thorough understanding of the problem and has the potential to produce original, holistic solutions (McNabb, 2017).

Synergies created from collaboration can develop more robust methodologies, richer data collection, and deeper data analysis, hence findings that will be more convincing and impactful on the target audience. It is, therefore, the highlighting of this collaborative nature of the proposed study that can be a very strong means of enhancing its overall persuasiveness and fundability.

CONCLUSION

In this study, the authors have presented the importance of elaborating research proposals in management science. They have demonstrated that this document is important for novice researchers that want to apply to universities that want to treat specific topics with the objective to advance management science. They also highlight the importance of research proposal in resolving practical problems of organizations.

The authors conducted a literature review across two databases, from which they selected 14 documents that were analyzed to have three main elements:

The first one extract insights from the literature about qualitative methods used in research proposals identified. These qualitative methods are literature review with its different types (narrative, systematic, comprehensive, …etc) ; case studies (single or multiple case studies) ; interviews (structured, semi-structured or even open interviews) ; focus groups ; documents searches ; content analysis ; observations ; and longitudinal analysis. The authors noticed researchers on the possibility to use one of these methods (or other ones that don't exist in our documents requested) or combine multiple methods to make the study as interesting as possible.

The second element was about the presentation of the different structures identified from literature. Moreover, it provides the main element that should be in a research proposal. The authors have proposed in this way to integrate 12 elements which are: title, abstract, introduction (including literature review /definition, justification of the aim of research, setting research objectives), methodology (including participants, methods, research tools and procedures and data analysis), expected results, discussion and conclusion, ethical considerations, suggested time framework, research limitations, possible implications and references.

The third element of finding is on the issue of how to craft persuasive research proposal in management science. From literature, eight elements have been brought out as:

- Coherence in the research proposal and rigor methodology.
- Transparency that clearly gives explanation for the selected methodology and data analysis techniques.
- Reflexivity, recognizing potential biases and limitations of a study and express self-awareness of a researcher.
- Contributions that show the capability of the researcher in coming up with theoretical and practical solutions towards the advancement of science.
- Transferability, which suggests that the study can be applied elsewhere.
- Simplicity, in which the research problem or gap needs to be clearly articulated, using visual aids such as graphs, charts, or visuals for enhanced comprehension and effective communication of important messages and information.
- Risking, that is a very vital issue in research funding because it might be biased against high-risk research by peer review.
- Scientific collaboration models emphasize the ability of developing more robust methodologies, richer data collection, deeper data analysis to get more convincing findings.

At all, we estimate that this study contributes to advance knowledge about research proposals in management science, especially those using qualitative methods to collect and analyze data for their studies.

CONTRIBUTIONS OF THE STUDY

This chapter has targeted the provision of comprehensive framework for the design and implementation of rigorous qualitative research proposals in the context of management science.

The contributions that this research shall make include:

- An examination of the basic constituent elements that compose a compelling qualitative research proposal, including coherence, transparency, reflexivity, and contributions.

- Practical guidance and best practice about designing qualitative research proposals that demonstrate methodological rigor, strategic alignment with research objectives.
- Insight into how the proposed qualitative inquiry can produce transferable findings and serve as a heuristic device into the conduction of high-impact qualitative studies across diverse management science contexts.

By this study, the authors hope to enhance knowledge and practice in management science. Furthermore, this study could be considered as a detailed framework of the best new practices for the development of impactful qualitative research proposals. It may well turn into a blueprint for scholars and students interested in designing and implementing qualitative studies in other domains within management science.

LIMITATIONS AND FUTURE PERSPECTIVES

One major limitation of this study is that it only relies on literature review to analyze its subject. This study could have been informed by a qualitative investigation involving key stakeholders responsible for financing research projects or an interaction directly with researchers experienced in the submission of successful research proposals.

This indicates a limitation of this study but points to further research where the underlying factors of persuasiveness in research proposals within the management discipline could be explored in-depth through studies using interviews, focus groups, or case studies with qualitative methodologies.

REFERENCES

Ajemba, M. N., Arene, E. C., Ajemba, M. N., & Arene, E. C. (2022). Research gaps for future research and their identification. *World Journal of Advanced Research and Reviews*, 16(1), 1. Advance online publication. 10.30574/wjarr.2022.16.1.1062

Aspers, P., & Corte, U. (2019). *What is Qualitative in Qualitative Research.* https://link.springer.com/article/10.1007/s11133-019-9413-7

Beck, S., Bercovitz, J., Bergenholtz, C., Brasseur, T.-M., D'Este, P., Dorn, A., Doser, M., Dosi, C., Effert, A., Furtuna, R., Goodyear, M., Grimpe, C., Haeussler, C., Hans, F., Heinisch, B., Katona, N., Kleinberger-Pierer, H., Kokshagina, O., LaFlamme, M., & Zyontz, S. (2021). Experimenting with Open Innovation in Science (OIS) practices : A novel approach to co-developing research proposals. *CERN ideaSquare Journal of Experimental Innovation*, 5(2), 28–49. 10.23726/cij.2021.1328

Bentalha, B. (2023a). A Framework for Exploring the Global Performance of Service Supply Chain Management : A Qualitative Study. In *Government Impact on Sustainable and Responsible Supply Chain Management* (pp. 106–129). IGI Global., 10.4018/978-1-6684-9062-4.ch007

Bentalha, B. (2023b). Consumer Perception of Robotic Mobile Fulfillment Systems : A Netnographic Case Study of Amazon. In *Innovation, Strategy, and Transformation Frameworks for the Modern Enterprise* (p. 243-264). IGI Global. 10.4018/979-8-3693-0458-7.ch010

Boddy, C. R. (2016). Sample size for qualitative research. *Qualitative Market Research*, 19(4), 426–432. 10.1108/QMR-06-2016-0053

Buffington, C., Foster, L., Jarmin, R. S., & Ohlmacher, S. (2016). *The Management and Organizational Practices Survey (MOPS) : An Overview* (SSRN Scholarly Paper 2793858). 10.2139/ssrn.2793858

Busse, C., & August, E. (2021). How to Write and Publish a Research Paper for a Peer-Reviewed Journal. *Journal of Cancer Education*, 36(5), 909–913. 10.1007/s13187-020-01751-z32356250

Cassell, C., & Bishop, V. (2019). Qualitative Data Analysis : Exploring Themes, Metaphors and Stories. *European Management Review*, 16(1), 195–207. 10.1111/emre.12176

Cavalcanti, M. F. R. (2017). Guidelines for Qualitative Research in Organization Studies : Controversy and Possibilities. *Administração: Ensino e Pesquisa*, 18(3), 457–488. 10.13058/raep.2017.v18n3.522

Charmaz, K. (2014). *Constructing Grounded Theory*. https://uk.sagepub.com/en-gb/eur/constructing-grounded-theory/book235960

Chigbu, U. E., Atiku, S. O., & Du Plessis, C. C. (2023). The Science of Literature Reviews : Searching, Identifying, Selecting, and Synthesising. *Publications / MDPI*, 11(1), 1. Advance online publication. 10.3390/publications11010002

Cristina García Vivar, Anne McQueen, & Navidad Canga. (2011). *Qualitative research proposal : A model to help novice researchers.*

da Silva, J. J., & Silva Cirani, C. B. (2020). The capability of organizational innovation : Systematic review of literature and research proposals. *Gestão & Produção*, 27(4), e4819. Advance online publication. 10.1590/0104-530x4819-20

Deb, D., Dey, R., & Balas, V. E. (2019). Literature Review and Technical Reading. In Deb, D., Dey, R., & Balas, V. E. (Eds.), *Engineering Research Methodology : A Practical Insight for Researchers* (pp. 9–21). Springer. 10.1007/978-981-13-2947-0_2

Derntl, M. (2014). Basics of research paper writing and publishing. *International Journal of Technology Enhanced Learning*, 6(2), 105–123. 10.1504/IJTEL.2014.066856

Dodgson, J. E. (2019). Reflexivity in Qualitative Research. *Journal of Human Lactation*, 35(2), 220–222. 10.1177/0890334419830990308492 72

Franzoni, C., & Stephan, P. (2023). Last Word on the Discussion regarding the paper : "Uncertainty and risk-taking in science : Meaning, measurement and management in peer review of research proposals". *Research Policy*, 52(3), 104710. Advance online publication. 10.1016/j.respol.2022.104710

Galdas, P. (2017). *Revisiting Bias in Qualitative Research : Reflections on Its Relationship With Funding and Impact.* https://journals.sagepub.com/doi/10.1177/1609406917748992

García-Arca, J., González-Portela Garrido, A. T., & Prado-Prado, J. C. (2024). Deploying Sustainability Through Employee Participation. An Action Research Proposal. *Systemic Practice and Action Research*, 37(1), 19–40. 10.1007/s11213-023-09639-x

Gaus, N. (2017). Selecting research approaches and research designs : A reflective essay. *Qualitative Research Journal*, 17(2), 99–112. 10.1108/QRJ-07-2016-0041

Glaser, B., & Strauss, A. (2017). *Discovery of Grounded Theory : Strategies for Qualitative Research*. Routledge., 10.4324/9780203793206

González-Sánchez, R., Alonso-Muñoz, S., & Medina-Salgado, M. S. (2023). Circularity in waste management : A research proposal to achieve the 2030 Agenda. *Operations Management Research : Advancing Practice Through Research*, 16(3), 1520–1540. 10.1007/s12063-023-00373-0

Grossoehme, D. H. (2014). Overview of Qualitative Research. *Journal of Health Care Chaplaincy*, 20(3), 109–122. 10.1080/08854726.2014.92566024926897

Jabar, M. A., Sidi, F., Selamat, M. H., Ghani, A. A. A., & Ibrahim, H. (2009). An Investigation into Methods and Concepts of Qualitative Research in Information System Research. *Computer and Information Science*, 2(4), 4. Advance online publication. 10.5539/cis.v2n4p47

Jacobs, T., De Maeyer, S., & Beck, M. (1999). Family Formation and the Divisions of Labour in Belgian Couple Households. *Quality & Quantity*, 33(3), 291–304. 10.1023/A:1004600921053

Javadian, G., Dobratz, C., Gupta, A., Gupta, V. K., & Martin, J. A. (2020, juillet). *Qualitative Research in Entrepreneurship Studies : A State-of-Science.* 10.1177/0971355720930564

Joseph, F. I., Earland, J., & Ahmed, M. A. (2021). Experience of Conducting Sensitive Qualitative Research as a Cultural Outsider : Formulation of a Guide for Reflexivity. *International Journal of Qualitative Methods*, 20, 16094069211058616. 10.1177/16094069211058616

Kapiszewski, D., & Karcher, S. (2021). Transparency in Practice in Qualitative Research. *PS, Political Science & Politics*, 54(2), 285–291. 10.1017/S1049096520000955

Kedia, B. L., Nordtvedt, R., & Pérez, L. M. (2002). International business strategies, decision-making theories, and leadership styles : An integrated framework. *Competitiveness Review*, 12(1), 38–52. 10.1108/eb046433

Klopper, H. (2008). The qualitative research proposal. *Curationis*, 31(4), 4. Advance online publication. 10.4102/curationis.v31i4.106219653539

Lamsiah, A., & Bentalha, B. (2022). Economic Diplomacy and Economic Intelligence Amalgamation : The Moroccan-Nigerian Economic Exchange as a Case Study. *Alternatives Managériales Economiques, 4*(4), Article 4. 10.48374/IMIST.PRSM/ame-v4i4.35571

Leo, S. (2019). Pitfalls of tourism graduate students in presenting the ingredients of research proposals. *Journal of Hospitality, Leisure, Sport and Tourism Education*, 24, 178–189. 10.1016/j.jhlste.2019.03.003

Lindgreen, A., Di Benedetto, C. A., Verdich, C., Vanhamme, J., Venkatraman, V., Pattinson, S., Clarke, A. H., & Khan, Z. (2019). How to write really good research funding applications. *Industrial Marketing Management*, 77, 232–239. 10.1016/j.indmarman.2019.02.015

Ma, S. Z. (2023). Considerations in the Area of Management : Directions and Currents. *East African Scholars Journal of Economics. Business and Management*, 6(08), 239–247. 10.36349/easjebm.2023.v06i08.001

Marican, N., Ali, A., Aziz, R., & Yusof, M. (2023). *Designing Value Proposition for Indigenous Inspired Spa Strategic Marketing : A Research Proposal | SpringerLink*. https://link.springer.com/chapter/10.1007/978-3-031-08093-7_22

Maxwell, J. (2012). *Qualitative research design : An interactive approach.*

McNabb, D. E. (2017, septembre). How to Write a Research Proposal. In *Routledge eBooks* (p. 100-108). 10.4324/9781315181158-9

Miles, A. (2017). *Research Methods and Strategies Workshop : A Taxonomy of Research Gaps : Identifying and Defining the Seven Research Gaps.* https://www.researchgate.net/publication/319244623_ARTICLE_Research_Methods_and_Strategies_Workshop_A_Taxonomy_of_Research_Gaps_Identifying_and_Defining_the_Seven_Research_Gaps

Miller, D., & Friesen, P. H. (1984). A Longitudinal Study of the Corporate Life Cycle. *Management Science*, 30(10), 1161–1183. 10.1287/mnsc.30.10.1161

Nanuru, R. F., & Utama, M. D. (2020). Dynamics of proposition and funding of lecturers' research proposal at private universities in higher education service institutes region XII – Indonesia. *International Journal of Scientific and Technology Research*, 9(1), 3085–3090. 10.4108/eai.18-7-2019.2289201

Nasution, N., Hasnati, H., Hasan, M. A., & Yandra, A. (2019). E-Service Application For Online Research Proposal Guidance. *International Journal of Scientific and Technology Research*, 8(7), 71–76.

Nowell, L. S., Norris, J. M., White, D. E., & Moules, N. J. (2017). Thematic Analysis : Striving to Meet the Trustworthiness Criteria. *International Journal of Qualitative Methods*, 16(1), 1609406917733847. 10.1177/1609406917733847

Oddershede, A. M., Watkins, F. J., & Quezada, L. E. (2018). Methodology proposal for appraising health service ICT system implication. *2018 7th International Conference on Computers Communications and Control (ICCCC)*, 58-62. 10.1109/ICCCC.2018.8390437

Pearce, P. F., & Ferguson, L. A. (2017). How to write abstracts for manuscripts, presentations, and grants : Maximizing information in a 30-s sound bite world. *Journal of the American Association of Nurse Practitioners*, 29(8), 452–460. 10.1002/2327-6924.1248628660736

Piperopoulos, P. (2010). Qualitative research in SMEs and entrepreneurship : A literature review of case study research. *International Journal of Economics and Business Research*, 2(6), 494. 10.1504/IJEBR.2010.035701

Ratan, S. K., Anand, T., & Ratan, J. (2019). Formulation of Research Question – Stepwise Approach. *Journal of Indian Association of Pediatric Surgeons*, 24(1), 15. 10.4103/jiaps.JIAPS_76_1830686882

Robertson, S. P. (2008). The Qualitative Research Process as a Journey - Mapping Your Course with Qualitative Research Software. *Qualitative Research Journal*, 8(2), 81–90. 10.3316/QRJ0802081

Rowley, J. (2012). Conducting research interviews. *Management Research Review*, 35(3/4), 260–271. 10.1108/01409171211210154

Silyn-Roberts, H. (2013, novembre). *A Research Proposal*. https://www.sciencedirect.com/science/article/pii/B9780080982854000054

Sousa, D. (2014). Validation in Qualitative Research : General Aspects and Specificities of the Descriptive Phenomenological Method. *Qualitative Research in Psychology*, 11(2), 211–227. 10.1080/14780887.2013.853855

Stake, R. E. (1995). *The art of case study research*. Sage Publications, Inc.

Suryani, A. W., & Utami, H. (2020). Rigour in Qualitative Studies : Are we on track? *Jurnal Akuntansi Dan Keuangan*, 22(2), 2. Advance online publication. 10.9744/jak.22.2.47-58

Suyadi, Husnaini, & Elvina. (2020). Undergraduate students' difficulties in writing a research proposal : A case study. *International Journal of Scientific and Technology Research*, 9(3), 838–843.

Taquette, S. R., & Borges da Matta Souza, L. M. (2022). Ethical Dilemmas in Qualitative Research : A Critical Literature Review. *International Journal of Qualitative Methods*, 21. 10.1177/16094069221078731

Tijan, E., Jović, M., Aksentijević, S., & Pucihar, A. (2021). Digital transformation in the maritime transport sector. *Technological Forecasting and Social Change*, 170, 120879. 10.1016/j.techfore.2021.120879

Tortora, A. M. R., Di Pasquale, V., Riemma, S., & Iannone, R. (2021). Towards a tool for SMEs to Maintenance Management process self-assessment and improvement : Literature overview and research proposal. *Proc. Summer Sch. Francesco Turco*. Proceedings of the Summer School Francesco Turco. https://www.scopus.com/inward/record.uri?eid=2-s2.0-85124673905&partnerID=40&md5=19b7115dddc283a3f4f6db1591876661

Tran, G., & Yeomans, J. S. (2009). Constructions of Strategies Toward Success in English Literacy in Children by Dual-Language Families : A Qualitative Research Proposal. *International Journal of Learning*, 16(1), 295–308. 10.18848/1447-9494/CGP/v16i01/45943

Vasanthakumari, S. (2021, avril). *Writing research proposal*. https://wjarr.com/sites/default/files/WJARR-2021-0138.pdf

Vollero, A., Siano, A., & Della Volpe, M. (2019). A Systems Perspective for Conceptualizing Sustainability in Long-lived Family Businesses. Research Proposals on Risk Taking and Innovativeness. *Systems Research and Behavioral Science*, 36(1), 111–127. 10.1002/sres.2548

Yadav, D. (2022). Criteria for Good Qualitative Research : A Comprehensive Review. *The Asia-Pacific Education Researcher*, 31(6), 679–689. 10.1007/s40299-021-00619-0

Yin, R. K. (2018). *Case Study Research and Applications* (6e éd.). https://uk.sagepub.com/en-gb/eur/case-study-research-and-applications/book250150

Zerbino, P., Stefanini, A., & Aloini, D. (2021). Process Science in Action : A Literature Review on Process Mining in Business Management. *Technological Forecasting and Social Change*, 172, 121021. 10.1016/j.techfore.2021.121021

Zhang, J. J. (2020). Ethical Issues in Research. In Kobayashi, A. (Ed.), *International Encyclopedia of Human Geography* (2nd ed., pp. 299–302). Elsevier. 10.1016/B978-0-08-102295-5.10176-3

Chapter 3
Case Study Methodology in Business Management:
A Systematic Review

Adil Cherkaoui
https://orcid.org/0000-0002-9857-3629
Faculty of Legal, Economic, and Social Sciences, Hassan II University of Casablanca, Morocco

Youness Oudrhough
https://orcid.org/0000-0002-3115-4112
Faculty of Legal, Economic, and Social Sciences, Hassan II University of Casablanca, Morocco

ABSTRACT

The case study is one of the most important forms of qualitative research in the field of organizational management and provides a rich analysis of modern phenomena and their context. This method entails the gathering of data from different sources with a view to critically analyzing management situations. It involves skepticism of typical theories and procedures with the objectives of determining critical variables and researching organizational change and development processes. This chapter is concerned with how case studies can be defined and used in relation to research questions. The method used in the research process is systematic with the R software to cover data gathering, analysis. The chapter is supported by qualitative case studies from the published literature to stress the practical applicability and importance of this research method in expanding the body of knowledge in organizational management.

DOI: 10.4018/979-8-3693-5543-5.ch003

INTRODUCTION

Case study is a basic vehicle of management sciences research, as few other methods provide the depth of analysis of the object under study in its natural context. Compared with quantitative research, case studies make efforts to identify general laws of the phenomenon, but offer in-depth, complex descriptions of a particular event, thus giving the reader or the researcher a wide and contextualized view of organizational behaviours and processes (Dyer & Wilkins, 1991). This approach is highly suitable in the management science as the contexts under which organisations operate are complex and cannot be easily simplified using other traditional approaches. Case studies are defined as empirical research which examines contemporary events and processes in their natural settings especially when there is no clear separation between phenomenon and context (Gerring, 2004). This methodology is particularly helpful when trying to answer the 'how' and 'why' questions about events in which the researcher has little control; this affords greater depth that may not be realized through other methods (Hollweck, 2016). The importance of case studies in the field of management science is that they offer rich descriptions, enable generation of new hypotheses, and allow for the examination of existing hypotheses in context (Eisenhardt, 1989; Gibbert et al., 2008). They provide a rich view of managerial processes and organizational events and can reveal patterns and connections that are not discernable by other means. This inductive approach is important to develop new perspectives and to build the theories regarding management with the help of empirical researches (Barratt et al., 2011).

Developing a case study presupposes defining the research question, choosing cases, and using various types of data to strengthen the reliability of the outcomes (Eisenhardt & Graebner, 2007). While developing the research design, it is crucial to use multiple data sources and types to ensure the triangulation and reliability of the data collected. To meet the rigor and validity, some techniques are used, including chain of evidence, member check, and replication logic in the multiple-case studies (Gummesson, 2007; Hollweck, 2016). Concisely, case studies are instrumental to managing science since they offer comprehensive and context-specific information on phenomena under investigation. They are critical in theory development and have significant managerial prescriptions. This chapter is aimed at providing the discussion and clarification of the case study approach in management science, its history, key methodological aspects, and practical use. The main questions we will address include:

What are the distinctive characteristics of the case study method and how does it support theory development and improvement in management science?

What are the principles of case method and what are the issues that may occur during the development and implementation of the case?

And how case studies help to identify the practical solutions to the actual management problems and improve the organizational decision making?

Such questions will help to structure the discussion concerning the role and applicability of case studies as one of the most effective means of enhancing theoretical knowledge and developing practical solutions in the sphere of management science.

EISENHARDT'S AND YIN'S THEORETICAL FRAMEWORK

The theoretical background of case study approach in business management is based on several principles, which help to explain and understand the considered phenomenon and its context. One of the basic tenets is context awareness, which deems it mandatory to study phenomena in the context where they exist to understand the environment in which they unfold (Burnard, 2023). This enables the analysis of the dynamics of the existence of relations and interactions that define organizational outcomes. Another important principle is the theoretical relevance that prescribes that case studies are based on theoretical concepts which offer a framework for analyzing the data (Eisenhardt et al., 2016). This grounding makes the study not only descriptive but also explanatory, which helps advance theory by refining, elaborating, or questioning theoretical concepts drawn from the literature. Thus, connecting the obtained results with the theory allows for arriving at reasonable conclusions that would contribute to the development of the problems studied in the field of business management. In addition, case study research is characterized by flexibility and adaptability, meaning that the researcher is able to modify the methods and topics investigated as new ideas occur during the study (Hoon, 2013). This iterative process helps to focus on the case more and identify new patterns and relations which were not noticeable at the first glance. Finally, the last principle is the emergent insights where the research findings are inductively developed as the case study research is conducted, and conclusions are made (Siggelkow, 2007) .This approach contributes to the recognition of new theories and concepts that have been derived from the empirical data, which enhances the theoretical basis of business management.

Eisenhardt's and Yin's works have been particularly useful in establishing case study research within the field of management science where they presented different but equally valuable perspectives. Eisenhardt's framework is recognized for its focus on the systematic process of developing theories from case studies, which stresses the cyclical nature of data collection and analysis for creating sound theoretical concepts. Yins on the other hand gives a detailed design-oriented framework that also recognizes validity and reliability and gives step by step procedures for data collection and analysis. Both theorists agree on the application of multiple sources of data and the need to have a well-defined research method to improve the credibility and validity of case study research. However, they diverge in their primary focus: Eisenhardt's approach is more modern and versatile, as it enables the constant elaboration of the theory in the process of research, while Yin's framework is more formalized and stricter, oriented on specific guidelines for the design of the research to ensure the methodological purity. Altogether, both their works provide a realistic and systematic synthesis that can help researchers to manage the multifaceted character of the case study research with methodological rigor. See the table (1).

Table 1. Eisenhardt's and Yin's Case Study Design

	Main steps
Eisenhardt's Framework	**Defining the research question**: Clear and focused research question is always at the center of the study. The question should be rather general and orienting, seeking to know details about the "how" and the "why" of occurrences. **Case selection** Cases are selected through theoretical sampling as opposed to the random sampling. This entails sorting out cases that are most suitable for shedding light on the research question and this may involve using extreme, special or sample cases. **Data Collection** Semi structured interviews, observations and document analysis are used to collect large amounts of qualitative data. The study's validity is boosted with multiple sources since a single source is likely to give a partial view of the phenomenon under investigation. **Data Analysis** Within-case and cross-case analyses are conducted to identify patterns, themes, and relationships. The usage of this process's cyclical structure means that theoretical concepts may be refined throughout the data collection and analysis process. **Theory Building** The theoretical frameworks that are being developed are constantly checked against the data to continuously build sound theories. This process involves a continuous back and forth between theory and data to make sure that the results are not a mere replication of theory but are instead theory informed by data.
Yin's Case Study Design	**Research Questions** Coming up with clear research questions that the case study will seek to answer. These questions should be specific, clear and should help in defining the objectives of the study. **Propositions** Developing propositions that direct attention to what should be examined within the scope of the study. These propositions provide a conceptual framework that guides the data collection and analysis. **Units of Analysis** Stating the case or cases under study. This includes defining what is being researched and the scope of the case. **Logic Linking Data to Propositions** Identifying a logical link between the collected data and the propositions made at the beginning of the research. This helps in making a systematic analysis of the data in reference to the research questions formulated. **Criteria for Interpreting Findings** Setting standards that will be applied when coming up with the study's conclusion. This relates to determining the criteria for measuring the outcomes of the strategies formulated.

Created by the authors.

ENSURING RIGOR AND VALIDITY

When it comes to the management science, the concerns related to the rigor and validity are crucial since the object of the study is inherently more complex and variable than in any physical sciences. Management cases frequently focus on complex problems like leadership, corporate behavior, and organizational culture, and these problems need a solid methodological framework to provide reliable information. It is crucial to choose cases that will be appropriate and typical in relation to the examined phenomenon. The analysis of the organizational context in its depth provides a better understanding of factors that affect the phenomena. It is useful to employ highly systematic research methods to build or extend theories within the field of management science, doing so enables scholars to create sound theoretical developments that are based in realistic organizational environments (Eisenhardt, 1989).

Neglecting the aspects of rigor and validity in the research seems to be a major drawback in generating reliable outcomes in the sphere of case study research, specifically management science. While the former relates to the tight conduct of the research process, the latter has to do with the truthfulness of the conclusions made. Both involve a systemized approach to different methodological issues that are related to the research process such as research design, data collection, data analysis and interpretation. Rigor and validity concerns are construct validity, which helps to identify the right operational measures for the constructs under study, internal validity, which helps to determine causality, external validity to determine the extent to which the findings of the study can be generalized and reliability which shows that the operations of the study can be repeated and will yield the same results (Hollweck, 2016; Yin, 2014). To minimize bias and increase credibility, triangulation is employed, which involves using multiple sources of data, methods, investigators, or theories to increase the reliability of the findings (N. K. Denzin, 2008; Farquhar et al., 2020). Triangulation methods cover data triangulation, methodological triangulation, investigator triangulation and theory triangulation that gives a rich and comprehensive understanding of the research problem.

MATERIALS AND METHODOLOGICAL PROCESSES

Analyzing the case study approach is both complex and crucial, and a systematic review is central to this investigation. Following a systematic method helps in the systematic collection and evaluation of current literature, which outlines the current state of knowledge in a given field (Broadus, 1987). Lastly, the systematic review acts as an effective tool for the academician as well as the practitioner providing a coherent and rigorous synthesis for the evidence based case study approach as well as charting the course for the future direction of this ever-evolving field of case study methodology (Rethlefsen et al., 2021; Sarkis-Onofre et al., 2021). Thus, it was tried to collect the relevant keywords for the formulation of research questions. The table 2 below provides the summary of all the keywords that have been used in the study. By researching in WOS and Scopus databases, 423 articles were reviewed by following the PRISMA guidelines. This systemic analysis outlines the major themes and directions, and evaluates the research methods and findings, as well as revealing the complex interdependencies of business management. Notably, it reveals the research voids and inconsistencies in the literature, enabling the next research to focus on the untapped areas (Donthu et al., 2021; Zupic & Čater, 2015). The PRISMA protocol structured our screening process into four distinct stages: identification, screening, eligibility, and inclusion. To collect all related articles with our research question, we searched for articles published from 1996 to June 2024. In the WOS database, we focused on two categories: Management, Business, and in relation to the other filters, table 2 summarizes the inclusion criteria we have mobilized. For the Scopus database, we considered two categories: Business, management. After a first reading of our database, we observed that a certain number of articles did not address our research question and did not directly touch the objectives of our research which gave us the possibility to exclude 25 in the eligibility phase, and in the final phase 60 duplicates were removed using the R studio software, as well as the bibliometrix library and biblioshiny access.

Table 2 PRISMA process and inclusion, exclusion criteria.

WOS research process

["Case study methodology" OR " Single case" OR " Multiple case" OR " qualitative research" and Business management] and 1996 or 1997 or 2001 or 2002 or 2004 or 2005 or 2006 or 2007 or 2008 or 2009 or 2010 or 2011 or 2012 or 2013 or 2014 or 2015 or 2016 or 2017 or 2018 or 2019 or 2020 or 2021 or 2022 or 2023 or 2024 (Publication Years) and Management (Citation Topics Meso) and Corporate Social Responsibility or Knowledge Management or Internationalization or Intellectual Capital or Entrepreneurship (Citation Topics Micro) and Article or Proceeding Paper or Early Access or Review Article (Document Types)

Scopus research process

(TITLE-ABS-KEY ("Case study methodology" OR " Single case" OR " Multiple case" OR " qualitative research") AND TITLE-ABS-KEY (business AND management)) AND PUBYEAR > 1996 AND PUBYEAR < 2024 AND (LIMIT-TO (DOCTYPE , "Article") OR LIMIT-TO (DOCTYPE , "Conference paper") OR LIMIT-TO (DOCTYPE , "Book chapter") OR LIMIT-TO (DOCTYPE , "Review"))

PRISMA process

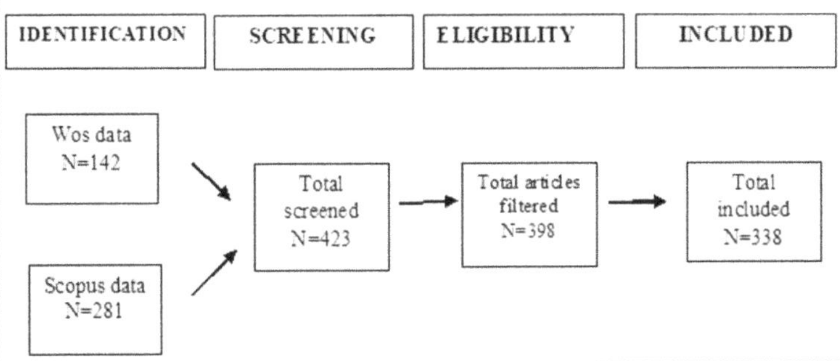

BIBLIOMETRIC RESULTS

In this section, we show bibliometric data such as countries' scientific output, word cloud, trend topics, and co-citation network to present the most productive countries in case study research in business management and the emerging issues and trends (Derviş, 2020). The discovery of the volume of countries' scientific output determines both the performance of countries and the promotion of further cooperation and investments (Aria & Cuccurullo, 2017), in the same order word cloud helps identify the most frequently used terms, which provide an idea of major fields and the possibility of identifying the gaps in the literature; trend topics depict the dynamics of the research interests to predict the future developments and make strategic decisions. The co-citation network is especially valuable as it indicates the positioning of the most often cited publications, which may reveal the key works and theories in the field, thus helping to comprehend the internal structure of knowledge (Aria & Cuccurullo, 2017). Bibliometric analyses are important for the identification of the intellectual background of the field, for the identification of the most influential works and authors, for the evaluation of the research productivity and performance and for providing guidance for the further development of the research and policy (Broadus, 1987; Cobo et al., 2011). They increased awareness and accessibility, helping scholars with the promotion of research outputs to achieve the greatest possible reach. Altogether, these considerations give a systemic view of the research orientations for providing the evidence based enhancements for the business management case studies (Waltman et al., 2010; Zupic & Čater, 2015) . All the outputs were produced in R Studio software and Biblioshiny packages.

COUNTRIES' SCIENTIFIC PRODUCTION

With the help of the map, we can get an idea of the most effective countries in terms of generating case studies in business management. The three countries that provided the most data towards our research question are the United States, China and the United Kingdom respectively. The above countries' leadership in case study production can be attributed to the following scientific and structural factors. The United States is the world leader in this area due to a large investment in higher learning and research as well as due to a strong tradition of academics. High density of business schools and research centers like Harvard Business School and Stanford Graduate School of Business ensures a steady production of high-quality case studies. Not only do these institutions focus on positive empirical research, but they also promote collaborative partnerships with business, which gives a practical slant to these case studies. Also, the funds from both the government and the private entities make it possible to support several research activities, and this prepares a high productivity in the business management research. China is fast developing economically which has seen its academic and research capacities grow thus making it a significant contributor. China has adopted a strategic educational policy to develop the higher education research infrastructure in the country in the past few decades. Chinese universities and research institutes are paying more and more attention to the field of business management, if the economy develops continuously, and the marketization and internationalization processes are proceeding in China. This has increased the overall research productivity and the research interest in case studies which probe the special features of Chinese firms and their management system and in the process has added new dimension to global knowledge base. The United Kingdom has done exceedingly well in this area, due to presence of several highly regarded universities and the culture of employing sound research methods. The nature of research in the UK is methodologically stringent and interdisciplinary which results in the production of good quality case studies. Furthermore, there are historical relations between the UK and other global markets which make it possible to come up with business management research cases that are both credible and relevant in the global context while at the same time following all the research methods properly.

Figure 1. Mapping of the most productive countries

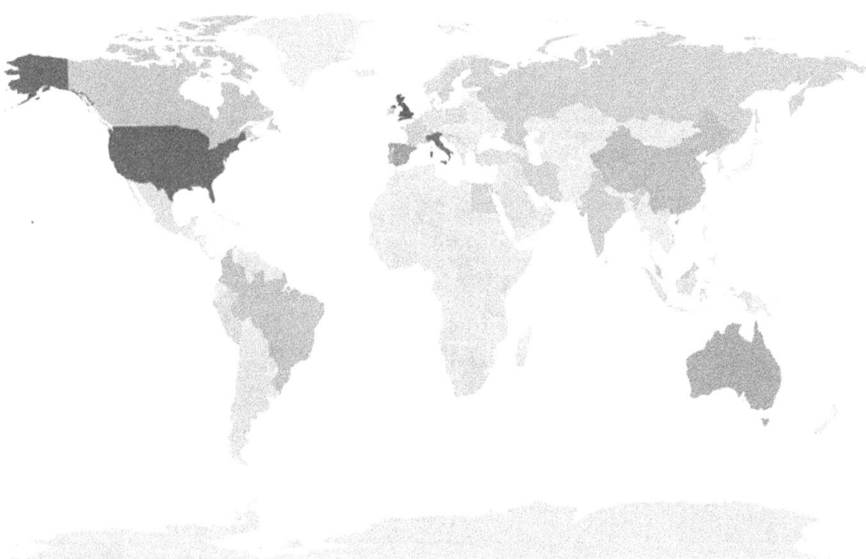

WORD CLOUD

The word cloud output reveals key themes within case study research in business management, highlighting the most attractive keywords: The most common terms were "case study," "corporate social responsibility (CSR)", "sustainability", "sustainable development", "supply chain management" and "innovation " .The frequent use of "case study" indicates a focus on the detailed analysis of specific business issues, stressing the methodological importance of examining particular phenomena in-depth. The focus on CSR and sustainability revealed a strong emphasis on ethical values and environmental concerns, which are essential for comprehending how organizations incorporate social responsibility into their management plans. The appearance of the supply chain management as one of the major keywords also indicates its significance for improving logistics and operations' effectiveness in the context of globalization challenges. Last of all, the focus on innovation shows the constant concern with the ways management encourages creativity and sustains competitive advantage through innovation. Altogether, these keywords indicate that there is a focus on the ethics of research, the need for sustainability, business operations, and new ideas, with the case study approach serving as a source of comprehensive information and best solutions for modern business problems.

Figure 2. Word cloud output

CO-CITATION NETWORK

The co-citation network analysis points out the authors Eisenhardt and Yin as the prominent figures in case study research area within business management context. It is important to note that all these authors are well-established for their methodological work on the case study approach, and their central role in the network illustrates the method's early and ongoing development. The high co-citation frequency of Eisenhardt and Yin reveals that the two sources are used in harmony, giving the researchers a balanced methodological toolbox that is both more rigid and more open. This ability mostly arises from the dual focus on methodological precision and theoretical depth, allowing researchers to develop detailed, context-sensitive understandings of organization processes and phenomena. The co-citation network therefore reveals that Eisenhardt and Yin have played an important part in defining the methodological direction of case study research and pointing researchers to sounder and more useful theoretical and practical developments.

Figure 3. Co-citation network

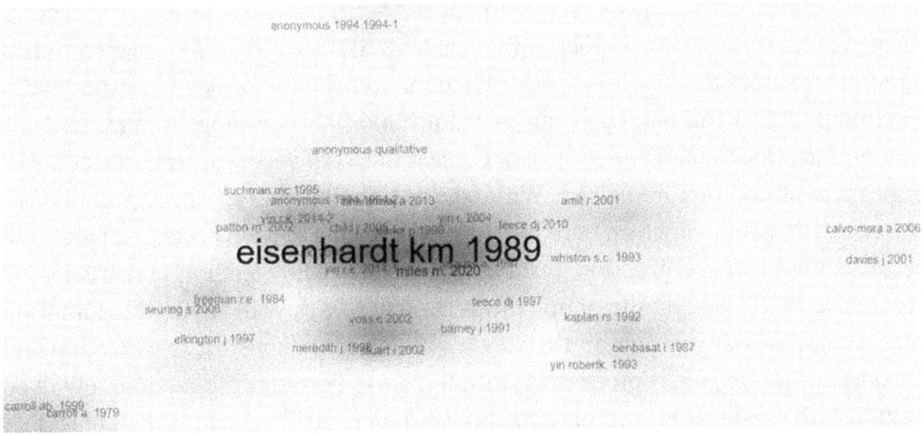

DISCUSSION

Interpretative and Constructivist Paradigms

The case study approach in management science has a strong theoretical foundation on which the analysis of organizational phenomena is based. This section focuses on the theoretical advancements and design philosophies by various authors in the field of management science. First, case studies are useful in that they close the gap between research constructs and their use in real-life situations (Voss et al., 2002). They are specially appropriate in case of exploratory research when the goal is to describe phenomena in context (Flyvbjerg, 2006; Gerring, 2004) . Eisenhardt (1989) stated that developing theories from case study research is a very systematic process of data collection and analysis that produces theories that are rich and grounded in context. Case study research method is rooted in the qualitative research paradigms and deals more with the interpretivism and constructivist approaches (Gummesson, 2007). It enables the researchers to look at the causal mechanisms and to formulate hypotheses for future investigations (Gibbert et al., 2008). Moreover, case studies offer deep insight into multifaceted realities and are thus said to be irreplaceable in management science (Dul & Hak, 2008).

Both the interpretative and constructivist paradigms are basic to qualitative research methods, including case study investigation in management science since they provide different ontological angles and epistemological stances on the creation and accumulation of knowledge, which are helpful when it comes to the analysis of complicated organizational phenomena (Takahashi & Araujo, 2019). The interpretivism paradigm assumes that reality is constructed via social interactions and experiences (Goldkuhl, 2012) and is focused on meanings and interpretations that individuals or groups ascribe to their experience or context of experience; subjectivity, context, and emergent findings are valued. While the positivist paradigm focuses on the objective reality about which an individual has no control, the constructivist paradigm stresses individual construction of reality stating that knowledge is constructed by experiences and through interactions with reality (Mir & Watson, 2000), admitting multiple realities depending on the view of reality and context it is located in and considering the research process as an interaction between the persons involved including the researcher and participants (Avenier, 2010). Within the context of case study research, the constructivist paradigm directs the researcher's attention to how participants interpret their roles and make meaning of their experiences within organizational settings (Creswell, 2008). Both paradigms are pertinent to case study research in management science because of the focus on context-saturated data (Hollweck, 2016), the capability of capturing and analyzing subjective perceptions and interpretation of reality, and the ability to be more responsive and adaptive in the research process (Eisenhardt, 1989). In this way, adopting these paradigms, case study researchers can find out rich, context sensitive insights into the nature of the organizational phenomena and thus better contribute into the development of the theoretical and practical knowledge (Yin, 2014).

Case Study Typologies

Case studies in management science are used for various types of research, and the categorization that serves different purposes accordingly contributes to both theoretical and practical advances. For a start, exploratory case studies are employed to research little understood phenomena or even wholly new kinds of problems to generate theories and the wonted direction for follow-up investigations (Yin, 2014). Moving on to descriptive case studies, these are intended for providing rich descriptions of an observed event, allowing a better understanding of the complexity and context that might have played (Yin, 2014). Exploratory work is more about finding out what there may be to know; in contrast, though, is descriptive work, with the emphasis being on how many and who. This sort of research is often used as a basis for other forms of analysis. Explanatory case studies, on the contrary, attempt to explain theoretical prediction-based models by comparing empirical patterns with theoretically predicted axes of change in order that a truly incremental explanation may follow (Verleye, 2019). Used especially to ask not what is happening, but why it happens by using the detailed context generated from descriptive studies as a basis for cause-and-effect relationships. Moreover, intrinsic case studies are intense investigations focused on cases of inherent interest with one goal directed at exhaustive comprehension well and not toward generalization (Stake, 1995). These are typically selected as the case is very special or its characteristics highlight broader ideas about what one can expect. Instrumental case studies promote specific examples to investigate wider themes or validate theoretical concepts, thereby contributing to theory building (Stake, 1995). Unlike intrinsic studies, which are chosen for their inherent value, instrumental studies employ the case as a means to an end, seeking to broaden understanding or theory. Lastly, longitudinal case studies also may be repeated observations of the same or similar research subjects at different time points and can help to improve their understanding about organization change. They are valuable because longitudinal studies can follow what happens over time, revealing temporal patterns and trajectories.

Grounded Theory and Case Study Research in Management Studies

The combination of grounded theory and case study research in management studies is a strong methodological approach to theory construction and applicable knowledge. Grounded theory, developed by (Glaser & Strauss, 2017), can be described as a systematic research approach that aims at developing theories from the data collected and analyzed during the research process. This approach is particularly suitable for case studies that are mainly exploratory and descriptive in nature since it seeks to gain a richer understanding of organizational realities as they exist within their contexts (Charmaz, 2006). According to Eisenhardt (1989), case studies are detailed research of recent events or phenomena in their natural setting, which helps to collect a vast amount of high-quality qualitative data. When applied to case studies, grounded theory enhances the methodological quality of scholars' works and, therefore, contributes to raising the quality of thinking grade. This is particularly so in the field of management, where it is crucial to appreciate the dynamics of behavior in organizations and the processes that characterize decisions. The integration of these approaches enables the formulation of pragmatic approaches to solving business problems using analytical methods which involves hypothesis making, testing and observation of results using case studies (Denzin & Lincoln, 2000). Also, both grounded theory and case study research focus on the context in which the behaviour and decisions of the organization. The method recognizes context in the construction of social reality, which is consistent with case study approaches to investigating phenomena in their natural environment (Guillemette, 2006). So, integrating these two types of research, investigators can generate theories which are contextualized and pertinent to certain settings in an organization. Moreover, both methodologies pay attention to methodological issues and validity concerning the use of theoretical sampling, saturation, triangulation, pattern matching, and explanation building in the study to ensure credibility and trustworthiness of the results (Charmaz, 2006). This integration of such approaches provides assurance that the research outcomes are credible and reliable. Therefore, it is evident that the integration of grounded theory and case study in the field of management research provides a strong methodological complementarity that strengthens both the theoretical and practical value of the studies. These methodologies when combined enable the researchers to develop comprehensive and informative management studies that go a long way in enhancing the body of knowledge in the field of management as well as in the provision of effective management solutions. Subsequent research should extend this methodological interaction and complement the development of the field of management science with more profound and sensible works.

Illustrative Case Studies

Application of Case studies is considered imperative in management research because the approach provides deep contextual understanding of the phenomenon under study. This approach is especially useful when studying the subtle interactions of HR practices in SMEs and family businesses, corporate governance choices, blockchain, supply chain, corporate decision-making, and branding of tourism destinations. In case study research, the main data collection instruments include face-to-face interviews, observation, and document reviews. There are two primary types of case studies used: tend to be single case studies (SCS) and multiple case studies (MCS). Single case studies center on an analysis of a single object or event, which provides crucial and detailed information about the examined issue. This approach is particularly useful in studying special or sensitive events that could offer much information. On the other hand, multiple case studies deal with several cases so that one can compare cases, similarities and differences in several situations. This method makes the results more general and offers a wider look at the research problem. Despite being subjected to strict validity and reliability standards, single and multiple case studies offer rich context-specific data that is crucial for the scholars, policymakers, and practitioners seeking to promote sustainable and effective business development strategies.

Moroccan Context

Concerning the Moroccan context, the analysed HRM practices are predominantly organically developed and based on the values and culture of managers of the SMEs (Hejaji & Cherkaoui, 2022; Cherkaoui & El Aabar, 2022). An exploratory case study to uncover dynamics shows that, in Moroccan SMEs, socially responsible HR practices focus on employees' health and sustainable development programs sensitive to local context. Besides, it also helps to build up the reputation of the organization and to strengthen stakeholders' trust. In addition, it also highlights the importance of SMEs as agents for change and the promoters of inclusive and sustainable growth in their local areas. These socially responsible practices are best explained through case study research because the methodology provides rich data on how managers incorporate CSR into their plans. Thus, based on the findings of specific cases, researchers can reveal the factors that determine CSR adoption in SMEs and offer useful recommendations to policymakers, practitioners, and academics to support sustainable business development in various contexts.

In Moroccan companies, the human resource management is evolving from the conception of employees as expenses to that of human capital (Cherkaoui & Belgaid, 2021). This shift entails devolving HR policies to the operating managers, who become major players in HR activities. This has shifted a new role of frontline managers in pushing for participative and decentralized HR policies to create more centralized organizational culture (Cherkaoui & Belgaid, 2018). These changes are explained by conducting a qualitative case study that includes interviews and observations with key practitioners who are frontline managers. This methodology focuses on the interpretations and enactments of HR policies by managers and illustrates the application and value of decentralization. The analysis of the case study enhances knowledge of the discussed initiatives and expands the literature about participatory HR practices and organizational flexibility.

In this way, the concept of CSR, its interpretation, and its use in Moroccan SMEs are logically unconscious, as they are based on the values and environment of managers. While CSR has become a more developed practice, it currently applies to both large businesses and small and medium-sized ones. As for the CSR adoption in SMEs, they still reflect its rationality and are closely connected to the managers' values. SMEs are originally grounded in local contexts, resulting in local managers adopting a proximity-based RSE that is a contextualized phenomenon. This goes to show how case study research is useful in studying such phenomena. Exploratory and comprehensive case studies as used by (Cherkaoui & Jallal, 2017) are crucial in understanding the responsible practices in SMEs more so the social ones. Some of these are In-depth interviews, direct observation, document analysis, these approaches allow the research to explore how RSE is implemented and translated in SMEs. These methods offer detailed and contextual understanding of the CSR phenomenon that is not always possible with quantitative methods, they capture the dynamics of CSR in SMEs. Thus, with the help of these examples, it is possible to understand how Moroccan SMEs implement CSR in their activities due to local conditions and managerial attitudes. These case studies help to advance knowledge on CSR in SMEs which in turn can aid in implementing better and culturally suitable CSR strategies and policies.

Other Contexts

When it comes to the subject of family businesses, one of the most valuable research approaches is the case study method that allows for the identification of the multifaceted factors that define the operations of such companies. It is worth noting that most of the literature on family businesses is mainly quantitative approach. Nevertheless, there is an emerging understanding of how interpretivism and critical realism can contribute to the development of this area. These provide more contextualized understanding that is not limited to the results that are obtainable from positivist epistemological frameworks. For example, a study that aimed at analyzing 75 qualitative family business case studies where researchers used a directed coding procedure that is both theoretical and empirical (De Massis & Kotlar, 2014). That way, highly rigorous approaches facilitate the investigation of the specific mechanisms by which family firms operate. The integration of these paradigms can help enhance the state of case study research of family businesses and lay a strong research foundation for developing theory in the field (De Massis & Kotlar, 2014).

Blockchain technology has the potential of revolutionizing the supply chain management by improving on the transparency, security, and efficiency within the different business fields. But the implementation of blockchain encounters some technical and economic issues that should be solved. Some of the technical challenges include designing adequate and efficient block chain frameworks, data privacy and security, and compatibility with current IT systems. Additionally, the adoption of blockchain in the worldwide market is hampered by legal issues, lack of norms and best practices, and organizational disapproval. To address these questions, a multi-case study research design is highly effective (Sivula et al., 2020). This approach entails the use of open-ended questions to obtain descriptive, qualitative information from the targeted respondents. Thus, this method enables researchers to obtain rich and context-specific details regarding applications of blockchain in SCM and challenges faced while implementing it. Case studies will allow to reveal specific ways of applying blockchain in supply chains, as well as to describe recommendations regarding potential challenges and the means of increasing the acceptance of the technology in the global market.

CONCLUSION

Studying the case study approach is a challenging yet critical process, and a systematic approach to the review is key to the study. Based on the PRISMA guidelines, this study selected 423 articles from the WOS and Scopus databases, based on the appropriate keywords. The systematic approach helped gather and assess the existing literature to establish the current state of knowledge and identify gaps for further research. The research trends mapped from the bibliometric analyses using R Studio and Biblioshiny packages revealed the United States, China, and the United Kingdom as the most productive countries because of their well-developed research frameworks and sound educational policies. The word cloud highlighted keywords like ‘case study’, ‘corporate social responsibility’, ‘sustainability’, ‘supply chain management’, ‘innovation’, which pointed at the ethics, sustainability, operations, and innovations. Trend topics illustrated the research dynamics to help with decision-making; the co-citation network using Eisenhardt and Yin as the basis of a balanced methodological tool kit. Therefore, it becomes important to be very rigorous and valid in the conduct of management science case studies to arrive at valid conclusions that can help in the understanding of complex organizational phenomena. All the methodological approaches described here put stress on the proper case selection, detailed examination, and compliance with the strict guidelines to increase the reliability and credibility of the results. Grounded theory when combined with case study research enhances theoretical credibility, which offers detailed, contextually specific data. Paradigm of interpretation and constructivism is the primary of case study research in the management science and provides insightful approaches to the knowledge production. These paradigms enable the investigation of the nature of people’s experiences and the meanings they attach to these experiences. Eisenhardt and Yin’s contributions are two of the most influential publications on the case study research method in the field of management science. Eisenhardt focuses on the recursive data collection and analysis, which is useful for the development of reliable theoretical frameworks from case study research. Yin’s design-oriented framework also lays emphasis on the validity and reliability of the study by laying down elaborate designing protocols. The two frameworks support the use of multiple data sources and the articulation of the research design hence increasing the reliability of case study research. Altogether, their works offer a reasonable mix of both the rigidity and the freedom necessary for carrying out case study investigations successfully and systematically. This way, case study research in the management science can substantially enhance the field by providing detailed and practical insights into various organizational contexts.

LIMITS AND FUTURE PERSPECTIVES

Nevertheless, several limitations of the case study approach should be considered in the management science field despite the significant benefits it provides. One major drawback is that cases selected may not be typical and therefore findings may not be generalized to other phenomena. Furthermore, the interpretive and constructivist paradigms, which are filled with contextual data, may pose questions of subjectivity to the study's objective findings and their ability to be replicated. The use of a large amount of qualitative data can also prevent researchers from developing numerical outcomes and using statistical analysis to make statistically valid conclusions. The suggested lines for future investigation regarding case study research in management science consist of the combination of qualitative and quantitative data alongside with mixed method designs to expand the area and relevance of the outcomes. Thus, there is a possibility that the use of longitudinal studies could provide more insight into the developmental of organizational phenomena. In addition, the use of other emerging technologies like machine learning and big data analysis can help in gaining more reliable data analysis and better prediction of the trends. It is also possible to enlarge the spectrum of data sources and to introduce non-traditional databases and interdisciplinary approaches into the given line of research. Thus, overcoming these limitations and discussing these future prospects will lead the field to the development of more extensive, accurate, and applied knowledge of organizational processes.

REFERENCES

Aria, M., & Cuccurullo, C. (2017). bibliometrix : An R-tool for comprehensive science mapping analysis. *Journal of Informetrics*, 11(4), 959–975. 10.1016/j.joi.2017.08.007

Avenier, M.-J. (2010). Shaping a Constructivist View of Organizational Design Science. *Organization Studies*, 31(9-10), 1229–1255. 10.1177/0170840610374395

Barratt, M., Choi, T. Y., & Li, M. (2011). Qualitative case studies in operations management : Trends, research outcomes, and future research implications. *Journal of Operations Management*, 29(4), 329–342. 10.1016/j.jom.2010.06.002

Bentalha, B. (2023). Consumer Perception of Robotic Mobile Fulfillment Systems: A Netnographic Case Study of Amazon. In Innovation, Strategy, and Transformation Frameworks for the Modern Enterprise (pp. 243-264). IGI Global. 10.4018/979-8-3693-0458-7.ch010

Bentalha, B. (2023). A Framework for Exploring the Global Performance of Service Supply Chain Management: A Qualitative Study. Government Impact on Sustainable and Responsible Supply Chain Management, 106-129.

Bentalha, B., & Hmioui, A. (2021). Smart service supply chain and Just Walk Out technology: a netnographic approach. In *The Proceedings of the International Conference on Smart City Applications* (pp. 223-236). Cham: Springer International Publishing

Broadus, R. N. (1987). Toward a definition of "bibliometrics". *Scientometrics*, 12(5), 373–379. 10.1007/BF02016680

Burnard, K. J. (2023). Developing a robust case study protocol. *Management Research Review*, 47(2), 204–225. 10.1108/MRR-11-2021-0821

Charmaz, K. (2006). Constructing Grounded Theory : A Practical Guide Through Qualitative Analysis. In *Introducing Qualitative Methods* (Vol. 1).

Cherkaoui, A. & Belgaid, A. (2018). Effectivité des pratiques RSE communiquées par les entreprises marocaines au prisme de la gestion de la diversité: enseignements d'une étude de cas. Revue internationale des sciences de l'organisation, 6, 141-181. 10.3917/riso.006.0141

Cherkaoui, A., & Belgaid, A. (2021). Les nouveaux défis des managers de proximité face à la décentralisation de la fonction RH au Maroc – Cas d'un groupe financier panafricain. *Question(s) de management, 32*(2), 43-68. 10.3917/qdm.212.0043

Cherkaoui, A. & El Aabar, A. (2022). Contribution à l'étude d'impact des pratiques RSE des entreprises au Maroc: vers une création de valeur durable et partagée – Cas de l'OCP. Question(s) de management, 39, 13-40. 10.3917/qdm.219.0013

Cherkaoui, A., & Jallal, R. (2017). Spécificités des pratiques RH socialement responsables au contexte des PME au Maroc : Les enseignements d'une étude de cas. *Question(s) de management, 18*(3), 63-77. 10.3917/qdm.173.0063

Cobo, M., López-Herrera, A., Herrera-Viedma, E., & Herrera, F. (2011). Science mapping software tools : Review, analysis, and cooperative study among tools. *Journal of the American Society for Information Science and Technology*, 62(7), 1382–1402. 10.1002/asi.21525

Creswell, J. W. (2008). *Research Design : Qualitative, Quantitative, and Mixed Methods Approaches* (3e édition). SAGE Publications Inc.

De Massis, A., & Kotlar, J. (2014). The case study method in family business research : Guidelines for qualitative scholarship. *Journal of Family Business Strategy*, 5(1), 15–29. 10.1016/j.jfbs.2014.01.007

Denzin, N., & Lincoln, Y. (2000). Grounded Theory : Objectivist and Constructivist Methods. In *Handbook of Qualitative Research* (p. 509-535).

Denzin, N. K. (2008). Strategies of Qualitative Inquiry. *Sage (Atlanta, Ga.).*

Derviş, H. (2020). Bibliometric Analysis using Bibliometrix an R Package. *Journal of Scientometric Research*, 8(3), 156–160. 10.5530/jscires.8.3.32

Donthu, N., Kumar, S., Mukherjee, D., Pandey, N., & Lim, W. M. (2021). How to conduct a bibliometric analysis : An overview and guidelines. *Journal of Business Research*, 133, 285–296. 10.1016/j.jbusres.2021.04.070

Dul, J., & Hak, T. (2008). *Case Study Methodology in Business Research.*

Dyer, W. G., & Wilkins, A. L. (1991). Better Stories, Not Better Constructs, to Generate Better Theory : A Rejoinder to Eisenhardt. *Academy of Management Review*, 16(3), 613–619. 10.2307/258920

Eisenhardt, K. M. (1989). Building Theories from Case Study Research. *Academy of Management Review*, 14(4), 532–550. 10.2307/258557

Eisenhardt, K. M., & Graebner, M. E. (2007). Theory building from cases : Opportunities and challenges. *Academy of Management Journal*, 50(1), 25–32. 10.5465/amj.2007.24160888

Eisenhardt, K. M., Graebner, M. E., & Sonenshein, S. (2016). Grand Challenges and Inductive Methods : Rigor Without Rigor Mortis. *Academy of Management Journal*, 59(4), 1113–1123. 10.5465/amj.2016.4004

Farquhar, J., Michels, N., & Robson, J. (2020). Triangulation in industrial qualitative case study research : Widening the scope. *Industrial Marketing Management*, 87, 160–170. 10.1016/j.indmarman.2020.02.001

Flyvbjerg, B. (2006). Five Misunderstandings About Case-Study Research. *Qualitative Inquiry*, 12(2), 219–245. Advance online publication. 10.1177/1077800405284363

Gerring, J. (2004). What Is a Case Study and What Is It Good for? *The American Political Science Review*, 98(2), 341–354. 10.1017/S0003055404001182

Gibbert, M., Ruigrok, W., & Wicki, B. (2008). What passes as a rigorous case study? *Strategic Management Journal*, 29(13), 1465–1474. 10.1002/smj.722

Glaser, B., & Strauss, A. (2017). *Discovery of Grounded Theory : Strategies for Qualitative Research.* Routledge., 10.4324/9780203793206

Goldkuhl, G. (2012). Pragmatism vs interpretivism in qualitative information systems research. *European Journal of Information Systems*, 21(2), 135–146. 10.1057/ejis.2011.54

Guillemette, F. (2006). L'approche de la Grounded Theory; pour innover? *Recherches Qualitatives*, 26(1), 32–50. 10.7202/1085397ar

Gummesson, E. (2007). Case study research and network theory : Birds of a feather. *Qualitative Research in Organizations and Management*, 2(3), 226–248. 10.1108/17465640710835373

Hejaji, I., & Cherkaoui, A. (2022). L'influence du profil du dirigeant-propriétaire sur les pratiques RH au sein des PME marocaines : Enseignements de deux études de cas. *Revue Management & Innovation*, 6(2), 46–69. 10.3917/rmi.206.0046

Hollweck, T. (2016). Robert K. Yin. (2014). Case Study Research Design and Methods (5th ed.). Thousand Oaks, CA : Sage. 282 pages. *The Canadian Journal of Program Evaluation, 30*. 10.3138/cjpe.30.1.108

Hoon, C. (2013). Meta-Synthesis of Qualitative Case Studies : An Approach to Theory Building. *Organizational Research Methods*, 16(4), 522–556. 10.1177/1094428113484969

Mir, R., & Watson, A. (2000). Strategic Management and the Philosophy of Science : The Case for a Constructivist Methodology. *Strategic Management Journal*, 21(9), 941–953. 10.1002/1097-0266(200009)21:9<941::AID-SMJ141>3.0.CO;2-D

Rethlefsen, M. L., Kirtley, S., Waffenschmidt, S., Ayala, A. P., Moher, D., Page, M. J., Koffel, J. B., Blunt, H., Brigham, T., Chang, S., Clark, J., Conway, A., Couban, R., de Kock, S., Farrah, K., Fehrmann, P., Foster, M., Fowler, S. A., Glanville, J., & Young, S. (2021). PRISMA-S : An extension to the PRISMA Statement for Reporting Literature Searches in Systematic Reviews. *Systematic Reviews*, 10(1), 39. 10.1186/s13643-020-01542-z33499930

Sarkis-Onofre, R., Catalá-López, F., Aromataris, E., & Lockwood, C. (2021). How to properly use the PRISMA Statement. *Systematic Reviews*, 10(1), 117. 10.1186/s13643-021-01671-z33875004

Siggelkow, N. (2007). Persuasion with case studies. *Academy of Management Journal*, 50(1), 20–24. 10.5465/amj.2007.24160882

Sivula, A., Shamsuzzoha, A., & Helo, P. (2020). Requirements for Blockchain Technology in Supply Chain Management : An Exploratory Case Study. *Operations and Supply Chain Management: An International Journal*, 14(1), 39–50. 10.31387/oscm0440284

Stake, R. E. (1995). *The art of case study research*. Sage Publications, Inc.

Sułkowski, Ł. (2009). Interpretative approach in management sciences. *Argumenta Oeconomica*, 23, 127–149.

Takahashi, A. R. W., & Araujo, L. (2019). Case study research : Opening up research opportunities. *RAUSP Management Journal*, 55(1), 100–111. 10.1108/RAUSP-05-2019-0109

Verleye, K. (2019). Designing, writing-up and reviewing case study research : An equifinality perspective. *Journal of Service Management*, 30(5), 549–576. 10.1108/JOSM-08-2019-0257

Voss, C., Tsikriktsis, N., & Frohlich, M. (2002). Case research in operations management. *International Journal of Operations & Production Management*, 22(2), 195–219. 10.1108/01443570210414329

Waltman, L., van Eck, N. J., & Noyons, E. C. M. (2010). A unified approach to mapping and clustering of bibliometric networks. *Journal of Informetrics*, 4(4), 629–635. 10.1016/j.joi.2010.07.002

Yin, R. K. (2014). *Case Study Research*. SAGE Publications.

Zupic, I., & Čater, T. (2015). Bibliometric Methods in Management and Organization. *Organizational Research Methods*, 18(3), 429–472. 10.1177/1094428114562629

Chapter 4

Participant Observation as a Qualitative Research Method in Management Sciences: Interests, Conditions, and Limitations

Azzouz Elhamma
Ibn Tofaîl University, Morocco

ABSTRACT

Certainly, selecting the appropriate research methodology remains a primary concern for all social science researchers. Among the less commonly used methods is participant observation (PO), considered a qualitative research approach. By using ethnography as a theoretical framework and adopting a systematic literature review approach, this chapter aims to discuss the significance and use of PO as a research method in management sciences. The main results indicate that the PO method is frequently considered in the literature as a research method to enhance analysis by providing access to the field and valuable insights into the studied objects. However, it also entails certain challenges and limitations that we have highlighted to ensure its utility, credibility, and optimize research results.

DOI: 10.4018/979-8-3693-5543-5.ch004

1. INTRODUCTION

Certainly, there has been a long-standing debate in the social sciences about the most appropriate methodology method for conducting research and obtaining data. Researchers have always had the choice between quantitative and qualitative methods to study different types of phenomena. On the one hand, according to Jennings (2005, p.219), "quantitative methodology Provides a framework for data collection and analysis that enables researchers to reduce, analyze, and report the collected data numerically and statistically". On the other hand, qualitative methodology "provides a framework for empirical material (data) collection and analysis that enables researchers to gather empirical information in its textual entirety. The empirical materials are not substantively reduced or represented numerically" Jennings (2005, p.219). While quantitative methods offer statistical rigor and the ability to generalize results, qualitative methods provide deeper, richer, and more comprehensive analysis.

For several researchers, the participant observation (PO) is considered as the classic form of qualitative research (Conrad, 2001) and the primary method used by anthropologists doing fieldwork (DeMunck and Sobo, 1998). According to DeWalt and DeWalt (2002), fieldwork involves "active looking, improving memory, informal interviewing, writing detailed field notes, and perhaps most importantly, patience". The PO is a method of social inquiry that "... aims to generate practical and theoretical truths about human life grounded in the realities of daily existence" (Jorgensen 1989, p. 14).

Unfortunately, despite being frequently referred to in the methodology manuals, the PO remains under-engaged by management researchers (Von Krogh, 2020). Vinten (1994, p.30) confirms that "qualitative methodology as a whole, of which participant observation is a sub-species, are approaches which are widely denigrated in the literature. This is unjust, since the method is not without its rigors, and the alternative, complementary and more highly regarded quantitative approach is not without its flaws, some of them fatal".

In this context, by using ethnography as a theoretical framework for PO involves a systematic approach to studying cultures and social phenomena through immersive fieldwork, this research aims to present a literature review on the main benefits and limitations of adopting participant observation in management science research. Unlike narrative reviews, which may not follow a standardized methodological protocol, in this research, we have chosen to adopt the systematic approach of literature reviews. Systematic literature reviews are very important in management science for their ability to provide an objective and comprehensive perspective on the current state of knowledge about a specific topic.

Through this article, we first present the conceptual framework of our research (§2). Secondly, we outline our methodological choices (§3). Thirdly, we highlight the main results of our study (§4). Finally, we provide a brief conclusion (§5).

2. CONCEPTUAL FRAMEWORK

This section is divided into two subsections. The first subsection is devoted to the presentation of the main qualitative research methods (2.1). The second subsection explains how participant observation can serve the research studies, including its definition, the main types, and examples of studies that have used participant observation as a research method in management sciences (2.2).

2.1. The Main Qualitative Research Methods

In management sciences, the most commonly used methods are quantitative methods. These methods are applied in several research areas including management control (Elhamma, 2012; 2013; 2015), financial accounting (Elhamma, 2023; 2024), etc. However, qualitative methods have also begun to be used in recent years (Alla et al., 2022; Bentalha, 2023, 2024; Bentalha & Hmioui, 2021). In management sciences in particular, and in social sciences in general, to understand in depth the studied phenomena, researchers generally use qualitative research methods. "Qualitative research methods can be useful in providing in-depth information to understand target behaviors" (Condon & Coulson, 2017, p. 27). According to Schonfeld and Dreyer (2008), qualitative research methods are "most appropriate in situations in which little is known about a phenomenon or when attempts are being made to generate new theories or revise preexisting theories". Qualitative research is defined "as an iterative process in which improved understanding to the scientific community is achieved by making new significant distinctions resulting from getting closer to the phenomenon studied" (Aspers & Corte 2019, p.139).

The main methods of qualitative research are summarized in the following table (table 1).

Table 1. The Main Qualitative Research Methods: Definitions, Purposes and Data Collection Methods

Types	Definitions	Purposes	Data Collection Methods
Narrative	Clandinin and Connelly (2000, p.20) defined narrative research as "a way of understanding experience [involving] collaboration between researcher and participants, over time, in a place or series of places, and in social interaction with milieus".	Explores situations, scenarios and processes	Interviews and documents
Case study	Yin (2009, p.18) defines case stud research as "an empirical inquiry that investigates a contemporary phenomenon with its real-life context, especially when the boundaries between phenomenon and context are not clearly evident".	Examination of episodic events with focus on answering "how" questions	Interviews, observation, document contents and physical inspections
Grounded theory	Martin and Turner (1986, p.143) define grounded theory as "an inductive, theory discovery methodology that allows the researcher to develop a theoretical account of the general features of a topic while simultaneously grounding the account in empirical observations or data".	Investigates procedures	Interviews and questionnaires
Historical	Borg (1963, p.188) defined historical research as "the systematic and objective location, evaluation, and synthesis of evidence in order to establish facts and draw conclusions about past events".	Description of past events	Interviews, surveys and documents
Phenomenological	Adams and van Manen (2008, p.615) define phenomenological research as, "the study of lived or experiential meaning and attempts to describe and interpret these meanings in the ways that they emerge and are shaped by consciousness, language, our cognitive and non-cognitive sensibilities, and by our preunderstandings and presuppositions'.	Understand or explain experiences	Interviews, surveys and observations
Ethnographic	Brewer (2000, p.9) explains ethnography as: "...a study of people in naturally occurring settings or 'fields' by means of methods which capture their social meanings and ordinary activities, involving the researcher participating directly in the setting, in order to collect data in a systematic manner but without meaning being imposed on them externally".	Describes and interprets social grouping or cultural situation	Interviews, observations and active participation

Source: Author

2.2. Participant Observation: Definitions, Types, and Use in Management Sciences

2.2.1. Definitions and Basic Principles

The participant observation (PO), developed by ethnologists, which has been transposed by sociologists, is a qualitative research methodology that is widely used by practitioner-scholars in many disciplines, including anthropology, sociology, communication studies, health studies, psychology, and management sciences, etc. In PO, the researcher observes the subject of the research study, either by participating directly in the action, as a member of the study population, or as a "pure" observer. In this latter case, the researcher does not participate in the action.

Schensul et al. (1999, p.91) define PO as "the process of learning through exposure to or involvement in the day-to-day or routine activities of participants in the researcher setting". Wilson (2005) defined the PO as follows: "Participant observation is

(i) a qualitative methodology (which can put quantitative data into context and/or give it greater meaning);
(ii) (ii) usually involving long-term research consisting of observing, interacting, conversing, sharing in work and leisure routines, and interviewing members of the community; and
(iii) (iii) that takes an insider's (emic) point of view—often to build etic as well as emic models—to understand routine, everyday life or unique phenomena occurring within the setting(s) occupied by members of the group or community".

According to Conard (2001), by using this methodological method, the researchers "collect observational data by observation in a situation or organization, after gaining access to their research sites". Therefore, the PO is considered a very important instrument for the study of social life over time in its most complex and hidden facets. Among the best examples of studies that have used this research method is that of the anthropologist and the ethnologist Malinowski (1922). In his work "Argonauts of the Western Pacific", Malinowski lived "as a native among the natives for many months together, watching them daily at work and at play, conversing with them in their own tongue and deriving all his information from the surest sources personal observation and statements made to him directly by natives in their own language without the intervention of an interpreter" (Malinowski, 1922, p.45).

2.2.2. The Main Types of PO

Generally, two main types of observations are identified by researchers. First, the participant observation that "involves being in the setting under study as both observer and participant" (Kawulich, 2012). Second, the direct observation that "involves observing without interacting with the objects or people under study in the setting" (Kawulich, 2012). Furthermore, another distinction between covert and overt observations can be highlighted. In Covert observation, the population subject of research are unaware that you are observing them. In overt observation, the participants are aware of being observed for research purposes.

The interaction between these two distinctions allowed Gold (1958) to highlight four positions of researchers in observations in a social context (table 2).

Table 2. The Four Positions of the Researcher in Observations in a Social Context According to Gold (1958)

Types	Definitions
Complete participant	"The researcher who is a member of the group of people under study; he/she is involved in the setting and in studying other group members without their knowledge" (Kawulich, 2012).
Complete observer	"The researcher is able to observe the setting and group under study without participating, but participants are unaware of being observed" (Kawulich, 2012).
Observer as participant	"The researcher who participates in the social setting under study, but is not a group member" (Kawulich, 2012).
Participant as observer	"The researcher who is a group member and who observes other group members with their knowledge" (Kawulich, 2012).

Source: Author

In the same sense, and according to the level of participation, Spradley (1980) suggests five different types of participant observations: non-participating; passive participation ; moderate participation ; active participation; and full participation.

Table 3. The Five Types of PO According to Spradley (1980)

Type of PO	Level of Involvement
Non-participating	No contact with the studied population or field.
Passive participation	The researcher is simply in the bystander role
Moderate participation	The researcher maintains a dual perspective, balancing between insider and outsider roles.
Active participation	The researcher integrates into the group by fully embracing their skills and customs for the sake of complete comprehension.

continued on following page

Table 3. Continued

Type of PO	Level of Involvement
Full participation	The researcher is fully integrated into the study population.

Source: author according to Spradley (1980)

2.2.3. Some Recent Studies in Management Sciences

In this point, we present three recent research studies in management sciences, which used the PO as a research methodology, in three different disciplines: Marketing, management control and audit (cited by Abarar et al. 2014).

Table 4. Some Recent Studies in Management Sciences using Participant Observation as a Methodological Method

Field	Study	Purposes	Methodology of Collecting Data
Marketing	Strategic marketing management (Valentina Kirova, 2011)	The objective of his study was to clarify the contributions of the marketing function to the general strategy of the company in order to understand how the link between the two functions – marketing and strategy – is achieved.	The research focused on a single case study based on PO as a method of access to the field. Its data collection can be summarized in two main phases. • The first was that of PO which took place over a period of one year. During this period, the researcher is immersed in the field and has effectively collected observation data. In this initial stage, the researcher has favored observation of behavior and informal interviews instead of semi-structured interviews. • The second phase of collection was conducted a few months after the end of the PO.

continued on following page

Table 4. Continued

Field	Study	Purposes	Methodology of Collecting Data
Management control	The budgetary slack (Prudence, 2010)	In this study, the author questioned the means of controlling budgetary slack. The aim of the study was to understand how budgetary slack is controlled through the longitudinal study of a case.	The general approach to collecting data was structured around four phases: • A phase of "exploratory" interviews (from May 2004 to February 2005): Data collection is based on semi-structured interviews with management controllers and sales managers. • A "complete participant" observation phase (from March 2004 to January 2005): Data collection is carried out from observation and collection of internal documents. • A "direct participant" observation phase (from January 2005 to September 2005): Data collection is carried out from: observation, collection of internal documents and interviews conducted internally. • A "non-participant" observation phase (from October 2005 to October 2008): Its collection of primary data is based on interviews and various internal documents.
Auditing	Social networks and the competence of the auditor (Razgani, 2011)	In this study, by using the theory of social networks, the author was interested in understanding the link between one of the foundations of audit quality, namely competence, and the audit firm as a social organization.	The general approach to collecting data is based on three phases: • A phase of "exploratory" interviews (from March to October 2005): Data collection is mainly based on exploratory interviews, mainly with financial auditors. • A "full participant" observation phase (from October 2005 to October 2007): Data collection is carried out from observation and the collection of internal documents. • A "direct participant" observation phase (from November 2007 to November 2009): Data collection is carried out from observation and semi-structured interviews.

Source: Author according to Abarar et al. (2014)

3. METHODOLOGICAL FRAMEWORK

In this section, we present our methodological choices for conducting the present study. Firstly, we highlight the chosen approach of literature review (3.1), and then we present our main sources of information (3.2).

3.1. The Selected Type of Review

Our purpose in this study is to present a literature review on the interests and challenges of PO in management sciences. A literature review can be generally defined as a methodical approach of collecting and synthesizing previous research findings. Generally, review studies do not present new data but rather assess existing publication on a specific topic. There are several types of literature reviews, which can be categorized into two main groups: narrative reviews (traditional or non-systematic reviews); and systematic reviews.

Narrative reviews are studies that offer an overview of knowledge on a subject. They are typically a type of literature that is easy to read, accessible, and authored by one or a few experts. They are sometimes referred to as "expert reviews" or "traditional reviews". Generally, this type of review do not require to cover all articles ever published on the studied topic. Narrative reviews do not typically have an initial research question or research strategy, and they are not necessarily supported by a standardized protocol. This lack of systematization in reporting is considered one of their major weaknesses, as these deficiencies can significantly increase the risk of bias in the author's interpretation and conclusions. In contrast, according to Baethge et al. (2019), narrative reviews are considered suitable for synthesizing diverse studies and literature using different methodologies, thereby providing a comprehensive overview of a specific topic. On the other hand, systematic review can be defined as a "research method and process for identifying and critically appraising relevant research, as well as for collecting and analyzing data from said research [...]. The aim of a systematic review is to identify all empirical evidence that fits the pre-specified inclusion criteria to answer a particular research question or hypothesis" (Hannah, 2019). This method is widely used in medical research but not as widely in business research (Hannah, 2019).

Certainly, systematic reviews "have strict requirements for search strategy and selecting articles for inclusion in the review, they are effective in synthesizing what the collection of studies are showing in a particular question and can provide evidence of effect that can inform policy and practice. However, systematic reviews are not always the best strategy" (Hannah, 2019). Given the diversity of research in our topic, we have chosen to use the systematic method. Unlike narrative reviews, which may not follow a standardized methodological protocol, systematic literature reviews adhere to a rigorous methodology aimed at minimizing bias of research.

3.2. Sources of Information

Nowadays, one of the most efficient ways to conduct a literature search study in management science is to utilize several electronic databases. In this study, we have chosen to search in several databases such as Google Scholar (www.scholar.google .com), Scopus (www.scopus.com), ScienceDirect (www.sciencedirect.com), etc.

4. RESULTS AND DISCUSSION

In this section, we present the main results of this study. Firstly, we discuss the benefits and interests of participant observation (4.1), and then we examine the main limitations of this research method (4.2).

4.1. Participant Observation: Interests and Conditions

In management sciences, the PO method allows to researchers to understand the mechanisms of the organization in depth and to explain them clearly. It makes the researcher an observer and actor in the life of the organization. The goal is to provide a new understanding of the objective studied, thus reducing the gap between a simplified discourse (interview) and the reality on the ground (Abarar et al. 2014). The PO "involves the active involvement of a researcher with members of a group or community being studied, typically as an equal-status member" (Fine, 2001). This way of gradually approaching the organization allows the establishment of a climate of trust between the researcher and the studied actors. It contributes to reduce the risks of artificialization of the phenomena observed, by leading their actors to rediscover their spontaneity and be faithful to their behavior and actions. In this case, the researcher can perceive the real meaning of the facts collected by a deeper acquaintance of the object studied and its actors.

By using PO, the researcher is required to use strategies of penetration and especially integration into the study environment. The most important principle is accessibility to others and interaction of the researcher with the studied target (Abarar et al., 2014). The PO promotes a more individualized and deeper relationship with society, the studied culture; it allows us to understand processes, events and relationships in their social context. This established mechanism, due to generally long cohabitation, facilitates all other data collection activities. It creates contact and reduces reactivity and allows us to imbibe the meaning that the actors themselves give to their practices, because it only allows us to understand these "inponderabilia" of which Malinowski wrote in The Argonauts of the Western Pacific: "there is a series of phenomena of great importance which cannot possibly be recorded by questioning or computing documents but have to be observed in their full actuality. Let us call them the inponderabilia of actual life" (Malinowski, 1922).

The prolonged cohabitation with the subject of study allows the researcher to share the lives, activities and feelings of people, in a face-to-face relationship. The participating observer therefore reflects the image of the social process of the life of the observed group. However, the observation must be objective: this is the most important principle. In this sense, Bernard (1865) confirms that "observers, we said, purely and simply note the phenomena before their eyes. They must be anxious only to forearm themselves against errors of observation which might make them incompletely see or poorly define a phenomenon. To this end they use every instrument which may help make their observations more complete. Observers, then, must be photographers of phenomena; their observations must accurately represent nature. We must observe without any preconceived idea; the observer's mind must be passive, that is, must hold its peace; it listens to nature and writes at nature's dictation".

Immersion in the research object, interaction with its environment and the establishment of a climate of trust with the actors observed, offer the researcher the opportunity to have access to and deploy different data collection techniques, especially interviews and observations, note-taking, etc. without this being seen as intrusive or doubtful by the actors (Abarar et al. 2014). In this context, PO is considered as "a methodology that depends upon the establishment of relationships. It relies upon sociability. One essential skill is the ability to recruit informants from the field" (Fine, 2001). According to Fine (2001), "the researcher who lacks the ability to make connections will have difficulty collecting credible research data".

4.2. Participant Observation: Main Limitations

The literature addresses several criticisms of the PO as a qualitative research method. Certainly, the first criticism is that of the lack of objectivity. The researcher cannot avoid any subjectivity and any biased perception of the studied. According to Ratner (2002) cited by Kawulich (2005) "some qualitative researchers believe that one cannot be both objective and subjective, while others believe that the two can coexist, that one's subjectivity can facilitate understanding the world of others".

Among the most important disadvantages of the PO is the low degree of reliability. According to Shrout and Lane (2012, p.643), reliability is defined as "the reproducibility of measurements, and this is the degree to which a measure produces the same values when applied repeatedly to a person or process that has not changed". Given the nature of the method, it would be impossible for another researcher to repeat the same study with the same results. The PO is based especially on the skills and personal characteristics of each researcher.

In addition, one of the most important difficulties is "the Hawthorne Effect" according to which people behave differently when they know that they are being observed (McCarney et al. 2007). According to Coombs and Smith (2003) cited by Oswald et al. (2014, p.53), "the Hawthorne effect is one of the greatest challenges research observers face when gathering data and has long been described as the 'Achilles heel' of participant research".

Finally, we can cite the limitation relating to the difficulty of the observer-researcher to access all facets of the study site and therefore his inability to identify all the possible interactions. In this regard, Jennings (2005) showed that "disadvantages of the method are associated with behaviors and patterns that may be influenced by the presence of the observer/participant, the difficulty in observing all interactions and events if only one observer is present in the business setting or study site, and with the fact that not all facets may be available to the observer because some contexts or interactions may be considered off-limits by the participants".

5. CONCLUSION

Our goal in this article is to highlight the participant observation as a qualitative research method. We have presented its main definitions, types, and some examples of studies that have used this tool as a research method in management sciences (marketing, management control and audit). We also highlighted its main advantages, conditions of use and limitations.

The results of this present research show that the PO involves several risks. These mainly concern the risk of provoking artificial behaviors by the presence of an observer, which would not have taken place or would have occurred differently. Likewise, actors may feel a certain reluctance to share their ideas in front of the researcher. Furthermore, observation could be accompanied by a risk of partial sight of reality. Indeed, the observer, by the obligation to integrate into the structure of the group, risks seeing his observation limited by this rooting and involvement in the life of the company and not being able to take an overall view. Finally, this integration can alter the objectivity of the researcher and modify the way we view the studied phenomenon (Abarar et al. 2014).

Despite the many disadvantages that could be directed towards this technique, the method of PO nevertheless remains the only method which makes it possible to understand the complexity of the studied object by offering the researcher a rich base of information necessary for the understanding and analysis of his or her research object.

Our current research offers several theoretical and methodological contributions. Theoretically, our work contributes to current knowledge on PO by providing a summary of this method, which is considered underexplored in management science. Methodologically, this study provides a targeted overview of the benefits and limitations of PO to guide researchers who wish to adopt this method, ensuring credibility in their results.

However, these results must be interpreted with caution due to the methodological limitations of our current research. Firstly, like other studies based on literature reviews, our work does not cover all articles using PO as a research method in management sciences. Secondly, literature reviews can be prone to researcher bias in the selection and interpretation of studies.

REFERENCES

Abarar, I., Elhamma, A., & Razgani, H. (2014). De l'observation participante comme méthode de recherché. *Revue Marocaine des Sciences de Management*, 3, 72–84.

Adams, C., & van Manen, M. (2008). Phenomenology. In Given, L. M. (Ed.), *The SAGE Encyclopedia of Qualitative Research Methods* (pp. 615–620). Sage Publications.

Alla, L., Bentalha, B., Megzari, G., & Hmioui, A. (2022). Les effets du COVID-19 sur les chaines logistiques: Une analyse qualitative. Reconfigurations du tourisme en contexte de la crise du Covid-19: Quelles démarches pour quelles résiliences au Maroc et ailleurs?, 1(1), 130-153.

Aspers, P., & Corte, U. (2019). What is qualitative in qualitative research. *Qualitative Sociology*, 42(2), 139–160. 10.1007/s11133-019-9413-731105362

Baethge C, Goldbeck-Wood S, Mertens S. (2019). SANRA-a scale for the quality assessment of narrative review articles. Res Integr Peer Rev. 26;4:5. .10.1186/s41073-019-0064-8

Bentalha, B. (2023). Consumer Perception of Robotic Mobile Fulfillment Systems: A Netnographic Case Study of Amazon. In Innovation, Strategy, and Transformation Frameworks for the Modern Enterprise (pp. 243-264). IGI Global. 10.4018/979-8-3693-0458-7.ch010

Bentalha, B. (2024). Employees and Robots in Amazon's Robotic Mobile Fulfillment Systems: A Netnographic Analysis of a Supply Chain Transformation. In Blockchain Applications for Smart Contract Technologies (pp. 188-207). IGI Global. 10.4018/979-8-3693-1511-8.ch009

Bentalha, B., & Hmioui, A. (2021). Smart service supply chain and Just Walk Out technology: a netnographic approach. In *The Proceedings of the International Conference on Smart City Applications* (pp. 223-236). Cham: Springer International Publishing.

Bernard, C. (1865). *An Introduction to the Study of Experimental Medicine* (Green, H. C., Trans.). Dover Publications.

Borg, W. R. (1963). *Educational research: An introduction.* David McKay Company, Inc.

Brewer, J. D. (2000). *Ethnography.* Open University Press.

Chigbu, U. E. (2019). Visually Hypothesising in Scientific Paper Writing:Confirming and Refuting Qualitative Research Hypotheses Using Diagrams. *Publications / MDPI*, 7(22), 22. Advance online publication. 10.3390/publications7010022

Clandinin, D. J., & Connelly, F. M. (2000). *Narrative Inquiry: Experience and Story in Qualitative Research*. Jossey-Bass.

Condon, L.A. & Coulson, N.S. (2017). Designing and Delivering Interventions for Health Behavior Change in Adolescents Using Multitechnology Systems: From Identification of Target Behaviors to Implementation. *Behavior Change Research and Theory Psychological and Technological Perspectives*, 27-45. 10.1016/B978-0-12-802690-8.00001-3

Conrad, P. (2001). Health Research, Qualitative, in International Encyclopedia of the Social &. *Behavioral Science*, 6608–6612. Advance online publication. 10.1016/B0-08-043076-7/03904-8

DeMunck, V. C., & Sobo, E. J. (1998). *Using methods in the field: a practical introduction and casebook*. AltaMira Press.

DeWalt, K. M., & DeWalt, B. R. (2002). *Participant observation: a guide for fieldworkers*. AltaMira Press.

Elhamma, A. (2012). The Relationship between Firm Size, Activity Based Costing and Performance: An Application on Moroccan Enterprises. *Journal of Accounting, Business & Management,* 19(1).

Elhamma, A. (2015). The relationship between budgetary evaluation, firm size and performance. *Journal of Management Development*, 34(8), 973–986. 10.1108/JMD-06-2014-0053

Elhamma, A. (2023). Impact of mandatory IFRS adoption on foreign direct investment: the moderating role of conflict of interest regulation. *Journal of Financial Reporting and Accounting*. 10.1108/JFRA-04-2022-0145

Elhamma, A. (2024). Determinants of national IFRS adoption: Evidence from the Middle East and North Africa region. *International Journal of Accounting. Auditing and Performance Evaluation*, 20(1-2), 69–90. 10.1504/IJAAPE.2024.135535

Elhamma, A., & Zhang, Y. F. (2013). The relationship between activity based costing, business strategy and performance in Moroccan enterprises. *Accounting and Management Information Systems*, 12(1), 22–38.

Fine, G. A. (2001). Participant Observation. in International Encyclopedia of the Social &. *Behavioral Science*, 2001, 11073–11078. 10.1016/B0-08-043076-7/00771-3

Gold, R. L. (1958). Roles in sociological field observations. *Social Forces*, 36(3), 217–223. 10.2307/2573808

Hannah, S. (2019). Literature review as a research methodology: An overview and guidelines. *Journal of Business Research*, 104, 333–339. 10.1016/j.jbusres.2019.07.039

Jennings, G. R. (2005). Business, Social Science Methods Used. Encyclopedia of Social Measurement, 219-230. 10.1016/B0-12-369398-5/00270-X

Jorgensen, D. (1989). The methodology of participant observation. In Pogrebin, M. (Ed.), *Qualitative Approaches to Criminal Justice: Perspectives from the Field* (pp. 12–26). Sage.

Kawulich, B. (2012). *Collecting data through observation, In book: Doing Social Research: A global context. McGraw Hill.*

Kawulich, B. B. (2005). Participant Observation as a Data Collection Method. *Forum Qualitative Social Research*, 6(2), 43. http://nbn-resolving.de/urn:nbn:de:0114-fqs0502430

Kirova, V. (2011). *La direction marketing stratégique comme vecteur de resserrement du lien marketing stratégie dans l'organisation: Etude du cas Laser-Confinoga.* Thèse Sciences de Gestion. Paris: Bordeaux – Montesquieu.

Malinowski, B. (1922). *Argonauts Of The Western Pacific*. Routledge & Kegan Paul.

Martin, P. Y., & Turner, B. A. (1986). Grounded Theory and Organizational Research. *The Journal of Applied Behavioral Science*, 22(2), 141–157. 10.1177/002188638602200207

McCarney, R., Warner, J., Iliffe, S., van Haselen, R., Griffin, M., & Fisher, P. (2007). The Hawthorne Effect: A randomised, controlled trial. *BMC Medical Research Methodology*, 7(1), 30. 10.1186/1471-2288-7-3017608932

Oswald, D., Sherratt, F., & Smmith, S. (2014). Handling the Hawthorne effect: The challenges surrounding a participant observer. *Review of Social Studies*, 1(1), 53–73. 10.21586/ross0000004

Prudence, J. (2010). *La maîtrise du slack budgétaire: Une analyse autour des théories de la déviance et de la criminologie: le cas d'une entreprise de distribution.* Thèse Sciences de Gestion. Paris: Université Paris Dauphine.

Razgani, H. (2011). *Les réseaux sociaux et la compétence de l'auditeur – Cas du cabinet Audit &Co.* Thèse Sciences de Gestion. Paris: Université Paris Dauphine.

Schensul, J., Schensul, S., & Lecompte, M. (1999). *Essential ethnographic methods: observations, interviews, and questionnaires (Book 2 in Ethnographer's Toolkit).* AltaMira Press.

Schonfeld D.J. & Dreyer B.P. (2008). Research Foundations, Methods, and Issues in Developmental-Behavioral Pediatrics, Developmental-Behavioral Pediatrics Evidence and Practice, 47-56. https://doi.org/10.1016/B978-0-323-04025-9.50006-4

Shrout, P. E., & Lane, S. P. (2012). Reliability. In Cooper, H., Camic, P. M., Long, D. L., Panter, A. T., Rindskopf, D., & Sher, K. J. (Eds.), *APA handbook of research methods in psychology, 1. Foundations, planning, measures, and psychometrics* (pp. 643–660). American Psychological Association. 10.1037/13619-034

Spradley, J. P. (1980). Participant Observation. New York: Holt, Rinehart and Winston.

Vinten, G. (1994). Participant Observation: A Model for Organizational Investigation? *Journal of Managerial Psychology*, 9(2), 30–38. 10.1108/02683949410059299

Von Krogh, G. (2020). Building capacity for empirical discovering in management and organization studies. *Academy of Management Discoveries*, 6(2), 159–164. 10.5465/amd.2020.0053

Wilson, T. D. (2005). Participant Observation. Encyclopedia of Social Measurement, 19-24. 10.1016/B0-12-369398-5/00398-4

Yin, R. K. (2009). *Case stud research: design and methods* (4th ed.). Sage Publications.

Chapter 5
Qualimatric Intervention Research:
Interest for Management Sciences Researchers

Abdessamad Dibi
ENSEM, Hassan 2 University, Morocco

Taoufiq El Moussaoui
Faculty of Sciences Dhar El Mehraz, Sidi Mohamed Ben Abdellah University, Morocco

Alaa Eddine El Moussaoui
Higher School of Technology, Hassan 2 University, Morocco

Chakir Loqman
Faculty of Sciences Dhar El Mehraz, Sidi Mohamed Ben Abdellah University, Morocco

ABSTRACT

Management sciences are a complex discipline that simultaneously raises technical issues, stemming from accounting and financial approaches, and human issues, of psychosocial inspiration. Therefore, they require appropriate methodologies to capture their full complexity. It is in this perspective that we can situate the development of qualitative field research methodologies. In this chapter, the authors propose to discuss, through a case of illustration of a field research conducted within a public non-profit organization, the relevance of the qualimetical intervention research method for researchers in management sciences, as a field research, engaged, collaborative, and transformative.

DOI: 10.4018/979-8-3693-5543-5.ch005

INTRODUCTION

Management science is a multifaceted discipline that intricately balances technical challenges, rooted in accounting and financial methodologies, with psychological and social dimensions. This field encompasses the integration of formal elements, such as costing methods, budgeting, and dashboard design, with informal dynamics like organizational learning, actor interactions, and resistance to change within organizations (Cappelletti & Khenniche, 2024). These dual aspects highlight the complexity of management science, where quantitative tools must be effectively applied in environments shaped by human behavior and social structures. Understanding and navigating this interplay is essential for developing robust management strategies that can adapt to both technical demands and the nuanced realities of organizational life.

From this, it can be deduced that management science is a highly complicated subject (Molloy and Barney, 2015), which requires the use of the most appropriate methodologies to understand its full intricacy (Lindsay et al., 2016; Pieter Jansen, 2018). In this context, we may consider the development of qualitative field research methodologies. Such methods are based on the collection of field data founded on the observation of professional practices through the researcher's involvement in the organisation studied (Jönsson & Lukka, 2005; Worley et al., 2015). Here, the researcher starts to explore a phenomenon in depth, to better understand its structure and the context in which it operates. Operationally, field research methods can be classified into the following categories: case studies (Yin 1989), experimental research, action research (Koenig 1997), engineering research (Martinet 1997) or even intervention research (David et al., 2000). Historically, field research in management sciences is fairly well developed, particularly through the studies conducted by Lewin (1946); Argyris (1970); Argyris et al. (1985). The intervention research approach began to emerge, particularly under the impetus of the Scandinavian school (Jönsson and Lukka, 2005; Lukka and Vinari, 2017) and the French school (David et al., 2000; Savall and Zardet, 2014; Cappelletti, 2009, 2010; Buono et al., 2018; Cappelletti and Khenniche, 2024).

Intervention research is a dynamic and practical approach aimed at designing and implementing effective management models and tools directly in the field, often within the context of a transformation project that may range from partially to fully defined (David et al., 2000). This methodology leverages a qualimetric character, which integrates qualitative, quantitative, and financial data, to create a comprehensive representation of the observed phenomena and generate robust theoretical insights (Savall and Zardet, 2014; Cappelletti et al., 2018). The qualimetric intervention research methodology is specifically geared towards managing organizational change, focusing on the identification of dysfunctions within the organization, evaluating their economic impacts through hidden costs, and subsequently implementing targeted actions to improve management quality and operational efficiency. By addressing these hidden costs and converting them into added value, this approach not only seeks to rectify immediate inefficiencies but also aims for long-term, sustainable improvements in the organization's performance. The combination of detailed data analysis and practical intervention makes this methodology particularly effective in producing actionable strategies that align with both the economic goals and the operational realities of the organization (Savall and Zardet, 2014; Cappelletti et al., 2018). This dual focus on theoretical rigor and practical application ensures that the interventions are not only scientifically grounded but also pragmatically viable, facilitating a smoother transition and greater acceptance of changes within the organizational structure. Furthermore, by emphasizing the integration of diverse data types, the qualimetric approach provides a nuanced understanding of organizational dynamics, which is crucial for devising interventions that are both effective and sustainable in the long run.

If, for a long time, the predominant epistemology in management sciences has been founded on a positivist epistemology based on the discovery of a pre-existing truth as well as the validation of knowledge by correspondence between artificial models and this reality, qualimetric intervention research on the other hand is founded on an integrated epistemology of 'generic constructivism' with its three pillars namely cognitive interactivity, contradictory intersubjectivity and generic contingency (Savall and Zardet, 2014; Buono et al, 2018; Cappelletti and Khenniche, 2024). Using an approach rooted in defining research problems based on the real-world context of the organization being studied and validating observations through rigorous, non-complacent discussions between researchers and practitioners, qualimetric intervention research addresses a significant research gap by bridging theory and practice. This methodology's three epistemological pillars link it fundamentally to engaged scholarship (Van de Ven and Johnson, 2006) and closely align it with evidence-based management (EBM) (Pfeffer and Sutton, 2006; Rousseau, 2006; Rousseau and Ten Have, 2022). Traditional management research often faces criticism for being overly theoretical and disconnected from practical application, leading to solutions that may not effectively address the complexities and nuances of real-world organizational challenges. In contrast, qualimetric intervention research emphasizes the importance of developing research questions that emerge directly from organizational issues, ensuring relevance and applicability. This approach fosters a dynamic interplay between researchers and practitioners, enabling them to co-create knowledge through continuous, critical dialogue. By validating observations through non-complacent discussions, the methodology mitigates biases and ensures that the findings are robust and reflective of actual organizational conditions. The consensus-building process not only enhances the credibility of the research outcomes but also facilitates the implementation of findings, as practitioners are more likely to adopt solutions they have co-developed. This integration of qualitative, quantitative, and financial data offers a holistic perspective that captures the multifaceted nature of organizational phenomena, addressing the gap in traditional research methodologies that often rely on single-dimensional data. By anchoring research in the lived experiences of organizations and maintaining a continuous feedback loop with practitioners, qualimetric intervention research not only contributes to the academic body of knowledge but also generates practical insights that drive meaningful organizational change. This approach underscores the necessity of bridging the gap between academia and practice, ensuring that management research not only advances theoretical understanding but also delivers actionable, evidence-based solutions that enhance organizational effectiveness and sustainability.

In this research paper, we propose to discuss the interest of the qualimetric intervention research method for management sciences researchers, based on a committed, collaborative and transformative field research. To this end, we will first present the method (I). We will then analyse the epistemological foundations of the method (II). Next, we will outline its implementation protocol (III). Finally, we will discuss the value of the method through an illustrative case of qualimetric intervention research carried out within a non-market public organisation (IV).

I. FOUNDING PRINCIPLES OF INTERVENTION RESEARCH

We first present the reference framework for intervention research as a field research. We then situate it against the other field research approaches in management sciences.

1.1 Reference Framework for Intervention Research

The concept of intervention is not new to the research community. It has been the subject of numerous studies, notably in social psychology (Lewin, 1951), anthropology (Bastide, 1971; Van Willigen, 1986) and operational research (Hatchuel, 1994; Landry, Banville and Oral, 1996; David, 1998). In management sciences, we can identify numerous research approaches that share the ambition of generating both practical knowledge useful for action and some general theoretical knowledge (Girin, 1990; David et al, 2000; Bartunek and Egri, 2012; Taylor, 2018). These include grounded theory (Glaser and Strauss, 1967), cooperative inquiry (Heron, 1971; Reason and Heron, 1986), action inquiry (Torbert, 1981), action science (Argyris, 1985), participatory action research (Fals-Borda and Rahman, 1991), engineering research (Chanal, Lesca and Martinet, 1997), diagnostic action research

(Koenig, 1997), and intervention research (Hatchuel, 1994; David, 1999) as well as qualimetric intervention research (Savall and Zardet, 2014, Bueno et al, 2018).

From a methodological perspective, the intervention research is structured around five founding principles: increased rationality, incompleteness, scientificity, isonomy and the principle of two levels of interaction (David, 2000).

According to the principle of increased rationality, the researcher must foster a cohesive relationship between empirical facts and the interactions among individuals (Hatchuel, 1994). This principle highlights the necessity of explicitly defining the researcher's role in relation to the organizational players they collaborate with, thereby diverging from the notion of universal rationality. Contrary to merely injecting external expert knowledge, increased rationality emphasizes the importance of creating new relationships and fostering novel understandings within the research context. This approach advocates for a collaborative and integrative method, where the generation of knowledge is co-constructed with the participants, enhancing the relevance and applicability of the interventions. Recent studies underline the efficacy of this approach in producing context-specific solutions and actionable insights, thereby amplifying the impact of intervention research (Johnson et al., 2021).

According to the principle of incompleteness, it is inherently impossible to predetermine the exact trajectory and outcomes of intervention research (Oliva, 2019). This principle recognizes the dynamic and emergent nature of the research process, where the primary aim is to generate new knowledge that can facilitate the evolution and adaptability of organizations (Argyris, 1985). This concept is intrinsic to the action science approach, which also emphasizes the iterative and responsive nature of knowledge creation within complex systems. Recent studies underscores the value of embracing this unpredictability as it allows for adaptive methodologies that respond to evolving circumstances and participant feedback, ultimately contributing to more effective and context-sensitive interventions. Such an approach enhances the relevance and applicability of research findings, ensuring they are not only theoretically robust but also practically viable in real-world settings (Burney et al, 2023).

According to the principle of scientificity, researchers must consistently adopt a critical stance toward the empirical facts they encounter, refraining from assuming the role of "expert of experts." Instead of uncritically accepting established knowledge, researchers are encouraged to scrutinize the conditions under which technical or sociological knowledge is validated within their interventions. This principle aligns with the broader ideal of truth in scientific inquiry and serves as a methodological extension of increased rationality (David, 2000). It mandates that researchers engage in reflexive practice, continually questioning and refining their methods and assumptions. Recent studies emphasize that this critical approach not only enhances the rigor and credibility of intervention research but also fosters more genuine and equitable collaboration with organizational stakeholders (Olmos-Vega et al, 2023). By maintaining this critical perspective, researchers ensure that their findings are robust, contextually relevant, and ethically sound.

According to the principle of isonomy, interventions should establish systems of exchange among stakeholders that uphold both the pursuit of truth and democratic values. This entails fostering an environment where all actors are equally engaged and their perspectives are considered with parity, ensuring that the effort to understand and address issues is applied uniformly across the board (Hatchuel, 1994). This principle is foundational in action research, as articulated by Fals-Borda and Rahman (1991), and in the action science approach championed by Argyris (1985). Recent scholarship underscores the significance of isonomy in creating equitable and participatory research environments that not only democratize knowledge production but also enhance the relevance and applicability of interventions by incorporating diverse viewpoints and expertise (Radaelli et al, 2014). By adhering to the principle of isonomy, researchers can facilitate more inclusive and balanced dialogues, ultimately leading to more effective and sustainable outcomes.

According to the principle of two levels of interaction, intervention research necessitates both an adaptive intervention system and a dynamic, knowledge-based approach. Within the intervention system, the researcher's relationship with other actors evolves as the intervention progresses, reflecting the fluid nature of real-world engagements and allowing for responsive adjustments based on emerging insights (Bradbury, 2015). Concurrently, the knowledge-based approach requires the researcher to foster a new dynamic of knowledge creation by actively comparing and integrating their own expertise with the insights and experiences of field actors. This dual approach not only enhances the depth and relevance of the research but also ensures that the knowledge generated is co-constructed, nuanced, and practically applicable. Such an approach aligns with contemporary understandings of participatory research, where the iterative process of engagement and knowledge exchange strengthens the intervention's impact and sustainability.

1.2 Position of Intervention Research in Field Research

Based on the general framework of field research in management sciences, we can see that the intervention research approach is indeed an extension of the other field research approaches, even though it differs from them in terms of the implementation of concrete changes within the organisations studied. In fact, the intervention research is based on the four characteristics common to other field research. The first is the objective of understanding in depth how the system works, helping it to define possible trajectories of change, choosing one, then implementing it and evaluating the results (Jönsson and Lukka, 2005; Worley et al., 2015). This first characteristic refers fundamentally to evidence-based management (EBM) (Pfeffer and Sutton, 2006; Rousseau, 2006, 2007; Cappelletti, 2009). The second is the production of knowledge through the interaction with the field (Bartunek and Egri, 2012; Taylor, 2018). The third one, is that the validation of knowledge implies a transversal validation at several levels of horizontal, vertical and oblique confrontation. These last two characteristics are fundamentally linked to the engaged scholarship (Van de Ven and Jonhson, 2006). The fourth characteristic is the prescriptive nature of the research. Above and beyond these characteristics which are common to the field research approaches, the intervention research helps to design and implement concrete changes within the organisations studied (Lukka, 2005; Lukka and Vinnari, 2017). It is exactly this transformative aim of the intervention research that clearly distinguishes it from the other field research approaches, in particular that of the grounded theory (Glaser and Strauss, 1967; Strauss and Corbin, 1990; Gurd, 2004).

The following table, based on the work of David (1999), situates the intervention research against the other field research in management sciences (Table 1).

Table 1. Research approaches in management sciences

		Objective	
		Mental construction of reality	**Concrete construction of reality**
Approach	Based on the observation of facts or on the group's work on its own behaviour.	**Observation, with or without participation (I):** developing a descriptive model of the functioning of the system studied.	**Action research (IIb):** Helping to transform the system through its own self-reflection, using a participatory approach.
	Starting from an idealised situation or a transformation concrete project.	**Designing management models and tools (IIa):** developing potential management tools and possible operating models, without any direct link to the field.	**Intervention research (III):** helping to design and implement the appropriate management models and tools in the field, following a more or less fully defined transformation project.

Source: David A. (1999)

According to David (1999), there is two cases: the first one indicates that the research may not use a formalised model (first line: Based on the observation of facts or on the group's work on its own behaviour). The second one shows that the research may use management tools and models as objectives or as exploration means (second line: Starting from an idealised situation or a transformation concrete project) ; The contribution of this research to the construction of reality can be reduced to the production of representations which have no direct consequences for action (left-hand column: mental construction of reality) or, on the contrary, it can represent a direct intervention in the transformation of reality (right-hand column: concrete construction of reality). In this way, each of the four squares represents an ideal type of research approach: observation, with or without participation (square

I), design of management models (square IIa), action research (square IIb) and intervention research (square III).

Following David (1999), Henriot and Khenniche (2019) articulate intervention research as a methodological process encompassing three stages within two distinct phases: the intelligibility phase and the actionability phase. During the intelligibility phase, researchers focus on observing and representing facts, aiming for a deep, comprehensive understanding of the phenomena under study. This involves systematic data collection, analysis, and interpretation to construct accurate representations of the context and issues at hand (Henriot & Khenniche, 2019). Transitioning to the actionability phase, the goal shifts towards developing actionable rules and strategies based on the representations established in the first phase. This phase emphasizes the practical application of insights gained, formulating concrete action plans and interventions that can effectively address the identified problems (Henriot & Khenniche, 2019). The dual-phase structure ensures that intervention research remains both theoretically sound and practically relevant, fostering a seamless integration of understanding and action. Recent references within the scholarly community support this structured approach, highlighting its efficacy in producing both insightful knowledge and tangible outcomes (Reason & Bradbury, 2008; Penuel et al., 2011).

The intervention research differs from other field research not only in the participation of researchers in the development and implementation of action rules, but also in the reflexive loop created by the transformation process. In fact, when we talk about intervention research, the researchers cooperate with the practitioners in the intelligibility phase and in the actionability phase. The two parts also interact in the reflexivity phase by discussing the results of the actions carried out jointly, as well as those of their effects on the organisation with regard to the initial situation (Cappelletti and Khenich, 2024).

Qualimetric intervention research distinguishes itself from other field research methodologies through its dual collaborative and transformative nature (Lewin, 1946; David et al., 2000; Lukka, 2005; Van de Ven & Johnson, 2006; Savall & Zardet, 2014; Buono et al., 2018). Its collaborative aspect involves the development of theoretical knowledge grounded in the rigorous observation of empirical facts, ensuring that this knowledge is directly applicable to enhancing managerial decisions (Pfeffer & Sutton, 2006; Rousseau & Ten Have, 2022). This method fosters a participatory environment where researchers and organizational stakeholders co-create insights, ensuring that the generated knowledge is both academically robust and practically relevant (Cappelletti & Khennich, 2024). The collaborative efforts enhance stakeholder engagement and foster a sense of co-ownership, which is crucial for the successful implementation and sustainability of research outcomes (Van de Ven & Johnson, 2006; Savall & Zardet, 2014). On the transformative side, qualimetric intervention research aims to co-produce knowledge by continuously interacting with the organization's stakeholders and transforming management situations to address specific organizational challenges (Cappelletti & Khennich, 2024; Buono et al., 2018). This transformative process leverages real-time feedback and iterative problem-solving to create adaptive and effective management solutions, making the research deeply embedded in the practical contexts it aims to improve (Lukka, 2005; Van de Ven & Johnson, 2006). By integrating these dual aspects, qualimetric intervention research ensures that the knowledge and interventions are not only theoretically sound but also practically impactful, thereby facilitating meaningful organizational change and development.

2. EPISTEMOLOGICAL PRINCIPLES OF INTERVENTION RESEARCH

We start with a presentation of the epistemological principles of qualimetric intervention research, followed by a discussion of its scientific legitimacy, in particular through the criteria of quality, validity and scientific rigour.

2.1 Qualimetric Intervention Research: An Integrated Research Epistemology

For a long time, the dominant epistemological framework in management sciences was rooted in positivist epistemology, which emphasizes the discovery of pre-existing truths and the validation of knowledge through a direct correspondence between artificial models and empirical reality. This approach typically relies on quantitative methods to test hypotheses and build predictive models, aiming for objectivity and generalizability. However, qualimetric intervention research diverges from this traditional paradigm by embracing a 'generic constructivist' epistemology characterized by three foundational pillars: cognitive interactivity, contradictory intersubjectivity, and generic contingency (Savall & Zardet, 2014; Buono et al., 2018; Cappelletti & Khenniche, 2024). Cognitive interactivity underscores the dynamic and reciprocal interactions between the researcher and participants, advocating for a co-constructive process of knowledge generation where continuous dialogue and mutual influence foster deeper insights and more relevant outcomes (Buono et al., 2018). In others words, it corresponds to the interactive process between the researcher and the actors in the organisation, designed to produce knowledge by successive iterations that respond to the permanent challenge of increasing the significance of the information processed in the scientific work. Contradictory intersubjectivity refers to the interactive process of confronting the subjective viewpoints of the different actors, allowing them, through successive iterations, to construct an acceptable common representation. It acknowledges and leverages the presence of diverse and often conflicting perspectives among stakeholders, considering these differences as opportunities to enrich understanding and produce more nuanced, robust conclusions (Savall & Zardet, 2014). The last pillar which is the generic contingency, constitues an epistemological framework that admits the presence of specificities in the functioning of organisations, but also posits the existence of regularities and invariants that are relatively stable over time, forming a hard core of knowledge and presenting a certain stability and a "certain universality" (Buono et al., 2018). In short, it emphasizes the significance of context-specificity and adaptability, recognizing that effective management interventions must be tailored to the unique conditions and evolving circumstances of each organizational setting (Cappelletti & Khenniche, 2024). Together, these pillars facilitate a more holistic, adaptive, and participatory approach to organizational research, challenging the static and reductionist tendencies of positivist methods. By integrating these elements, qualimetric intervention research not only broadens the scope of knowledge creation but also enhances the practical relevance and transformative potential of its findings, fostering a deeper engagement with the complex realities of organizational life.

These three epistemological pillars also allow qualimetric intervention research to be fundamentally linked to the Engaged Scholarship approach (Van de Ven and Johnson, 2006) and to the Evidence Based Management (EBM) approach (Pfeffer and Sutton 2006; Rousseau 2006; Rousseau and Ten Have, 2022). Indeed, through the principles of 'the co-production of knowledge' between researchers and practitioners and that of 'contradictory arbitration', in other words the validation of observations by non-complacent discussions between researchers and practitioners leading to a consensus on reality, the qualimetric intervention research converges resolutely with 'Engaged Scholarship approach'. By defining research issues rooted in the reality of the organisation being studied, the design of a research project following a collaborative learning approach, the adoption of a longitudinal research project and the alternation of phases of immersion in the field with those of distancing, allowing the accumulated research hypotheses to be regularly revised. The qualimetric intervention research is basically driven by 'Evidence Based Management' (Pfeffer, Sutton, 2006; Henri, 2006; Rousseau and ten Have, 2022).

Research intervention is characterized by an iterative process that alternates between immersive fieldwork and reflective analysis, encapsulating what scholars describe as the immersion and distancing phases. During the immersion phase, researchers dive deeply into the organizational context, closely observing and engaging with real-world practices and stakeholders to gather rich, empirical data (David et al., 2000; Savall & Zardet, 2005). This hands-on involvement enables the researcher to gain a nuanced understanding of the complexities and dynamics within the field. Following this, the distancing phase involves stepping back from the immediate context to critically analyze and interpret the collected data. This reflective analysis is crucial for identifying patterns, generating insights, and validating observations, allowing researchers to formulate and refine theoretical constructs from the lived experiences and practices observed in the field (Cappelletti, 2009; Buono et al., 2018). The key strength of this methodological approach lies in its iterative nature, fostering a continuous dialogue between theory and practice. Each cycle of immersion and distancing not only deepens the understanding of the context but also enhances the robustness and applicability of the emerging theories. This back-and-forth movement ensures that theoretical developments are firmly grounded in empirical reality while also being subjected to rigorous critical scrutiny. The result is a dynamic and evolving body of knowledge that remains relevant and actionable for practitioners (David et al., 2000; Savall & Zardet, 2005; Cappelletti, 2009; Buono et al., 2018).

2.2 Legitimacy Criteria for Qualimetric Intervention Research

The scientific legitimacy of qualimetric intervention research can be discussed when research hypotheses are being formulated, when results are being evaluated and when these results are being communicated.

The research's legitimacy, in terms of quality, validity and scientific rigour, is prepared and discussed during the phase of formulating hypotheses using field observations (Van de Ven and Johnson, 2006; Buono et al., 2018). Thus, the criterion of scientific quality can be assessed through the soundness of the hypotheses (positioning of the problem and the hypotheses); the research project (quality of the proposed solutions); the implementation of the research (substantial results and monitoring); and the evaluation of the research (criticism and adjustment of tested solutions). The criterion of scientific validity can be assessed from the three angles of internal, external and ecological validity (Lee, 1999). Internal validity corresponds to the judgement that a research process is suitable for testing the proposed hypotheses. External validity indicates that the results of the research can be generalised. Ecological validity refers to the fact that research integrates and describes the characteristics of the environment studied. For its part, the criterion of scientific rigour depends on two fundamental variables: the research's quality and its scientific interest. Both are assessed through the expression of the problem raised by the research and the hypotheses formulated to answer it.

The results obtained during the intervention research are subjected to careful evaluation and monitoring in the distancing phase, an essential step to balance the practitioner's perspective that might dominate during the immersion phases. The discussion initiated in the immersion phase is augmented with additional arguments and critical analysis, ensuring a holistic view. This approach counteracts the potential bias introduced by the practitioner's immediate, context-driven insights.

Intervention research fundamentally relies on the cooperation between researchers and practitioners in both the actionability phase, where practical interventions are designed and implemented, and the reflexivity phase, where findings are critically discussed and analyzed (Argyris & Schön, 1989; Buono et al., 2018). For the intervention researcher, this means meticulously measuring the degree of validation of the proposed hypotheses, distinguishing between contingent insights specific to the particular context and more generic theories that have broader applicability (Reason & Bradbury, 2008; Coghlan & Brannick, 2014). This critical evaluation process enables researchers to construct more robust and comprehensive responses to the studied problems, ensuring that the solutions are not only empirically grounded but also theoretically sound.

Finally, communicating the research results to the academic community serves as a crucial step in avoiding external controversy over the findings and in finalizing their validation. The scientific validity of intervention research is significantly enhanced by adhering to a rigorous, structured process that includes formulating the research problem, alternating between phases of immersion and distancing in the field, and effectively disseminating the results (Popper, 1959). This comprehensive approach ensures that the research is thoroughly examined and critiqued by peers, which helps in identifying any potential biases or methodological flaws. By fostering a transparent and open dialogue with the academic community, researchers can refine their theories and methodologies, contributing to the overall reliability and robustness of their findings (Popper, 1959). Furthermore, public dissemination of research results encourages broader scrutiny and replication, which are essential components of the scientific method, thereby bolstering the credibility and generalizability of the research (Wacheux, 1996). In intervention research, where practical applicability is paramount, this dual focus on rigorous methodology and transparent communication ensures that the knowledge produced is not only valid but also actionable and beneficial for both academic and practitioner audiences (Maxwell, 1996). Thus, the iterative process of problem formulation, field immersion and distancing, and result dissemination creates a robust framework for producing scientifically and practically valuable insights.

3. PROTOCOL FOR IMPLEMENTING THE QUALIMETRIC INTERVENTION RESEARCH

We present the various stages in the implementation of intervention research, and then discuss its interest for management sciences researchers as a field research approach.

3.1 Stages Involving the Implementation of Qualimetric Intervention Research

Intervention research is a unique field-based methodology designed to create and implement suitable management models and tools, often grounded in a transformation project that varies in its level of definition (David et al., 2000). The distinctive qualimetric character of intervention research, as established by H. Savall and his team at the ISEOR (Institute of Socioeconomics of Enterprises and Organizations), integrates qualitative, quantitative, and financial data. This multidimensional approach enhances the representation of observed phenomena and strengthens the theoretical statements produced (Savall & Zardet, 2014; Cappelletti et al., 2018). By leveraging this combination of data types, qualimetric intervention research offers a comprehensive view that captures the complexity of organizational dynamics, making the findings more nuanced and actionable. Recent studies have shown that this integrated approach leads to more effective managerial interventions by providing a holistic understanding that combines empirical rigor with practical relevance (Nielsen & Miraglia, 2017; Bracci et al., 2019). For instance, incorporating financial metrics alongside qualitative insights allows researchers and practitioners to assess both the economic and human impact of management practices, thereby facilitating more balanced decision-making processes (Buono et al., 2018). Furthermore, the collaborative nature of this research methodology encourages active engagement from organizational stakeholders, ensuring that the resulting management tools and models are not only theoretically sound but also practically viable (Coghlan & Shani, 2021). This participatory element is crucial for the successful implementation and sustainability of changes within organizations, as it fosters a sense of ownership and alignment among all involved parties. By combining diverse data sources and engaging stakeholders in the research process, the qualimetric approach in intervention research exemplifies a robust and adaptive framework that addresses the multifaceted challenges of contemporary organizational life, aligning well with modern calls for more integrated and context-sensitive research practices (Bradbury, 2022; Cappelletti et al., 2018).

Based on the principles of Engaged Scholarship (Van de Ven and Johnson, 2006) and Evidence-Based Management (EBM) (Pfeffer and Sutton, 2006; Rousseau, 2006, 2007; Ten Have, 2022), the methodological framework for intervention research has evolved to provide researchers with five essential directions for robust inquiry. These include formulating research problems that are deeply rooted in practical realities, designing research projects through a collaborative learning approach to ensure diverse stakeholder engagement, and planning long-term research endeavours that allow for sustained and in-depth analysis. Additionally, the framework emphasizes the importance of utilizing a variety of theoretical perspectives and research methods to enrich the study and foster comprehensive understanding. Crucially, it advocates for the regular revision of accumulated research hypotheses to stay aligned with new evidence and theoretical advancements (Cappelletti, 2010; Cappelletti and Khenich, 2024). Recent studies corroborate these strategies, suggesting that such a dynamic and collaborative approach not only enhances the relevance and applicability of research findings but also bridges the gap between academic inquiry and practical application (Brown & Duguid, 2022; Choudhury et al., 2023).

Using the methodology adopted by Savall and Zardet (2004, 2021), this research protocol consists of five successive stages: conducting the diagnosis, forming the project group, implementing the project, evaluating the actions and finally disseminating the management tools.

During the diagnosis stage, the researcher undertakes a comprehensive evaluation of the organization's hidden cost indicators—absenteeism, workplace accidents, staff turnover, quality defects, and direct productivity gaps—by conducting interviews with the organization's managers and operational staff. The diagnostic process is structured into three key modules that collectively provide a thorough analysis of dysfunctions and hidden costs within the micro-space. The first, a qualitative module, involves identifying dysfunctions through in-depth qualitative interviews with managers and at least 30% of the operational staff within the diagnosed micro-space. This allows for a nuanced understanding of the underlying issues from those directly involved or impacted. The second module is financial in nature, in which hidden costs are quantified using a robust general calculation model developed from new quantitative and financial interviews conducted with the managers. This step is crucial for translating qualitative insights into concrete financial terms, thereby providing a tangible measure of the economic impact of identified dysfunctions. Finally, the third module, known as the "mirror effect," involves presenting the diagnostic results to the micro-space members in a session led by the management controller. During this session, the collected data and findings are discussed and reflected back to the operational staff and managers, allowing them to see a 'mirror image' of the organization's inefficiencies and hidden costs. This collaborative validation process ensures that the dysfunctions and hidden costs are acknowledged and accepted by consensus, fostering a shared understanding and commitment to addressing the identified issues. This structured, iterative approach not only ensures a comprehensive and accurate diagnosis but also engages the organizational members in a meaningful dialogue aimed at fostering organizational improvement.

The second stage of the process involves forming project groups tasked with developing solutions to mitigate the identified dysfunctions and analyzing their socio-economic consequences. These groups play a crucial role in translating diagnostic insights into actionable strategies aimed at organizational improvement. Each project group is managed by a project leader, typically the hierarchical manager of the diagnosed micro-space, and is supported by the involved researcher. Structurally, the project group is divided into two distinct bodies: a small group and a plenary group. The small group consists of the project leader and a representative of the executive committee, who are kept apprised of the plenary group's progress and ensure that proposed solutions align with the organization's overall strategy. The plenary group, on the other hand, includes the project leader and the members of the diagnosed micro-space. This larger group is responsible for generating specific solutions and thoroughly studying their socio-economic implications to ensure practicality and efficacy. By having these dual bodies within the project group, the organization can effectively balance strategic oversight with detailed, ground-level input and analysis. The small group's strategic perspective ensures that proposed solutions are viable within the broader organizational context, while the plenary group's more inclusive composition ensures that the solutions are grounded in practical realities and have the buy-in of those directly affected. This collaborative approach fosters a comprehensive evaluation of potential solutions, enhancing the likelihood of successful implementation and leading to sustainable organizational improvements.

The third stage of the project involves its implementation, a critical phase that is divided into two essential sub-phases: planning and budgeting, and the effective execution of actions. During the planning and budgeting sub-phase, detailed plans are developed outlining the various actions necessary to address the identified dysfunctions. This includes allocating resources, defining timelines, and setting specific goals to guide the project. Budgeting is a key component, as it ensures that adequate financial resources are allocated to each planned action, taking into account both direct and indirect costs that might arise during implementation. Once the planning and budgeting are finalized and approved, the project moves into the effective implementation phase, where the planned actions are executed according to the established timelines and resource allocations. This phase is characterized by ongoing monitoring and guidance through a permanent audit of the process. The audit serves multiple functions: it tracks the progress of each action against the planned objectives, identifies any deviations or emerging issues, and facilitates timely interventions to keep the project on track. Regular audits also ensure compliance with both internal and external standards and help in maintaining the alignment of the project with the overall strategic goals of the organization. Furthermore, continuous feedback loops are established, allowing for adaptive changes in response to real-time data and evolving circumstances. This rigorous approach not only promotes transparency and accountability but also enhances the efficiency and effectiveness of the project implementation, ultimately leading to successful outcomes and sustainable improvements within the organization.

The fourth stage, evaluation, is crucial for assessing the effectiveness of the implemented solutions in reducing dysfunctions and hidden costs. Once the solutions are in place, the researcher conducts a thorough evaluation to measure their impact. This process involves compiling a comparative assessment of both the economic and social performance resulting from the actions taken. To achieve a comprehensive evaluation, various methods are employed. Qualitative methods include conducting in-depth interviews with various actors in the field, such as managers, operational staff, and other stakeholders involved in the project. These interviews provide nuanced insights into the perceived changes and improvements resulting from the implementation. Quantitative methods focus on analyzing the evolution of the five hidden cost indicators: absenteeism, workplace accidents, staff turnover, quality defects, and direct productivity gaps. By comparing pre-implementation and post-implementation data, the researcher can quantify the extent to which these indicators have improved, thereby providing a clear measure of the solutions' effectiveness. This dual approach—combining qualitative insights with quantitative data—ensures a robust and multi-faceted evaluation. The findings from this evaluation are critical not only for validating the success of the implemented actions but also for identifying areas that may require further attention or adjustment. Moreover, these insights contribute to the broader body of knowledge, informing future projects and organizational strategies aimed at reducing hidden costs and improving overall efficiency and effectiveness. This iterative process of implementation followed by rigorous evaluation underscores the importance of continuous improvement and adaptive learning within organizations, helping them to sustain positive changes and achieve long-term success.

The fifth and final stage of the project involves disseminating management tools to all the organization's managers through structured training sessions organized by the researcher. This training is vital for equipping managers with six specific tools, each designed to address particular organizational needs and enhance overall management efficiency. These tools are introduced through comprehensive sessions that include theoretical instruction and practical demonstrations, ensuring that managers grasp both the conceptual framework and the practical applications of each tool. Interactive elements such as workshops, case studies, and role-playing exercises facilitate experiential learning, allowing managers to practice using the tools in a controlled environment before integrating them into their daily operations. The researcher also provides follow-up support, including user manuals and access to a helpdesk, to assist managers in overcoming initial implementation challenges. This approach ensures not only the introduction but also the institutionalization of these tools within the organizational culture. By systematically training managers and embedding these tools into regular managerial processes, the organization promotes sustained improvements and fosters an environment of continuous operational excellence and adaptive learning. A summary is presented in the table below (Table 2).

Table 2. Management tools used in intervention research

Requirements	Tools	Contents	Intake
Need medium and long term vision	**The internal-external strategic action plan (INTEXSAP)**	Tool for formalising the strategy, which is aimed at external targets (the product-market couple, customers, suppliers, etc.) as well as internal targets (tangible and intangible investments, the adequacy of training/employment, etc.)	Developed over five years, based on the internal forces of the organization and the opportunities offered by the environment, INTEXSAP is a real tool for steering the establishment project
Need for prioritisation of actions	**The priority action plans (PAP)**	Tool corresponding to the individual implementation of the strategic objectives and actions in the INTEXSAP	Allows a multiplication of the INTEXSAP strategic actions in priority actions to be carried out by the teams to achieve the strategic objectives.
Need to emulate actors	**The activity contract negotiable periodically (ACNP)**	Tool corresponding to the individual implementation of strategic objectives and actions in the PAP	Evaluated on the basis of qualitative, quantitative and financial indicators, it allows managers to negotiate with their subordinates the objectives to be achieved, ensuring their feasibility over a six-month period.
Need to have performance monitoring indicators on hand	**Balanced ScoreCard (BSC)**	Groups together the indicators needed to manage the organisation's activities	Allows management to monitor and evaluate the implementation of the internal and external strategic action plan in a qualitative, quantitative and financial manner
Need for skills matching	**The skills grid (SG)**	Representing one of the tools of the BSC, it allows to visualize the versatility and vulnerabilities in terms of skills within a team	Develops and implements training plans tailored to the needs of the institution
Need to optimize time and activity management	**The time self analysis grid (TSAG)**	A coherent set of time management instrumentation to improve planning and scheduling of activities	Allows to identify the tasks carried out over a given period in order to structure efficiently the use of time of the different actors, to develop the individual and collective programming of activities, and allow for a concerted delegation of certain activities.

Source: Authors

3.2 Discussion of the Interest of Qualimetric Intervention Research for Management Science Researchers

According to the theoretical framework of qualimetric intervention research, the latter is deployed on the basis of an integrated epistemology (Savall and Zardet, 2004, 2021; Buono et al., 2018), which means that research risks leaving the scientific consultancy field and moving towards the consultation and expertise fields (Cappelletti and Khennich, 2024). This is exactly the point on which the interest of this approach for management science researchers can be discussed. In fact, in qualimetric intervention research, a correlation between the different phases of the research and the various epistemological pillars of the approach can be observed. In the diagnostic phase, the principle of cognitive interactivity enables observation to be formalized on the basis of interactions with the field actors. This interactive engagement serves as a cornerstone of qualimetric intervention research, where empirical data collection is intricately tied to direct, real-time interactions with stakeholders in the field. Such a hands-on approach ensures not only the integration of diverse perspectives but also the validation of findings through a continuous feedback loop with practitioners. Consequently, while this methodology might appear to migrate from purely scientific consultancy to more practitioner-oriented domains, it simultaneously enriches theoretical constructs with practical insights. This dual focus provides management science researchers with a holistic toolkit that is both rigorously theoretical and pragmatically applicable. The epistemological underpinnings, particularly during the diagnostic phase, emphasize cognitive interactivity—where the co-construction of knowledge between researchers and field actors allows for a more grounded and nuanced understanding of complex management issues. By fostering such interactions, the research process becomes inherently reflexive, challenging researchers to remain adaptable and responsive to the dynamics of the field. The qualimetric intervention approach, with its emphasis on cognitive interactivity during the diagnostic phase, exemplifies a sophisticated and nuanced methodology that can significantly advance the rigor and impact of management science research.

The knowledge relating to the management situation is then co-produced during the various interviews between the researcher and the actors in the field. This knowledge clarifies, enriches or contradicts the hypotheses adopted by the researcher in approaching the research field. The observation therefore confirms a hypothesis or enables a new one to be formed for future research. In the phase of steering groups, the knowledge generated in the diagnostic phase will be validated or invalidated following a comparison of the viewpoints of the various actors involved. At the end of this stage, an initial level of knowledge is stabilised. The principle of contradictory intersubjectivity therefore allows, through contradictory discussions between the researcher and the field actors, to ensure that the collected observation acquires the status of contingent knowledge and valid at the field level.

In the phases of evaluating and communicating research results, researchers must carefully determine whether their findings are specific to the studied field or have broader applicability, while also assessing their academic impact. This evaluation process can be significantly enhanced through active engagement with the scientific community, such as participating in research seminars and conferences. These interactions facilitate critical feedback and diverse perspectives, which helps in discerning the scope and significance of the results (Garcia-Morales et al., 2018; Hellekson et al., 2020). Employing the principle of generic contingency allows researchers to elevate contingent knowledge to a more generic status by rigorously comparing field observations against existing literature and findings from other domains. This synthesis not only ensures a robust validation of the results but also aids in identifying universal patterns and principles applicable beyond the specific research context (Hammersley, 2017; Bryman, 2016). Such methodology reinforces the relevance and academic rigor of research, ultimately contributing to the advancement of knowledge across disciplines (Maxwell & Mittapalli, 2019).

This shows that, in the context of qualimetric intervention research, the field immersion phase consists of constructing and implementing an integrated system of scientific observations. This system is deployed through the experimental implementation procedures. These procedures are incarnated in the tools, methods and systems, such as dashboards and steering groups, which form the basis of the intervention technology. Finally, they are implemented in accordance with the protocol formalised at the beginning of the qualimetric intervention research. This protocol defines the scheduling, timing and the actors involved in the various procedures, taking into account the context of the field studied and the research problem addressed.

On the basis of this reflexive analysis, Cappelletti and Khennich (2024) propose a model based on the triptych of epistemological principles/research phases/procedures, as shown in the figure below (Figure 1).

Figure 1. Proposed model for intervention research in management sciences

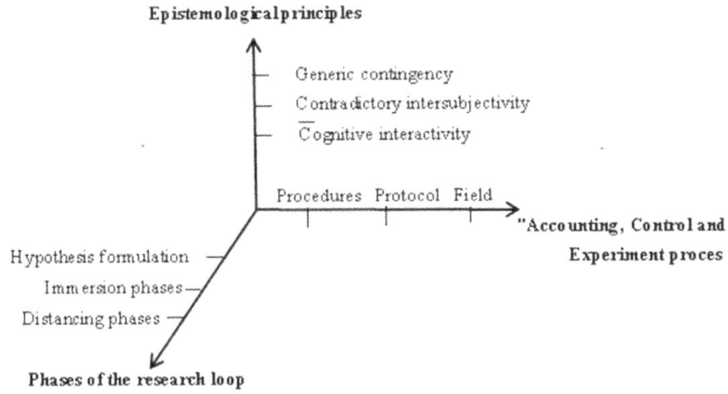

Source: Cappelleti and Khennich, 2024

Three substantial observations can be made about this model. The first one is that the model is based on a cyclical dynamic. The intervention process is first initiated in response to a demand arising from a need expressed by the actors in the research field and shaping the research problem. Next, it is founded on an implementation protocol adapted to the field context and to the treatment of the problem, making it possible to integrate the three epistemological principles of the intervention research. Finally, it is deployed according to the three phases of the research loop: formulation of hypotheses, immersion and distancing (Cappelletti and Khennich, 2024). The second point concerns the dynamic way in which the knowledge is processed, moving along a continuum of professional practices/contingent knowledge and theories/generic knowledge. This cooperation in the intelligibility and actionability phases of the intervention research is particularly conducive to the generation of shared reflexivity and, ultimately, to the creation of knowledge (Cappelletti and Khennich, 2024). The third point indicates that, if the two axes of epistemology and research phases are common to all intervention research in management sciences, whatever the discipline in which it is based, the procedures and protocols are on the other hand specific to each discipline. Thus, an intervention research in human resources management will mobilise procedures and protocols specific to the observation of objects characterizing this discipline, such as recruitment, the loyalty of talents, the leadership of HR managers, etc., while an intervention research in accounting should mobilise procedures and protocols relating to accounting, control and audit, such as management charts, cost calculation methods or budgetary approaches, etc.

It should be remembered that the three epistemological pillars of qualimetric intervention research allow this approach to be fundamentally crossed with that of Engaged Scholarship (Van de Ven and Johnson, 2006) and to be brought closer to the Evidence Based Management (EBM) approach (Pfeffer and Sutton 2006; Rousseau 2006, Rousseau and Ten Have, 2022). Indeed, through the principles of 'co-production of knowledge' between researchers and practitioners and that of 'contradictory arbitration', in other words the validation of observations through non-complacent discussions between researchers and practitioners leading to a consensus on reality, the qualimetric intervention research is resolutely 'engaged scholarship'. Through its approach based on the definition of research issues rooted in the reality of the organisation being studied, the design of a research project using a collaborative learning approach, the adoption of a longitudinal research project and the alternation of field immersion phases with those of distancing, enabling the accumulated research hypotheses to be regularly revised. This makes the qualimetric intervention research fundamentally "Evidence Based Management".

4. ILLUSTRATIVE CASE

We begin by presenting the context, the protocol and the process of the qualimetric intervention research carried out within a non-market public organisation, and then we analyse and discuss this research as committed, collaborative and transformative field research.

4.1. Presentation of the Research

Research Context: An Expressed Need for Organisational Development

Our research field is a public higher education establishment located in Casablanca, Morocco. It covers an area of 25 hectares and has 66 administrative, pedagogical and technical personnel as well as a team of 208 professors. The annual enrolment planned for 2020 was 6,000 trainee professors, 190 agrégatifs (candidates preparing for the agrégation competitive exam) and 240 trainee administrative staff.

Since its creation in 2012, this institution has been operating in a general environment of perpetual mutations, with consequences for the institution's strategy, internal organisation and social climate (Dibi, 2022). Indeed, as a public administrative organisation, its functioning is based on specific statutes and procedures that the executive team is obliged to know and respect. The members of this team are therefore assigned the role of "rule applicator", who is supposed to know the basics of public law, rather than that of “manager”, who is supposed to be better attuned to the methods and practices of management sciences. This “bureaucratic” model, which still dominates the educational establishments in Morocco, hinders organizational development and constitutes a source of dysfunctions that negatively affect the organization's economic and social performance (Dibi, 2022).

According to Savall and Zardet (2020), the historical origin of several organizational dysfunctions can be traced back to what they term the Taylor-Fayol-Weber (TFW) virus. This metaphorical virus encapsulates the imbalanced principles derived from early 20th-century theories on organizational management. Firstly, Taylor's model of the scientific organization of work (1911) propagated the idea of maximal division of labor, which emphasized task specialization to enhance efficiency but inadvertently led to fragmented workflows and diminished employee engagement. Secondly, Fayol's administrative organization model (1918) introduced a dichotomy between the design, decision-making, and implementation of activities, creating a stratified structure where the executive layer makes decisions, and the subordinates carry them out without room for input or flexibility. This hierarchical division stifles innovation and responsiveness within organizations, as critical feedback and frontline insights are often overlooked. Finally, Weber's bureaucratic organization model (1934) contributed to the depersonalization of jobs, advocating for a rules-based, impersonal approach to management to ensure consistency and fairness, but in practice leading to rigid structures where employee individuality and initiative are undervalued. The cumulative effect of these historical practices manifests in contemporary organizational dysfunctions such as siloed departments, inefficient communication, disengaged workforce, and an overall lack of adaptability. Recognizing these origins is crucial for modern management science as it seeks to diagnose and remedy these inherited issues. Addressing the TFW virus involves rethinking the principles of labor division, decision-making processes, and the role of personal agency within bureaucratic frameworks. Essentially, the path forward requires integrating flexibility, collaborative decision-making, and personalized job roles into the modern organizational design to mitigate the lingering impacts of these historically embedded practices. This holistic re-evaluation and transformation are vital for evolving organizational structures to be more adaptive, innovative, and human-centered in today's dynamic business environment.

Based on the context presented above, a series of meetings were held with the director of the establishment to define the strategic direction of our intervention, aiming to mobilize the members of the executive team around a project focused on organizational transformation and change management to enhance overall performance. During these discussions, we concentrated on the key objectives, the operational procedure, the intervention schedule, and the anticipated challenges. We agreed on the following objectives: contributing to the establishment's organizational development by implementing new management tools and methods designed to improve both economic and social performance; developing shared principles, tools, and methods to enhance organizational functioning and management; and achieving immediate and sustainable economic and financial results by converting hidden performance costs into added value for the organization. Together, these goals were designed to foster a cohesive and effective transformation process aimed at significantly boosting the establishment's performance and operational efficiency.

Research Protocol: A Plan of Operations Adapted to the Field Context and the Problem to Be Addressed

Our qualimetric intervention research within this educational establishment was founded on an integrative approach that combined qualitative, quantitative, and financial data (Savall & Zardet, 2004). Conducted over a longitudinal period from January 2019 to June 2020, this prolonged engagement enabled us to immerse deeply into the research field, facilitating extensive interactions with various stakeholders. The research process was methodically structured into five distinct stages, which included initial field research and immersion phases followed by periods of hindsight and analysis for critical reflection and distancing (Buono et al., 2018). This cyclic alternation between in-depth, hands-on involvement and analytical reflection allowed for a global and holistic understanding of the establishment's organizational dynamics. During the immersion phases, we closely observed and interacted with faculty, staff, and administration, collecting rich qualitative data that were then juxtaposed with quantitative metrics and financial performance indicators to form a comprehensive dataset. These multifaceted data streams provided a robust foundation for diagnosing the establishment's strengths and weaknesses. The distancing phases allowed us to step back and synthesize the findings, drawing insights that informed subsequent stages of intervention. This iterative process ensured that each phase of the research was constantly informed by real-time feedback and empirical evidence, enhancing the reliability and depth of our conclusions. The ultimate objective was to develop actionable strategies aimed at organizational improvement, leveraging the combined insights from qualitative narratives, quantitative analyses, and financial evaluations. This structured yet flexible approach underscores the efficacy of qualimetric intervention in capturing the complexity of educational environments and devising tailored solutions that are both evidence-based and contextually relevant. The longitudinal aspect of the study was particularly instrumental in tracking the evolution of the intervention's impact over time, thus ensuring that the strategies proposed were not only effective but also sustainable in the long run. In essence, our qualimetric intervention demonstrated the transformative power of an interdisciplinary, data-driven approach in fostering significant improvements within educational organizations. The process was carried out in five stages, as shown below (Figure 2).

Figure 2. Five stages of the research process

Source: Authors

The first stage of our intervention was the carrying out of a dysfunctional diagnosis to identify all or some dysfunctions and hidden costs in the organisation. By definition, the hidden costs are not found in the formal information system of companies and organisations. However, they are at least partially memorised by the actors in the field. Assessing hidden performance costs therefore means reconstructing data that has not been entered at source in an information or monitoring system, through the memory and cooperation of the actors involved, in particular via interviews conducted with them. This data provides an indication of what the final result might have been without the malfunctions, or at least due to their control (Khachan et al., 2023). The hidden costs were then quantified after a series of interviews with the managers of the educational establishment. Based on a set of sample sentences, we matched the dysfunctions identified with the necessary regulatory actions and the corresponding consequences, and finally calculated the corresponding hidden costs. This operation enabled us to make a qualimetric correspondence between the qualitative data (the dysfunctions), the quantitative data (the acts of regulation) and the financial data (the value of the hidden costs in monetary units). At the end of this phase, we were able to identify the five components that illustrate the economic impact of dysfunction regulation, namely extra wages, extra time, extra consumption, non-production, non-creation of potential and risks.This calculation of hidden costs was based on the hourly contribution to the added value of variable costs. It enabled us to highlight the performance level of the establishment and then to compare it with that expected. Subsequently, all the actors in the organisation were mobilised to set up project groups working on concrete solutions for improving the quality of operations and the way they are managed. The next stage consisted of implementing the various solutions proposed, in particular by planning and budgeting the actions selected. These actions have affected both the organisation's structures and the behaviour of those working in the field.The final phase of the intervention was that of evaluation, in which we compared the state of the organisation's economic and social performance before and after the implementation of the organisational development action plan.

Research Process: Linking Policy, Process, and Tools

In this qualimetric intervention research, we experimented the management control of hidden performance costs through three axes: policies, change process and management control tools.

The "policy" axis has been the catalyst for change within the organisation. It is through this strategic impetus that policy decisions, in terms of optimising processes and involving local actors in the process of change, have contributed to overcoming weaknesses and ensuring the survival and the development of the establishment. On the one hand, it facilitated the cultivation of a complex vision of the organisation by addressing the root causes of dysfunctions and generating observable data about them (dysfunctions and hidden costs). On the other hand, it was originally responsible for setting up the organisational change process and implementing new management tools. Next, the "change process" axis enabled the actors in the establishment studied to identify the various organisational dysfunctions, in particular those relating to working conditions, work organisation, communication-coordination-consultation, time management, and integrated training as well as the strategic implementation. It invited them to analyse the causes of these various dysfunctions (qualitative, quantitative and financial diagnosis), to implement corrective actions to reduce them (managerial innovation project) and to measure the financial impact of the corrective action plan, in particular by assessing the reduction of hidden costs and organisational vulnerabilities. Through this "change process" axis, the actors in the field were given the opportunity to define and implement a project that would address the lack of good working conditions, work disorganisation or even the vulnerability of skills observed in the actors of the establishment. Lastly, the "tools for change" axis proposed the tools needed to concretely implement the actions for change within the organisation. The use of these tools enables us to keep a close eye on organisational dysfunctions, which in turn improves the control of hidden performance costs.

4.2. Analysis-Discussion

Observation of Epistemological Principles in the Process of Creating Knowledge

Our qualimetric intervention research conducted within this establishment was extended over a fairly longitudinal period, allowing us a prolonged presence in our research field, a significant interaction with the actors and an alternation of field research with periods of hindsight and analysis (Buono & al. 2018). Thus, to build and maintain a climate of trust with the actors in the field, while preserving our research subject and its problematic, we were regularly led to renegotiate the architecture and/or planning of our intervention with the organisation's decision-makers. In fact, this renegotiation enabled us to avoid the hierarchy's direction when it sought to distance us from the treatment of the research problem (Baskerville and Wood-Harper, 1996) or even to amend the results obtained (Cappelletti and Khennich, 2024).

In our research, the knowledge creation process followed the epistemology of generic constructivism, which is based on three pillars: cognitive interactivity, contradictory intersubjectivity and generic contingency (Buono & al. 2018).

To ensure adherence to the principle of cognitive interactivity within our research, we employed semi-directive interviews in which notes were meticulously written by hand rather than recorded, a practice designed to foster the trust necessary for the project's successful execution (Weick, 1995; Buchanan et al., 1988). This approach was pivotal in creating a comfortable and open environment for participants, encouraging candid and unfiltered communication. The dysfunction verbatims gathered during these interviews were analyzed systematically, categorized by actor type and dysfunction domain to ensure a thorough and nuanced understanding of the issues at hand. This granular analysis provided a foundational basis for subsequent stages of the research, allowing for targeted and precise interventions. To further ensure the integrity and quality of the data, we employed the "mirror-effect" mechanism (Savall & Zardet, 2004), wherein the analyzed dysfunctions were presented back to the interviewees for validation. This process not only served to verify the accuracy of our interpretations but also reinforced the collaborative nature of the research, as participants were actively involved in verifying and refining the findings. By engaging in this iterative validation, we ensured that the insights drawn were deeply rooted in the participants' lived experiences and perceptions, thereby enhancing the overall authenticity and reliability of the data. Additionally, this method facilitated continual reflection and adjustment, ensuring that the dynamic and evolving nature of organizational contexts was captured and addressed. The backing of the mirror-effect mechanism, coupled with a rigorous semi-structured interview methodology, underscored our commitment to a high-quality, interactive research process (Château-Terisse et al. 2016). Consequently, this meticulous and participatory approach not only strengthened the validity of our findings but also cultivated a sense of ownership and alignment among the stakeholders, thereby laying a robust groundwork for effective and sustainable organizational change. Through the strategic intertwining of cognitive interactivity with meticulous data validation processes, our research framework exemplified a rigorous, yet empathetic approach to capturing and addressing organizational dysfunctions, fostering a cooperative and productive research environment.

The principle of contradictory intersubjectivity was meticulously implemented through a "divergence-convergence" analysis (Savall & Zardet, 2004), a sophisticated approach that involves confronting the diverse viewpoints of field actors to enable a cross-pollination of perspectives. This methodology facilitates a dynamic exchange of ideas, leading participants to collectively develop a shared vision of organizational phenomena. By embracing this pluralistic dialogue, the process inherently promotes a deeper understanding of complex organizational dynamics, ensuring that improvement actions are not only well-informed but also broadly supported by stakeholders. The iterative cycle of divergence—where differing opinions are aired, debated, and critically examined—followed by convergence—where consensus is sought and synthesized—ensures that the decisions related to proposed improvements are robust, inclusive, and grounded in the multifaceted realities of the organization. This principle of contradictory intersubjectivity was also prominently observed during discussions between the intervening researcher and actors within project groups, further aligning the research process with the concrete realities of the organization under study. By directly engaging with a broad spectrum of voices within the project groups, the researcher could capture a more authentic and nuanced picture of the organizational landscape, bridging the gap between theoretical constructs and practical implementation. This ongoing dialogical interaction ensured that the research was continually calibrated to reflect the lived experiences and operational realities of the participants, thereby increasing the relevance and applicability of the findings. The convergence phase, in particular, allowed for the distillation of diverse insights into cohesive, actionable strategies, fostering a sense of shared ownership and commitment to the proposed changes. The inclusion of contradictory perspectives not only enriched the analytical process but also empowered participants by validating their experiences and contributions. Consequently, this approach harnessed the collective intelligence of the group, transforming potential conflicts into opportunities for creativity and innovation.

To ensure the necessary distancing from the results expressed by field actors and maintain an objective perspective, we employed the technique of "expert opinion" (Savall & Zardet, 2004). This approach entailed conducting our own comprehensive analysis to elucidate the root causes of the identified dysfunctions. By stepping back and applying our expertise, we aimed to transcend the immediate, subjective viewpoints of individual actors and achieve a more holistic understanding of the organizational issues at play. Through this process, we systematically identified the key ideas surrounding each theme of dysfunction, allowing us to detect underlying patterns and systemic issues that may not be immediately apparent to those immersed in daily operations. The expert analysis provided a structured framework to sift through the complexities and contradictions often associated with organizational dysfunctions, thereby facilitating a clearer articulation of problems and potential solutions. On the basis of this "expert opinion," we formulated actionable insights and strategic interventions tailored to address the specific dysfunctions identified. This methodology was crucial in translating the empirical data and qualitative narratives gathered from the field into coherent, actionable plans aimed at enhancing the organization's overall performance. The distancing afforded by this expert analysis not only ensured a more objective evaluation but also added a layer of rigor and depth to our findings, reinforcing their credibility and utility. By distinguishing immediate symptoms from root causes, we were able to propose targeted, effective actions designed to foster sustainable improvement. Additionally, this expert-led analysis helped in prioritizing interventions, ensuring that efforts were concentrated on high-impact areas likely to yield significant performance gains. Thus, the technique of "expert opinion" was instrumental in bridging the gap between the subjective experiences of field actors and the objective requirements of organizational change, providing a robust foundation for informed decision-making and strategic planning. This balanced, expert-informed approach ultimately enabled an effective response to organizational challenges, driving meaningful and lasting improvements in performance.

This succession of 'mirror-effect', 'divergence-specificity' and 'expert opinion' analyses enabled us, ultimately, to build our knowledge of our research subject following a generic constructivism whose three epistemological pillars converge with those of Evidence-Based Management (EBM) (Pfeffer, Sutton, 2006; Rousseaua and Ten Have, 2022).

Validity of the Research Conducted

In any intervention research, the researcher aims both to create relevant insights for the practitioners involved in the intervention and to develop rigorous knowledge (Starkey et al., 2009).

Our qualimetric intervention research conducted within this establishment fundamentally enabled us to generate valuable knowledge instrumental for organizational management (Asselineau et al., 2022). Through a meticulous evaluation of the improvement action plan implemented during our intervention, we observed a marked increase in both the economic and social performance of the organization. This positive outcome was chiefly attributed to the significant reduction in organizational dysfunctions and the mitigation of previously unrecognized hidden costs. By systematically addressing these dysfunctions, we were able to streamline processes, enhance operational efficiency, and promote a more cohesive and motivated workforce. The data collected and analyzed throughout the intervention provided actionable insights that informed better decision-making and strategic planning. This approach not only highlighted specific areas needing immediate attention but also fostered a culture of continuous improvement. The comprehensive analysis done during the research uncovered latent inefficiencies, which, once addressed, resulted in considerable cost savings and resource optimization. Enhancing social performance was another critical achievement, as it involved improving employee well-being, job satisfaction, and organizational culture, all of which contributed to a more harmonious and productive work environment. The intervention demonstrated that tackling both economic and social dimensions holistically can lead to sustainable organizational growth. The knowledge gained from this research extended beyond mere diagnostics; it provided a roadmap for ongoing improvement and adaptability in the face of future challenges. Moreover, by deploying this qualimetric approach, we ensured that the interventions were deeply contextual, reflecting the unique dynamics and needs of the establishment. This tailored approach maximized the relevance and impact of the improvement measures implemented. As a result, our research not only achieved immediate performance gains but also laid the groundwork for enduring organizational health and resilience. Through this intervention, we have shown that using an evidence-based, qualimetric approach can generate significant, multifaceted improvements, offering a robust model for other organizations seeking to enhance their performance systematically.

The validity of our research was further substantiated by the scientific community, as evidenced by the opportunity to present approximately ten papers at academic conferences and publish several articles in esteemed, peer-reviewed journals. This recognition underscores the rigorous nature of our qualimetric intervention research, which seamlessly integrated immersion and distancing phases to create robust scientific knowledge. During the immersion phase, we engaged closely with field actors, fostering an environment conducive to generating rich, contextually relevant insights through direct interactions and observations. This phase was instrumental in capturing the nuanced realities of the organizational context, thereby ensuring that the knowledge generated was both grounded and practically applicable. Conversely, the distancing phase afforded us the necessary objectivity to critically analyze and synthesize the data collected, transforming field interactions into structured, scientific contributions. This phase also enabled us to position our findings within the broader academic discourse, inviting scrutiny, validation, and feedback from the scholarly community (Asselineau et al., 2022). The alternation between these phases not only strengthened the internal validity of our research but also facilitated the creation of knowledge that was both empirically robust and theoretically significant. By subjecting our findings to the rigorous assessment of academic peers, we ensured that our research adhered to high standards of scientific rigor and methodological soundness. The dissemination of our work through conferences and journal publications not only validated the quality of our research but also contributed to the ongoing dialogue within the field, offering new insights and methodologies for organizational improvement. This dual-phase approach, characterized by deep immersion in practical contexts followed by scholarly distancing and analysis, exemplifies a comprehensive strategy for generating impactful, credible research. Ultimately, the recognition from the scientific community highlights the efficacy of our qualimetric approach in bridging the gap between theory and practice, fostering advancements in both academic knowledge and practical organizational interventions. Through this rigorous methodological framework, our research has made significant contributions to the field, illustrating the power of an integrated, evidence-based approach to understanding and enhancing organizational performance.

Our intervention research can therefore be seen as an extension of organisational development approaches, in particular clinical research (Schein, 1987), participative inquiry (Reason & Bradbury, 2001, 2008) and Appreciative Inquiry (Cooperrider & al., 2008). These approaches facilitate the expression of the actors by involving them in the elaboration of organisational development projects. The qualimetric aspect of our intervention research also enabled us to help the actors to translate their qualitative expressions into quantitative and financial data. (Roche & al, 2023).

CONCLUSION

Given the various crises—climatic, health, energetic, and geopolitical—that currently affect the functioning of companies and organizations, researchers in management sciences are on the lookout for research methods that produce rigorous observations of reality to build robust and useful prescriptions and anticipations. In line with "Evidence Based Management," "Engaged Scholarship," and "Transformative Research" principles, qualimetric intervention research presents a compelling field research approach for generating actionable insights. This method is grounded in an integrated research epistemology that seamlessly combines hypothesis formulation, immersion, and distancing phases, emphasizing procedural rigor and field protocols. Key to its value is its foundation in defining research problems rooted in the organization's real conditions, designing research projects through a collaborative learning approach, adopting longitudinal studies, and alternating immersion in the field with periods of reflective distancing. This ensures a regular revision and refining of accumulated research hypotheses, making findings both dynamic and relevant. For managers, the implications are profound: qualimetric intervention research fosters the 'co-production of knowledge' between researchers and practitioners, ensuring that insights are validated through non-complacent discussions that lead to a consensus grounded in reality. This participatory and iterative process results in richer, more nuanced understandings that are immediately applicable for strategic decision-making and operational improvements. Managers can thus leverage these thoroughly vetted and contextually relevant insights to craft evidence-based strategies that enhance organizational resilience and adaptability in the face of ongoing and future crises. By engaging with this rigorous and collaborative research method, they can drive sustainable transformations, ensuring that their organizations are better equipped to navigate complexity and uncertainty.

However, the qualimetric intervention research approach is not without criticism. Primarily, it risks veering closer to consultancy rather than maintaining the rigor expected of academic research. Researchers must navigate the fine line between practical application and academic rigor, often facing political and cultural constraints from the governance of the organization being studied or even time and career pressures. Such constraints may push researchers to prioritize the pragmatic and immediate needs of the organization over the scientific integrity of the research. For intervening researchers, it is therefore crucial to deploy consultancy tools and methods as research technologies aimed at building a scientific consultancy foundation. Future research could explore methodologies to balance this dual role effectively, ensuring that the practical contributions of consultancy do not eclipse the generation of rigorous scientific knowledge. Investigating frameworks to measure and validate the scientific output in qualimetric intervention research could provide clarity and guidelines for maintaining academic standards. Additionally, examining the impact of political and cultural organizational constraints on research outcomes could yield insights into how these factors influence the trajectory and focus of intervention studies. By understanding these dynamics, new strategies can be developed to mitigate the influence of such constraints, ensuring a balance between practical relevance and scientific rigor. Longitudinal studies focusing on the career progression of researchers engaged in qualimetric interventions might also identify the pressures and incentives that shape research priorities, offering a pathway to harmonize career aspirations with the production of robust scientific findings. Moreover, comparative studies investigating the differences between qualimetric intervention research and traditional consultancy could illuminate best practices for maintaining a strong scientific orientation while delivering practical solutions. Such future research endeavors could ultimately contribute to refining the qualimetric approach, fortifying its position as a rigorous and valuable methodology within the broader landscape of organizational research.

Secondly, the intervening researcher can often be 'contaminated' or 'polluted' by the prevailing mental structures of the organization being studied. This contamination occurs when the researcher becomes influenced by the organization's stable elements such as the ideology of the managers or the prevailing opinions of employee groups. Such influences can compromise the objectivity and analytical rigor of the research. To mitigate this risk, it is crucial for the intervening researcher to develop systems for regular 'decontamination.' These systems could include arranging phases of distancing that facilitate frequent exchanges with external scholars and practitioners, thereby refreshing the research perspective and diminishing the risk of bias. Additionally, researchers must cultivate a robust research ethic that guards the foundational principles of objectivity and rigor. This can be achieved by adhering to established methodological standards and ethical guidelines, which act as safeguards against the pressures exerted by the organization under study. Ensuring the development of necessary skills to resist organizational pressures, such as critical thinking, resilience, and ethical decision-making, is also vital. Future research could explore the efficacy of various 'decontamination' techniques and their role in preserving the integrity of qualimetric intervention research. Longitudinal studies could examine how regular interactions with external peers and participation in academic communities impact the objectivity and quality of the research outcomes. Another avenue for future research could involve developing and testing frameworks for ethical guidelines specifically tailored to qualimetric intervention research, ensuring they address the unique challenges posed by organizational influences. Investigating the impact of researcher training programs that emphasize critical reflection and ethical conduct could also provide insights into best practices for maintaining objectivity. Comparative studies that examine the outcomes of intervention research with and without structured 'decontamination' phases could highlight the importance of these practices in safeguarding scientific rigor. By addressing these areas, future research can provide a clearer roadmap for intervening researchers to navigate the complex interplay between organizational influences and research integrity, thereby enhancing the overall quality and reliability of qualimetric intervention research.

The transformative nature of qualimetric intervention research often disrupts organizational routines, sometimes leading managers to resist or retreat from the change process. This presents a significant risk for researchers, as it can jeopardize the organizational transformation project. To mitigate this risk, it is essential for researchers to continually negotiate and refine the specifications of their research, ensuring that the study subject and research questions remain intact without compromising the relevance and rigor of the transformation effort. Future research could explore methods for effective negotiation and stakeholder management in intervention research, examining how researchers can balance fidelity to their research objectives with the need for organizational buy-in. Investigating strategies for maintaining consistent communication and demonstrating the value of the transformation to resistant managers could provide practical insights. Additionally, longitudinal studies could examine the long-term impacts of these negotiations on the success and sustainability of organizational changes. By understanding these dynamics, future research can offer frameworks and best practices to help researchers navigate the complexities of transformative interventions while maintaining scientific rigor.

Despite these limitations, the qualimetric intervention research methodology retains the advantage of offering management science researchers the possibility of theorising professional practices, by capturing phenomena that would not be accessible, or would be very difficult to access, using other methodologies (Cappelletti and Khennich, 2024).

REFERENCES

Alla, L., Bentalha, B., Megzari, G., & Hmioui, A. (2022). Les effets du COVID-19 sur les chaines logistiques: Une analyse qualitative. Reconfigurations du tourisme en contexte de la crise du Covid-19: Quelles démarches pour quelles résiliences au Maroc et ailleurs?, 1(1), 130-153.

Alla, L., Jaouhari, E. L., & Bentalha, B. (2022). Repenser l'entrepreneuriat au sein du secteur informel au Maroc à l'ère du post Covid-19: Pour approche prospective de la formalisation intégrative. *Alternatives Managériales et Economiques*, 4(1), 180–202.

Argyris, C. (1985). *Action Science*. Jossey-Bass.

Asselineau, A., Combaudon, C., & Fouré-Joopen, H. (2022). In Validation scientifiqued'une nouvelle approche du management par la recherche-intervention:Cas de 12 entreprises « libérées ». Recherche en Sciences de Gestion-ManagementSciences-Ciencias de Gestión, 150, 31-55.

Bentalha, B. (2023). Consumer Perception of Robotic Mobile Fulfillment Systems: A Netnographic Case Study of Amazon. In Innovation, Strategy, and Transformation Frameworks for the Modern Enterprise (pp. 243-264). IGI Global. 10.4018/979-8-3693-0458-7.ch010

Bentalha, B. (2023). A Framework for Exploring the Global Performance of Service Supply Chain Management: A Qualitative Study. Government Impact on Sustainable and Responsible Supply Chain Management, 106-129.

Bentalha, B. (2024). Employees and Robots in Amazon's Robotic Mobile Fulfillment Systems: A Netnographic Analysis of a Supply Chain Transformation. In Blockchain Applications for Smart Contract Technologies (pp. 188-207). IGI Global. 10.4018/979-8-3693-1511-8.ch009

Bradbury, H. (2015). *The SAGE Handbook of Action Research*. SAGE Publications. 10.4135/9781473921290

Buchanan, D. A., Boddy, D., & McCalman, J. (1988). Getting. In *Getting Out, Getting Back: the art of the possible. In A. Bryman (dir.), Doing Research inOrganisations*. Routledge.

Buono, A., Savall, H., & Cappelletti, L. (dir.) (2018). Intervention-Research: from-Conceptualization to Publication. Charlotte: IAP.

Burney, V., Arnold-Saritepe, A., & McCann, C. M. (2023). Rethinking the place of qualitative methods in behavior analysis. *Perspectives on Behavior Science*, 46(1), 185–200. 10.1007/s40614-022-00362-x37006599

Cappelletti, L. (2009). Performing an internal control function to sustain SOX 404and improve risk management: Evidence from Europe. Management AccountingQuaterly, été, 10(4), 17-27.

Cappelletti, L. (2010). *La recherche-intervention en contrôle de gestion. Actes duCongrès de l'AFC*. CD-Rom.

Cappelletti, L., & Baker, R. (2010). Measuring and developing human capital througha pragmatic action research: A French case study. *Action Research*, 8(2), 211–232. 10.1177/1476750309349976

Cappelletti, L., Heurtel, P., & Lefebvre, P. (Dir.). (2022). Crise de la connaissance etconnaissance de la crise. Les points de vue du CNAM. E. *Ms*.

Cappelletti, L., Voyant, O., Savall, H., & Noguera, F. (2018). 40 Years of Socio-Economic Approach of Management (SEAM): What We Know and Where WeGo? Academy of Management Annual Meeting, Chicago.

Château-Terrisse, P., Codello, P., Béji-Bécheur, A., Jougleux, M., Chevrier,S. et-Vandangeon-Derumez, I. (2016). Réflexivité et éthique du chercheur dans laconduite d'une recherche-intervention. La revue des sciences de gestion, 277, 45-56.

Chiapello, E., & Gilbert, P. (2019). *Management Tools. A Social Science Perspective*. Cambridge University Press. 10.1017/9781108553858

David, A. (1999). Intervention methodologies in management research. Track:Collaborative Management / Research Approaches.

Dibi, A. (2022). Le pilotage de la performance globale des établissements éducatifs: Des pistes pour faire mieux avec peu. *ACCRA*, 15(3), 61–91. 10.3917/accra.015.0061

Rousseau, D. M., & ten Have, S. (2022). Evidence-based change management. *Organizational Dynamics*, 51(3), 100899. 10.1016/j.orgdyn.2022.100899

Savall, H. et Zardet, V. (2014). Recherche qualimétrique. Observer l'objet complexe. Economica.

Strauss, A., & Corbin, J. (1990). *Basics of qualitative research: Grounded theory proceduresand techniques*. Sage.

Van de Ven, A. H., & Johnson, P. E. (2006). Knowledge for Theory and Practice. *TheAcademy of Management Review*, 31(4), 802–821.

Wacheux, F. (1996). Méthodes qualitatives et recherche en gestion. *Economica.*

Weick, K. (1995). *Sensemaking in Organizations.* Sage.

Worley, C. G., Zardet, V., Bonnet, M., & Savall, A. (2015). *Becoming Agile - How the-SEAM Approach to Management Builds Adaptability.* Wiley. 10.1002/9781119182351

Yin, R. K. (1990). Case Study Research - Design and methods. *Sage (Atlanta, Ga.).*

Chapter 6
Understanding Entrepreneurial Motivation Through Qualitative Study

Khalid Ouintejgal
https://orcid.org/0009-0009-4749-4275
Ibn Zohr University, Morocco & National School of Business and Management, Agadir, Morocco

Laila Cherradi
Ibnou Zohr University, Morocco

ABSTRACT

This chapter aims to cover entrepreneurial motivation as a key component of the emerging field of entrepreneurship, particularly through studies developed using qualitative methods. It indicates how qualitative methods particularly contribute to the understanding of entrepreneurial motivation through examples drawn from previous publications. A synthesis of seven papers on this topic, based on qualitative research methods is presented. The chapter subsequently presents some implications and contributions of qualitative research methods on entrepreneurial motivation study.

DOI: 10.4018/979-8-3693-5543-5.ch006

INTRODUCTION

A key factor that has contributed to the development of the entrepreneurial field is methodological pluralism of studies covering each of its main components (Shepherd, 2015).

There are several features to the greater methodological diversity in the field, but the one that will be the focus of this chapter is the growing popularity of qualitative research meeting current challenges in entrepreneurship research (Hlady-Rispal et al., 2021).

This chapter focuses on one of the main components of entrepreneurship, namely entrepreneurial motivation, which is analyzed, first and foremost, using qualitative methods.

In addition to the traditional distinction between opportunity and necessity for the survival or success of the business, the motivations can be classified according to the particular importance attached to need of achievement, autonomy, risk aversion, financial aspects and the influence of family and friends. (Drews, 2015).

The aim of this chapter is therefore to examine the singular contribution of specific qualitative methods, through the analysis of seven documents. How qualitative research makes a distinctive contribution to understanding entrepreneurial motivation. In addition, this work aims to assess the best features of these studies for entrepreneurship studying, and to inform on the strategies and techniques deployed to apprehend entrepreneurial motivation.

The path chosen to complete the understanding of the scope of entrepreneurial motivation could not take shape and succeed without a detailed analysis of the concepts of entrepreneurship and entrepreneurial motivation. A second focal point would be to present the preponderant theories that have transcended this field. Thirdly, the presentation of a literature review makes it in a position to identify the main lines taken to study entrepreneurial motivation and the research methods applied to this end especially qualitative ones.

The contents of this chapter are set out as follows. It, first, identifies entrepreneurship and entrepreneurial motivation from a conceptual viewpoint. The three most predominant theories used in the study of entrepreneurial motivation are presented in support. Then, the chapter presents a literature review on entrepreneurial motivation over the last thirty years (1983-2023) whose purpose is to understand the areas of research and the methods applied. Finally, the chapter shows the nature of qualitative research in management and entrepreneurship. After emphasizing the importance of the qualitative methods in the entrepreneurial motivation research area, the chapter aims to cover seven contributions to understand entrepreneurial motivation. While emphasis is mainly placed on qualitative methods, the publications analyzed are based on a variety of empirical approaches and epistemological postures. In terms of contributions addressing the theoretical and practical implications of this chapter, some comments and observations on the qualitative research methods used as part of the entrepreneurial motivation study are presented and supported by implications for the study of the general field of entrepreneurship applied to the Moroccan context.

1. ENTREPRENEURSHIP AND ENTREPRENEURIAL MOTIVATION: CONCEPTUAL CONCERNS

This section presents the definition of entrepreneurship, entrepreneurial motivation and the main theories underlying the study of this entrepreneurship main component.

1.1. Defining Entrepreneurship

"Entrepreneurship is a multidimensional and multidisciplinary phenomenon" (Hlady-Rispal et al., 2021). Research in this field did not stop attracting researchers from different disciplines. For this purpose, three main approaches have been adopted in entrepreneurship research (Stevenson & Jarillo, 1990) ; (Danjou, 2002): First, context-based studies that analyze the conditions or effects of entrepreneurial action in a socio-economic environment. From this point of view, disciplines such as economics (Filion, 1997) and sociology examine the impact of entrepreneurial activities on the economic context and, conversely, on the context conducive to the development of entrepreneurship. Secondly, the actor-focusing studies that analyze the psychological and demographic characteristics of the entrepreneur. From this perspective, disciplines such as psychology (Low & MacMillan, 1988) investigate the entrepreneur, his personal characteristics, motivations and personality traits. Third, the action-oriented studies that analyze the entrepreneur's actions. From this point of view, disciplines such as management (Gartner, 1988) simultaneously examine entrepreneurial management and the entrepreneurial process.

The challenge to develop entrepreneurship lies in the limited theoretical basis and absent consensus on a scientific definition covering whole area's conceptualization (Low & MacMillan, 1988) ; (Cunningham & Lischeron, 2001). This knowledge gap is the result of the phenomenon's inherent nature qualified as a complex, heterogeneous and multidimensional (Bruyat & Julien, 2001).

Entrepreneurship, in a "mainstream" sense, has been defined in a range of approaches as research on the field has developed. Here, we adopt the original definition by Professor Howard Stevenson, sponsor of entrepreneurship studies at Harvard Business School-HBS. As defined by Stevenson: Entrepreneurship is the opportunities seeking above managed resources (Stevenson, 1984).

The term "pursuit" calls a single-minded and unrelenting determination. Entrepreneurs frequently perceive a tight and limited window of opportunity and have a sense of urgency rarely found in companies where each new opportunity is part of a larger overall project, and where resources are more available (Thomas R. Eisenmann, 2013).

The term "opportunity" has been defined in several ways. For (Kirzner, 1973) "opportunity" is equivalent to "profit opportunity". He defines profit opportunity as the presence in the economy of a resource misallocation that exists when there is a demand for which individuals want to pay and for which the required resources are available in concert (Chelly, 2004).

Following the definition of entrepreneurship as a general notion, we present the definition of the specific notion of entrepreneurial motivation.

1.2. Defining Entrepreneurial Motivation

Entrepreneurial motivation is a key component of entrepreneurship (Gabarret & Vedel, 2015). Numerous definitions of this concept have been developed, in view of general and applied perspectives.

In general terms, motivation is defined as per the oxford dictionary "the conscious or unconscious stimulus for action towards a desired goal" This research line has been a "core topic in psychological science and organizational behavior for over a century" (Kanfer Ruth, 1990).

According to Levy-Leboyer (1984), motivation can be defined as "that which triggers activity, directs it towards certain goals and prolongs it until these goals are achieved" as cited in (Louche, 2019). This definition establishes the characteristic field of motivation: it is at the origin of behaviors of varying intensity, which are directed towards certain goals whose attainment is conditional on persistence over time.

Another definition presents motivation as "the study of the factors that determine the direction, intensity and duration of behavior" as cited in (Borman et al., 2003).

Under an operational conception, Naylor & Pritchard (1980) define motivation by studying the way behaviors are emitted (direction) and the intensity (amplitude) with which these behaviors are launched. Accordingly, Motivation is the process of allocating personal resources, in terms of time and energy, to diverse acts in order to ensure the greatest possible anticipated affect from them (Naylor et al., 1980).

From this perspective, entrepreneurial motivation refers to people's underlying thoughts and expectations about the potential personal returns from the pursuit of entrepreneurship (Sørheim & Landström, 2001) ; (Stefanovic et al., 2010). It corresponds to the intent or target that entrepreneurs are looking to meet through entrepreneurial property ownership (Solymossy, 1998).

In their work, other authors note that an actor's motivation to start a business is often both complex and multi-faceted (Shane et al., 1991).

In order to better grasp the scope of the definition of entrepreneurial motivation, we present below the dominant theories relating to this concept.

1.3. Leading Theories in the Study of Entrepreneurial Motivation

Initially, with a view to understanding motivational theories and grasping motivation as a concept defined on its own foundation, various authors proposed categorizations (Fenouillet, 2009). Below, we focus on the three most predominant theories used in the study of entrepreneurial motivation.

Content and Process Theories of Entrepreneurial Motivation

Two main types of motivational theories can be identified: content theories and process theories (Campbell, J., Dunnette, M., Lawler, E., & Weick, 1970).

-Content Theories of motivation focus on the "who?" & "what?" of motivation, or what people need in their lives (i.e. what motivates them). Among the best-known content theories: Maslow's Hierarchy of Needs, Herzberg's Two Factors Theory, McClelland's Theory of Needs, Alderfer's ERG-Existence Relatedness Grow-Theory.

-Process theories focus on the "why?" & "How?" of motivation, and are concerned with the psychological and behavioral processes that affect an individual's motivation. Some of the best-known motivational theories include: Skinner's reinforcement theory, Vroom's Expectancy Theory, Adam's Equity Theory, Locke and Latham's Goal Setting Theory.

Motivation theories are based on a set of assumptions about the nature of people and the factors that stimulate action, as well as on psychological and socio-psychological elements, economic value and cognitive theories. These hypotheses, and the theories derived from them, can be seen as part of a descriptive continuum ranging from the mechanistic to the organismic, which views development as a process by which human beings internalize, develop, perfect and integrate their internal structures or representations of themselves and the world around them. (Deci, Edward L and Ryan, 1985) (Hessels et al., 2008).

Push-Pull Theory

"Push and Pull" theory presents the reasons for setting up a business under two categories of motivation: "push motivations" and "pull motivations" (Bellanca & Colot, 2014).

This basic model in the literature on entrepreneurial motivation has two extensions, called "push" and "pull" respectively. According to this approach, the creation of a company can be chosen or undergone. It may be triggered by the presence of a business opportunity, or by the need for retribution in the event of unemployment (Gabarret & Vedel, 2015).

-The "PUSH" approach or incentive factors: necessity and dissatisfaction

The factors attributed to this first approach are coupled with negative understandings (Kirkwood & Campbell-Hunt, 2007). There are two "push" drivers of entrepreneurial motivation: (i) necessity (lack of employability, prolonged unemployment), and (ii) dissatisfaction (unsatisfactory work). These factors correspond to necessity, precarious or flexible employment, lack of work, redundancy, corporate restructuring (Hughes, 2003) (Mcmullen et al., 2008).

-The "pull" approach: opportunity and independence

At the other extreme, the "pull" theory presents business creation as the translation of an entrepreneurial culture following recognition of a business opportunity to be developed, motivated by a quest for independence and autonomy (Hughes, 2003).

The factors imputed to this second approach are coupled with positive understandings reflecting an individual's voluntary resolve to create a business. There are two dimensions to the "pull" approach to entrepreneurial motivation: (i) business opportunity and (ii) the desire for independence (Kirkwood & Campbell-Hunt, 2007).

Deci and Ryan's Theory of Intrinsic and Extrinsic Motivation

This theory, originally presented by Richard Deci in 1975 and enriched by Deci and Ryan (1985, 2002), presents originality compared to previous theories. A person aspiring to an end is said to be motivated, while another with no intention of action is said to be unmotivated. As a result, motivation and related concerns are a unitary phenomenon, varying on a scale of intensity from very little to a great deal of motivation (Deci & Ryan, 2000).

According to the theory of self-determination[1] different reasons or objectives are at the origin of actions. A distinction is made between intrinsic motivation, which drives people to carry out an action whose effect is intrinsically beneficial, and extrinsic motivation, which links action to the expectation of a dissociable result (Deci & Ryan, 1985).

This environmental approach to entrepreneurial motivation is based on two main theories: (i) the "drive theory", which presents motivation as a response to an internal need (e.g. achievement); (ii) the "incentive theory", which presents motivation as a response to external rewards (e.g. income) (Carsrud, A., 2009).

In entrepreneurship, motivation can be intrinsic or extrinsic, or both. Intrinsic and extrinsic motivation are not mutually exclusive. People can be motivated by both intrinsic and extrinsic motivations to perform an entrepreneurial act.

- Intrinsic motivation (IM):

Intrinsic motivation analyzes the performance of an activity for its intrinsic contributions, before referring to the resulting effect. By being intrinsically motivated, a person's action is not conditioned by the expectation of external rewards, but remains driven by the related pleasure or challenge (Deci & Ryan, 2000).

- Extrinsic motivation (EM):

Extrinsic motivation is a construct that considers the outcome dissociable from the activity. Extrinsic motivation is perceived when behavior responds to instrumental ends and values (gaining gratification or escaping punishment). Thus, there is considerable variation in the degree of autonomy of extrinsic motivation (Deci & Ryan, 2000).

Upon overview of the dominant theories of entrepreneurial motivation, we initiate a narrative literature review that presents the perimeter of the field of study and lines of research that have transcended the notion of entrepreneurial motivation over the last thirty years.

2. LITERATURE REVIEW ON ENTREPRENEURIAL MOTIVATION RESEARCH

This section presents a literature review on entrepreneurial motivation over the last thirty years (1983-2023). It outlines the selection criteria for selected papers, the main figures and the results of this narrative literature review.

2.1. Literature Review: Main Criteria for Selected Papers

A prerequisite for any literature review is to specify the selection criteria adopted. The main paper selection criteria for the present literature review are as follows: (1) be closely related to the object of study, i.e. entrepreneurial motivation; (2) be published between 1983 and 2023; (3) present a theoretical text or empirical study of reputed quality (Author, Institution, publication or editing journal, research interest). To this end, articles were selected and classified following a methodological selection sequence based on an initial electronic search carried out on the Scopus (Elsevier) / Web of Science (Clarivate) / JSTOR / CAIRN databases. The identification of articles matching the indicated keywords or their synonyms, i.e. "Entrepreneurial motivation" (the term most frequently used in the literature to designate the subject of our study), "entrepreneurial motives", "entrepreneurship motivational factors", "entrepreneurship motivation" and "motivation entrepreneuriale". Our review is then supplemented by a detailed manual analysis of the references cited in the articles, to compensate for the shortcomings of electronic searches.

It should be pointed out that our literature review includes a salient range of the dominant academic production on entrepreneurial motivation over the last thirty years. It cannot, therefore, claim to be exhaustive.

2.2. Literature Review: Some Key Figures

A numerical analysis of the selected publications has enabled us to classify them according to publication chronology, main authors and publication journals.

Table 1. Chronology of Published Articles on Entrepreneurial Motivation (source: authors)

Years	Paper's count
From 1983 to 2000	11
From 2001 to 2010	29
From 2011 to 2020	35

continued on following page

Table 1. Continued

Years	Paper's count
From 2021 to 2023	28
Total	**103**

The trend in the number of publications has been growing over the last thirteen years. Table 1 shows a significant number of articles published in the last three years, from 2021 to 2023.

Table 2. Main Authors of Papers on Entrepreneurial Motivation (Source: authors)

Main authors	Authors count	Contributions count
Robichaud Yves	3	2
Grilo Isabel & Thurik Roy	2	2
Sayagh Ilham	2	2
Carsrud Alan & Brännback Malin	2	1
Hessels Jolanda & Van Gelderen Marco & Thurik Roy	3	1
Kirkwood Jodyanne & Walton Sara	2	1
Collins Christopher J. & Hanges Paul J. & Locke Edwin A.	3	1
Shane Scott, Locke Edwin A., Collins Christopher J.	3	1
Murnieks Charles Y. & Klotz Anthony C. & Shepherd Dean A.	3	1
Lloyd Warren	1	1
Others	one or more authors	90
Total		**103**

Research into entrepreneurial motivation is dominated by a community of mainly Anglo-Saxon authors who have focused their papers on the theoretical production initiating understanding of the phenomenon. Currently, researchers are focusing on building the field of study through empirical applications.

Table 3. Main Publications on Entrepreneurial Motivation (Source: authors)

Publication media	Papers count
Entrepreneurship Theory and Practice	6
Journal of Business Venturing	5
Journal of Small Business Management	5
Sustainability (Switzerland)	3

continued on following page

Table 3. Continued

Publication media	Papers count
Small Business Economics	3
Innovations	3
Frontiers in Psychology	3
SSRN Electronic Journal	2
International Journal of Gender and Entrepreneurship	2
SAGE Open	2
International Journal of Entrepreneurial Behaviour and Research	2
other (67 items)	1
Total	**103**

Three journals dominated publications on entrepreneurial motivation during the period under study "Entrepreneurship Theory and Practice", "Journal of Business Venturing" and "Journal of Small Business Management". The presentation and evaluation of the three journals, taken from SCImago Journal Rank (SJR Indicator) is as follows:

- Entrepreneurship Theory and Practice - ETP: an interdisciplinary scientific journal devoted to conceptual and empirical research that develops, tests or extends theory relating to entrepreneurship, broadly defined.

Country: United States / Publisher: Wiley-Blackwell / H Index: 185

- Journal of Business Venturing - JBV: a multi-disciplinary, multi-functional and multi-contextual journal that aims to deepen our understanding of the entrepreneurial phenomenon in its many forms.

Country: United States / Publisher: Elsevier Inc / H Index: 211

- Journal of Small Business Management -JSBM: Official journal of the International Council for Small Business (ICSB), the journal's main objective is to publish scientific research articles in the fields of small business management and entrepreneurship.

Country: United Kingdom / Publisher: Taylor and Francis Ltd / H Index: 127

2.3. Results: Main Research Directions of Entrepreneurial Motivation

By visualizing the grouping of the main lines of research, we can identify previous approaches to the subject. There are four main lines of research on entrepreneurial motivation:

- Study of entrepreneurial motivation within or between countries:

A first category studies entrepreneurial motivation within a country.

A second category of studies focuses on the global level of entrepreneurial motivation, opting for comparative studies between countries.

- Study of entrepreneurial motivation in relation to company size or phases of development:

A third category of studies analyzes entrepreneurial motivation in relation to company size.

A fourth category of studies looks at entrepreneurial motivation in relation to the phases of the company's development process from start-up, growth and exit.

- Gender and ethno-cultural studies of entrepreneurial motivation:

A fifth category of studies approaches entrepreneurial motivation from a gender perspective.

A sixth category of studies analyzes entrepreneurial motivation in relation to ethnocultural references.

- Study of entrepreneurial motivation through various levels of analysis:

A seventh category of studies analyzes entrepreneurial motivation on individual levels or through meta-analyses.

- Study of entrepreneurial motivation applied to specific business models:

An eighth category of studies analyzes entrepreneurial motivation applied to specific economic models, such as the circular economy, agritourism, or specific business sectors.

A ninth line of research analyzes entrepreneurial motivation as applied to a typology of entrepreneurship: sustainable, green, social or eco-preneurship.

The methodology used in empirical studies is mainly quantitative (Anghel & Anghel, 2022) ; (Al Shammre et al., 2023) ; (Sulastiningsih & Wikaningtyas, 2023) ; (Prabowo et al., 2022) ; (Mia et al., 2022) ; (Wang et al., 2021) ; (Li et al., 2023), and is based on the use of measurement scales such as GET (*General measure of Enterprising Tendency*)[2] (Diop, 2012) (Estay et al., 2012). Qualitative studies are nonetheless evolving. we cite some of the groundbreaking works guiding this approach (Kirkwood & Walton, 2010) ; (Tjitalaksana, 2014) ; (Turegun, 2020) (Flagstad & Johnsen, 2022).

Next to the presentation of the results of the literature review, the following section highlights the contribution of qualitative methods to the study of the general field of entrepreneurship, management and the specific one of entrepreneurial motivation.

3. QUALITATIVE RESEARCH IN ENTREPRENEURIAL MOTIVATION

This section presents the nature of qualitative research in management and entrepreneurship in general terms. Secondly, an overview of the contributions of qualitative research methods on entrepreneurship and entrepreneurial motivation study is given through case studies of qualitative research applied to entrepreneurship main component.

3.1. Nature of Qualitative Research in Management and Entrepreneurship

Despite the lack of a consensual definition for qualitative research, some authors consider it to be research that describes phenomena in words, rather than numbers or measurements (Wiersma, 2000).

A number of authors tend to use a more generic expression: qualitative evaluation which is a «broad designation for a variety of approaches to evaluating social and educational programs, policies, projects, and technologies that make use of typically 'qualitative' methods for generating data and nonstatistical means of analyzing and interpreting those data» (Schwandt, 2007).

The purpose of qualitative research is to build knowledge that will give further understanding of social phenomena in natural contexts, focusing on the meanings, experiences and insights of all participants (Pope & Mays, 1995). This class of research approaches is a particularly valuable tool in fields where theory is less widely developed (Edmondson & Mcmanus, 2007). Researchers exploit these materials to investigate social and cultural issues in depth, with a strong textual focus (Myers, 2009).

Qualitative research makes use of techniques that enable the researcher to provide elaborate interpretations and uncover the proven meanings of observed phenomena without solid attachment to numerical measures (Bentalha, 2023) (Javadian et al., 2020) ; (Tjitalaksana, 2014) ; (Bajpai, 2018) ; (Zikmund et al., 2010).

In business and management, (Meyers, 2009) identifies four main qualitative research methods that are most commonly used as strategies of enquiry and means of gathering empirical data: action research "aims to solve practical problems while contributing to research in a particular academic discipline"; case study research "aims to explore or explain contemporary real-life situations; in business and management the focus is normally on one organization"; ethnographic research "relies on participant observation and fieldwork in order to provide a deep understanding of people within their social and cultural context"; and grounded theory "seeks to develop theory that is grounded in data that have been systematically gathered and analyzed". This author presents in support qualitative data collection techniques such as interviews "data gathering technique that involves questioning a subject", focus groups "the researcher interviews a group of people to get collective views on a defined topic", observation "data gathering technique where the researcher watches people from the outside with little, if any, interaction", content analysis "A systematic approach to qualitative data analysis that looks for structures and patterned regularities in the text" and discourse analysis "An approach to qualitative data analysis that focuses on the way that texts are socially constructed.". These techniques can be used either in single or in combination configuration (Myers, 2009) ; (Thiétart, 2014).

As applied to entrepreneurial motivation, qualitative research methodology aims to interpret and make an in-depth understanding of this topic (Javadian et al., 2020) ; (Hlady Rispal & Jouison-Laffitte, 2015).

Furthermore, qualitative scholars assert that the qualitative research is one of the most effective paths to make an in depth and contextualized understanding of people's motivations, reasons, beliefs and actions (Myers, 2009).

3.2. Case Studies of Qualitative Research Applied to Entrepreneurial Motivation

Our effort to consolidate qualitative research in entrepreneurial motivation is focusing on a selection of seven publications summarized in table 4.

The guiding principle in the selection of the sample in a qualitative study is that it should be very informative (Yin, 1994) ; (Miles & Huberman, 1994). The researcher selects all the groups that will allow to generate as many properties of a category as possible, in order to associate these categories and their properties with each other (Glaser & Strauss, 1967).

Our effort to consolidate qualitative research in entrepreneurial motivation is focusing on a selection of seven publications corresponding, respectively, to a specific research method. The cases are summarized in table 4.

Table 4. Summary of the Seven Publications Analyzed (source: authors)

Author(s)	**Qualitative method(s) extended**	**Research question(s)**	**Additions to entrepreneurial motivation understanding**	**Supplies for applying qualitative methods**
(Siswanto et al., 2013)	phenomenology of Schutz's	**What** are the "because motives that drive business and entrepreneurial activity in Pondok Pesantren Sidogiri (PPS) Pasuruan, East Java, Indonesia" context?	-Exploring entrepreneurial motivation by grasping "the actions of management in the business and entrepreneurial units to find the meanings". -Identifying how "external and internal environment encourages entrepreneurial development"	"Understanding the meaning of human action, where the action is a "project" through phenomenology"
(Tjitalaksana, 2014)	Single Case Study	**What** are the "motivational factors of ecopreneurship for start-up business"?	-Identifying five "motivational factors of ecopreneurship for start-up business".	Using "the case study is an investigation and analysis of a single case, intended to capture the complexity of an object of study"
(Kirkwood & Walton, 2010)	Multi-case study	**What** motivates ecopreneurs to start businesses?	-Comparing "the findings with results of extant literature on entrepreneurial motivations" -Using a sociological theoretical framework	-Making operational links through case studies are appropriate. -Using "fieldwork in the form of semi-structured interviews" -Using "(Eisenhardt, 1989) method and process for building theory" from cases that is "highly iterative" and "tightly linked to data".

continued on following page

Table 4. Continued

Author(s)	Qualitative method(s) extended	Research question(s)	Additions to entrepreneurial motivation understanding	Supplies for applying qualitative methods
(Flagstad & Johnsen, 2022)	Focus group interviews	**What** is "the role of the founder in forming a green organizational climate in small-scale manufacturing companies"?	-Identifying four motivational categories to the founders' environmental: opposition, frugality, activism and idealism	Using in-depth study in compagnies with a green profile, in a specific context: Norwegian manufacturing.
(Drews et al., 2015)	In-depth qualitative interviews	**What** are the different reasons "motivating entrepreneurs to start a business, their relationships amongst each other and with growth expectations"?	-"Motivations for starting a business are complex and that motivations other than the traditional opportunity-driven and necessity-driven distinction are more closely related to business survival and success."	Using In-depth qualitative interviews combined with Entrepreneurial Motivation Survey.
(Azam et al., 2023)	Ethnographic research design	**What** are the "motivational factors boosting female entrepreneurship in patriarchal society"?	"Themes which provide motivations for females to be an entrepreneur; personal motivation, financial stability, social support, flexible timings, and empowerment through business"	-Using ethnographic research design including the cultural contexts of society -Using "an emic viewpoint that enables its researchers to be from the same culture and setting" -Conducting in depth interviews in patriarchal society using Audio recording and field notes

continued on following page

Table 4. Continued

Author(s)	Qualitative method(s) extended	Research question(s)	Additions to entrepreneurial motivation understanding	Supplies for applying qualitative methods
(Donath Olomi, 2001)	Grounded theory approach	**How** does transition in primary motive for being in business occur?	-Contributing to the understanding of the evolution of entrepreneurial motivation through an attempt to describe the process of transition from "economic necessity" to "entrepreneurship" -Identifying conditions that trigger or favor and those that disfavor successful transitions at different levels of the development of the enterprise and the entrepreneur in the Tanzanian context	-Exploring the entrepreneurial motivation's process, develop some empirically grounded relevant concepts and a preliminary understanding of this process.

3.3. The Contributions of Qualitative Research Methods on Entrepreneurial Motivation Study

With the aim of providing a better understanding of entrepreneurship and entrepreneurial motivation, we present the contribution of qualitative research to the development of this discipline, with specific implications for entrepreneurial motivation, but also for the broader field of entrepreneurship in the Moroccan context.

The Contributions of Qualitative Research Methods on Entrepreneurial Motivation Study

From a reading of the cases presented, we can observe that qualitative researchers used different strategies to understand entrepreneurial motivation: phenomenology, single case study, multi-case study, focus group interviews, in-depth qualitative interviews and ethnographic research design.

The types of questions asked by qualitative research dealing with entrepreneurial motivation are: how, why, what and when: How do entrepreneurial motives occur? Why these motivations occur? What are the motivational factors in a specific context? When did it happen? The main questions of the analyzed cases are "What" and "how" question oriented. These questions are designed to help researchers understand entrepreneurial motivation by using open-ended questions to gather comprehensive, in-depth answers.

In terms of contributions to scientific research, two aspects are presented: the first presents advance in the understanding of entrepreneurial motivation and the second presents progress in qualitative research applied to this theme.

For this purpose, we can state that there is no shortage of advantages to be gained from using the qualitative research in entrepreneurship generally and entrepreneurial motivation research specifically. First, this type of research enables us to get the signification of human action and capture the complexity of the studied object namely entrepreneurial motivation by examining how this process changes gradually and how entrepreneurs constantly reorient and adapt to new circumstances and opportunities. Second, qualitative approach provides us making operational links as it allows us to pinpoint easily overlooked details in quantitative data especially when participants have their distinctive perceptions, readings and affective responses. Third, qualitative concerns help us building theory including the cultural contexts of society from cases with high iteration and close connection to data. Fourthly, the qualitative data are often rich and more refined and there's no need to pre-determine precise concepts and measures when collecting qualitative data.

It should be recalled, according to the results of our literature review, that the main approaches used in empirical studies on entrepreneurial motivation are quantitative (Anghel G.A. & Anghel M.A., 2022) ; (Al Shammre et al., 2023) ; (Sulastiningsih & Wikaningtyas, 2023) ; (Prabowo et al., 2022) ; (Mia et al., 2022) ; (Wang et al., 2021) ; (Li et al., 2023), and are based on the use of measurement scales such as GET (General measure of Enterprising Tendency) (Diop, 2012) ; (Estay et al., 2012). Qualitative studies are nonetheless evolving (Turegun, 2020) ; (Flagstad & Johnsen, 2022) and are in line with the aim of exploring the in depth, granular and extensive data provided by qualitative research.

All these findings suggest that qualitative approaches can be very valuable in the study and understanding of entrepreneurial motivation. In particular configurations and dynamic contexts, they can fill in the gaps of quantitative studies in order to enhance the validity and reliability required for both rigorous and relevant research. The articulation of quantitative and qualitative methods ensures a reasonable degree of external and internal validity. The two approaches (qualitative and quantitative) are complementary more than competitive, and the adoption of one should be based on the nature of the problem to be studied and its adequacy and applicability to the research questions (Silva & Sousa, 2016).

Implications for the Study of Entrepreneurship: Analysis of the Obstacles and Conditions of Development in the Moroccan Context

Entrepreneurship in Morocco is one of the mechanisms contributing to economic and social development. Its growth is supported by a range of national and regional incentives, but it faces a number of internal and external obstacles to promoting this priority lever for the national economy.

- ***Barriers to entrepreneurship in Morocco***

The Moroccan entrepreneurial ecosystem is facing a variety of obstacles that hinder its development. The manifestation of these obstacles covers administrative aspects (bureaucratic burden), financial limitations (access difficulties due to lack of guarantees), commercial constraints (economic volatility) and human aspects (lack of skills) (Touzani, 2024).

- ***Morocco's efforts to promote entrepreneurship***

As a corollary, in order to counteract these challenges in an integrated manner, various actions are being implemented by Morocco to promote entrepreneurship. Indeed, the promotion of innovation and entrepreneurship involves the launch of specific programs (e.g. Forsa Program), the simplification of administrative procedures by setting up dedicated one-stop shops (Regional Investment Centers), the facilitation of access to financing (e.g. the Damane Express program), the expansion of commercial opportunities, and the improvement of human capital.

The purpose of this package of measures is to boost innovation, job creation and economic growth (Aftiss, 2019).

Another area of action in the promotion of entrepreneurship in Morocco is the adoption of an integrative formalization approach to the informal sector, which accounts for up to 30% of GDP and employs 60% to 80% of the working population (CESE, 2021) (El Rhaz & Bouzineb, 2021). To this end, a range of programs directly or indirectly promoting the integration of the informal sector have been launched, and a series of flagship measures are planned (Alla et al., 2022).

- ***Contribution of qualitative contextualization analysis in the study of entrepreneurship***

The contribution of academic research is therefore appreciable in the analysis and understanding of the extent of the obstacles and conditions for the development of entrepreneurship in Morocco. Indeed, qualitative contextualization analysis that takes into account ethical, religious, political and social variables is of great use in the in-depth understanding of the Moroccan entrepreneurial profile and the entrepreneurial motivations behind the actions of entrepreneurs. In this regard, we reiterate the call of several authors who recall that management research in Morocco must integrate cultural specificities in order to arrive at actionable models (Louitri & Sahraoui, 2014).

CONCLUSION

Overall, the elements of this chapter provide a general framework that defines the scope of studies devoted to the entrepreneurial motivation that releases the creation of a new business venture. The framework developed in this chapter suggests some ideas for the development of this specific field.

Qualitative research is, undoubtedly, the source of a multitude of benefits for understanding a complex and multidimensional phenomenon such as entrepreneurship, through the exploration of the meanings, perceptions and experiences of human subjects by taking into account multiple and contextual factors that influence entrepreneurial behavior and outcomes.

This type of research can provide rich and relevant information about personal motivations, the institutional framework that punctuates the entrepreneurial ecosystem.

However, qualitative research remains subject to certain challenges, mainly the validity and reliability of the results. To this end, strategies such as triangulation can be adopted. In addition, the difficulties of managing the volume of data generated and the dissemination of results in this type of research can be an additional challenge.

In our opinion, qualitative research is a method that can provide academic support for a better understanding of entrepreneurship provided that the related limits are taken into account and invested in overcoming them in order to contribute to the development of the study of management sciences in general. This observation can be transposed to the current teaching context, which is subject to evolutionary adaptation, particularly with the advent of the digital era and the use of new technologies exploited within the framework of massive open online courses (Hmioui et al., 2019).

This chapter, which examines motivation in the entrepreneurial process, sought to understand its outlets through qualitative research methods. This is significant for the following two reasons. First, the qualitative research helps generating knowledge in the field of entrepreneurship. Second, qualitative research in this field provides knowledge that is applied to the key components of this process.

In view of the contributions covered, we aimed to provide a framework for highlighting the methods that illustrate the practices applied to entrepreneurial motivation in qualitative research. Despite the multitude of qualitative research methods, we cannot claim to be exhaustive. We have tried to present as representative a sample as possible. Finally, we report on the request made by entrepreneurship researchers using qualitative methods to encourage future research into the complex and innovative questions surrounding this field (Hlady-Rispal et al., 2021).

REFERENCES

Aftiss, A. (2019). L'entrepreneuriat au Maroc : Les contraintes et les conditions de développement. *Revue de l'Entrepreneuriat et de l'Innovation*, II(8), 1–9.

Al Shammre, A. S., Alshebami, A. S., Ali Seraj, A. H., Elshaer, I. A., & Al Marri, S. H. (2023). Unleashing environmental performance: The impact of green entrepreneurial motivation on small enterprises. *Frontiers in Environmental Science*, 11(April), 1–12. 10.3389/fenvs.2023.1176804

Alla, L., El Jaouhari, S., & Bentalha, B. (2022). Repenser l'entrepreneuriat au sein du secteur informel au Maroc à l'ère du post Covid-19 : Pour approche prospective de la formalisation intégrative. *Alternatives Managériales et Economiques, 4*(1), 180–202. 10.48374/IMIST.PRSM/ame-v4i1.30472

Anghel, G. A., & Anghel, M. A. (2022). Green Entrepreneurship among Students—Social and Behavioral Motivation. *Sustainability (Basel)*, 14(14), 1–19. 10.3390/su14148730

Azam, O., Zulfiqar, H., & Lillah, Z. I. (2023). An Ethnographic Study on Female Entrepreneurship in Patriarchal Structure. *Motivation and Challenges An Ethnographic Study on Female Entrepreneurship in Patriarchal Structure : Motivation and Challenges*, XIII(April), 108–118.

Bajpai, N. (2018). Business Research Methods. *Pearson Education*, 362. https://www.amazon.com/Business-Research-Methods-2nd-Edition/dp/1741032539

Bellanca, S., & Colot, O. (2014). Motivations et typologie d'entrepreneurs. *La Revue Des Sciences de Gestion*, 266(2), 63. 10.3917/rsg.266.0063

Bentalha, B. (2023). Influence des actions sociales de la RSE sur les valeurs perçues des employés : Étude qualitative des œuvres sociales de l'entreprise Taajmouati. *Repères et Perspectives Economiques, 7*(2), 183–204. 10.34874/IMIST.PRSM/RPE/43586

Borman, W. C., Ilgen, D. R., & Klimoski, R. J. (2003). *Handbook of psychology, volume twelve : Industrial and organizational psychology* (New York:).

Bruyat, C., & Julien, P. A. (2001). Defining the field of research in entrepreneurship. *Journal of Business Venturing*, 16(2), 165–180. 10.1016/S0883-9026(99)00043-9

Campbell, J., Dunnette, M., Lawler, E., & Weick, K. (1970). *Managerial Behavior Performance And Effectiveness.*

Carsrud, A., and M. B. (2009). *Understanding the Entrepreneurial Mind: Opening the Black Box* (H. Springer (ed.)).

CESE. (2021). *Une approche intégrée pour résorber l'économie informelle au MarocConseil Economique Social e Environnemental.*

Chelly, A. (2004). *L'identification et développement d'opportunités entrepreneuriales.*

da Silva, O. D. L., & de Sousa, Á. S. T. (2016). Gestion de la qualité dans la recherche en sciences sociales et humaines : articulation entre les méthodes qualitatives et quantitatives. *Recherches Qualitatives, 18*, 171–181. http://www.recherche-qualitative.qc.ca/Revue.html

Danjou, I. (2002). l'entrepreneuriat : Un champ fertile à la recherche de son Unité. *Revue Française de Gestion*, 28(138), 109–125.

Deci, E. L., & Ryan, R. M. (1985). The general causality orientations scale: Self-determination in personality. *Journal of Research in Personality*, 19(2), 109–134. 10.1016/0092-6566(85)90023-6

Deci, E. L., & Ryan, R. M. (2000). The "What" and "Why" of Goal Pursuits: Human Needs and the Self-Determination of Behavior. *Psychological Inquiry*, 11(4), 227–268. 10.1207/S15327965PLI1104_01

Deci & Ryan. (1985). intrinsic motivation and self- determinaton in human behaviour. In *Plenum Press* (Springer S).

Diop, P. M. (2012). *La motivation entrepreneuriale dans le contexte sub-saharien francophone.*

Donath Olomi, B. R. (2001). Evolution Of Entrepreneurial Motivation: The Transition From Economic Necessity to Entrepreneurship. *Business Management Review*, 7(2), 107–136.

Drews, C. (2015). *Understanding Motivations for Entrepreneurship A Review of Recent Research Evidence Ute Stephan, Mark Hart and Cord-. February*, 1–54. 10.13140/RG.2.1.3343.2165

Drews, C., Stephan, U., & Hart, M. (2015). Understanding Motivations for Entrepreneurship: A Review of Recent Research Evidence. *Enterprise Research Centre, February*, 1–54. 10.13140/RG.2.1.3343.2165

Edmondson, A. C., & Mcmanus, S. E. (2007). Methodological fit in management field research. *Academy of Management Review*, 32(4), 1155–1179. 10.5465/amr.2007.26586086

Eisenhardt, K. M. (1989). Building Theories from Case Study Research Published by : Academy of Management Stable. *Academy of Management Review*, 14(4), 532–550. 10.2307/258557

El Rhaz, L., & Bouzineb, Y. (2021). Le secteur informel au Maroc: Principales caractéristiques et tendances générales d'évolution. *Les Brefs Du Plan*, 212(0), 1–5.

Estay, C., Durrieu, F., & Diop, P. M. (2012). Motivation entrepreneuriale et logique d'action du créateur. *Revue Internationale P.M.E (Norwalk, Conn.)*, 24(1), 135–165. 10.7202/1012554ar

Fenouillet, F. (2009). Chapitre 16. Vers une approche intégrative des théories de la motivation. In *Traité de psychologie de la motivation* (pp. 305–338). 10.3917/dunod.carre.2019.01.0305

Filion, L. J. (1997). Le champ de l ' entrepreneuriat : historique, évolution, tendances Le champ de l ' entrepreneuriat : historique, évolution, tendances. *Revue Internationale P.M.E. Économie et Gestion de La Petite et Moyenne Entreprise, 10*(2), 129–172. https://id.erudit.org/iderudit/1009026ar

Flagstad, I., & Johnsen, S. Å. K. (2022). The psychology of green entrepreneurship: Founder-driven development of green climate in small-scale companies. *Cogent Business and Management*, 9(1), 2079245. Advance online publication. 10.1080/23311975.2022.2079245

Gabarret, I., & Vedel, B. (2015). Pour une nouvelle approche de la motivation entrepreneuriale. *La Revue Des Sciences de Gestion*, 271(1), 13. 10.3917/rsg.271.0013

Gartner, W. B. (1988). "Who is an entrepreneur?" is (still) the wrong question. *American Journal of Small Business*, 12(4), 11–32. 10.1177/104225878801200401

Glaser, B. G., & Strauss, A. L. (1967). *The Discovery of Grounded Theory: Strategies for Qualitative Research*. Aldine Tra., 10.1037/h0083159

Hessels, J., Van Gelderen, M., & Thurik, R. (2008). Entrepreneurial aspirations, motivations, and their drivers. *Small Business Economics*, 31(3), 323–339. 10.1007/s11187-008-9134-x

Hlady-Rispal, M., Fayolle, A., & Gartner, W. B. (2021). In search of creative qualitative methods to capture current entrepreneurship research challenges. *Journal of Small Business Management*, 59(5), 887–912. 10.1080/00472778.2020.1865541

Hlady Rispal, M., & Jouison-Laffitte, E. (2015). La contribution des méthodes qualitatives au développement du champ de l'entrepreneuriat. *Revue de l'Entrepreneuriat*, 14(1), 15–40. 10.3917/entre.141.0015

Hmioui, A., Alla, L., & Bentalha, B. (2019). Évaluation des formations en ligne : Etude qualitative des perceptions des utilisateurs des MOOC au Maroc. *ICT Day's19*.

Hughes, K. D. (2003). Pushed or pulled? Women's entry into self-employment and small business ownership. *Gender, Work and Organization*, 10(4), 433–454. 10.1111/1468-0432.00205

Javadian, G., Dobratz, C., Gupta, A., Gupta, V. K., & Martin, J. A. (2020). Qualitative Research in Entrepreneurship Studies: A State-of-Science. *The Journal of Entrepreneurship*, 29(2), 223–258. 10.1177/0971355720930564

Kirkwood, J., & Campbell-Hunt, C. (2007). Using Multiple Paradigm Research Methodologies To Gain New Insights Into Entrepreneurial Motivations. *Journal of Enterprising Culture*, 15(03), 219–241. 10.1142/S0218495807000137

Kirkwood, J., & Walton, S. (2010). What motivates ecopreneurs to start businesses? *International Journal of Entrepreneurial Behaviour & Research*, 16(3), 204–228. 10.1108/13552551011042799

Kirzner, I. (1973). *Competition and Entrepreneurship*.

Li, Y., Mohd Nordin, N. R., Akhter, S., Kumar, T., & Shaheen, M. (2023). Does green entrepreneurial behavior enhance through entrepreneurship education, perceived-ability to use technology, and commitment to environment? Understanding the contribution of entrepreneurial motivation and university support. *Ekonomska Istrazivanja*, 36(3), 2153375. Advance online publication. 10.1080/1331677X.2022.2153375

Louche, C. (2019). Chapitre 6. La motivation au travail. In *Psychologie sociale des organisations*. Paris: Dun.

Louitri, A., & Sahraoui, D. (2014). Contexte et contextualisation dans les recherches en Management au Maroc. *Revue Marocaine Des Sciences De Management*, 85–100.

Low, M. B., & MacMillan, I. C. (1988). Ent past research and future challenges. *Journal of Management*, 14(2), 139–161. 10.1177/014920638801400202

Mcmullen, J. S., Bagby, D. R., & Palich, L. E. (2008). Economic freedom and the motivation to engage in entrepreneurial action. *Entrepreneurship Theory and Practice*, 32(5), 875–895. 10.1111/j.1540-6520.2008.00260.x

Mia, M. M., Rizwan, S., Zayed, N. M., Nitsenko, V., Miroshnyk, O., Kryshtal, H., & Ostapenko, R. (2022). The Impact of Green Entrepreneurship on Social Change and Factors Influencing AMO Theory. *Systems*, 10(5), 132. Advance online publication. 10.3390/systems10050132

Miles, M. B., & Huberman, M. A. (1994). *Qualitative Data Analysis*. Sage Publi.

Myers, M. D. (2009). Qualitative Research in Bsiness & Management. In *Revista Brasileira de Linguística Aplicada* (SAGE Publi, Vol. 5, Issue 1). https://revistas.ufrj.br/index.php/rce/article/download/1659/1508%0Ahttp://hipatiapress.com/hpjournals/index.php/qre/article/view/1348%5Cnhttp://www.tandfonline.com/doi/abs/10.1080/09500799708666915%5Cnhttps://mckinseyonsociety.com/downloads/reports/Educa

Naylor, J. C., Pritchard, R. D., & Ilgen, D. R. (1980). Theory of Behavior in Organizations. In *Organizational Behavior 6*. Academic P. 10.4324/9781315701967-22

Pope, C., & Mays, N. (1995). Qualitative Research: Reaching The parts other methods cannot reach: An introduction to qualitative methods in health and health services research. *BMJ (Clinical Research Ed.)*, 311(6996), 42–45. 10.1136/bmj.311.6996.427613329

Prabowo, H., Ikhsan, R. B., & Yuniarty, Y. (2022). Drivers of Green Entrepreneurial Intention: Why Does Sustainability Awareness Matter Among University Students? *Frontiers in Psychology*, 13(March), 1–17. 10.3389/fpsyg.2022.87314035418917

Ruth, K. (1990). Motivation theory and industrial and organizational psychology. *Handbook of Industrial and Organizational Psychology, 1*(2), 75–130.

Schwandt, T. A. (2007). The SAGE Dictionary of Qualitative Inquiry. In *The SAGE Dictionary of Qualitative Inquiry*. 10.4135/9781412986281

Shane, S., Kolvereid, L., & Westhead, P. (1991). An exploratory examination of the reasons leading to new firm formation across country and gender. *Journal of Business Venturing*, 6(6), 431–446. 10.1016/0883-9026(91)90029-D

Shepherd, D. A. (2015). Party On! A call for entrepreneurship research that is more interactive, activity based, cognitively hot, compassionate, and prosocial. *Journal of Business Venturing*, 30(4), 489–507. 10.1016/j.jbusvent.2015.02.001

Siswanto, Armanu, Setiawan, M., & Nimran, U. (2013). Entrepreneurial Motivation in Pondok Pesantren. *International Journal of Business and Behavioral Sciences*, 3(2), 42–54.

Solymossy, E. (1998). *Entrepreneurial dimensions : The relationship of individual, venture, and environmental factors to success: Vol.* Academia.e.

Sørheim, R., & Landström, H. (2001). Informal investors - A categorization, with policy implications. *Entrepreneurship and Regional Development*, 13(4), 351–370. 10.1080/08985620110067511

Stefanovic, I., Prokic, S., & Rankovic, L. (2010). Motivational and success factors of entrepreneurs: The evidence from a developing country. *Zbornik Radova Ekonomskog Fakultet Au Rijeci*, 28(2), 251–269.

Stevenson, H. H. (1984). A Perspective on Entrepreneurship. *Harvard Business School, 9*(384–131), 85–94. https://books.google.de/books?hl=de&lr=&id=jqWuD7dF3GcC&oi=fnd&pg=PA7&dq=A+Perspective+On+Entrepreneurship&ots=IfMZFRw4e4&sig=NaodtCg2v8fPel2eVhspBHlGmmM

Stevenson, H. H., & Jarillo, J. C. (1990). A paradigm of entrepreneurship: Entrepreneurial management. *Strategic Management Journal, 11*, 17–27. 10.1007/978-3-540-48543-8_7

Sulastiningsih & Wikaningtyas, S. U. (2023). Green Entrepreneurship Motivation Model. *Academic Journal of Interdisciplinary Studies, 12*(2), 262. 10.36941/ajis-2023-0047

Thiétart, R.-A. (2014). Méthodes de recherche en management. Dunod. 10.3917/dunod.thiet.2014.01

Tjitalaksana, S. M. (2014). Identification of Five Motivational Factors of Ecopreneurship for Start-Up Business. *Universitas Telkom*, 1–8. https://openlibrary.telkomuniversity.ac.id/pustaka/files/100142/jurnal_eproc/identification-of-five-motivational-factors-of-ecopreneurship-for-start-up-business-a-case-study-of-enigami-papers-.pdf

Touzani, A. (2024). L ' Entrepreneuriat et développement économique au Maroc : Défis et Perspectives. *Revue Franaise d'Economie et de Gestion*, 5(5), 218–235.

Turegun, O. (2020). Motivational Factors of Ecopreneurs. *Uluslararası Akademik Yönetim Bilimleri Dergisi*, 6(9), 10–19.

Wang, W., Cao, Q., Zhuo, C., Mou, Y., Pu, Z., & Zhou, Y. (2021). COVID-19 to Green Entrepreneurial Intention: Role of Green Entrepreneurial Self-Efficacy, Optimism, Ecological Values, Social Responsibility, and Green Entrepreneurial Motivation. *Frontiers in Psychology*, 12(October), 732904. Advance online publication. 10.3389/fpsyg.2021.73290434721199

Wiersma, W. (2000). Research Methods in Education: An Introduction. Allyn and Bacon.

Yin, R. K. (1994). Case study research and applications: Design and methods. In *Journal of Hospitality & Tourism Research.* Sage Publi. 10.1177/109634809702100108

Zikmund, W., Babin, B., Carr, J., & Griffin, M. (2010). Business Research Methods Eight Edition. *South Western Educ Pub*, 668.

ENDNOTES

1 This theory proposes an "organismic conception of intrinsic and extrinsic motivation based on the dialectic between the environment and the individual in his or her quest to satisfy psychological needs" (Fenouillet, 2009, p.49)

2 Questionnaire that brings together the needs that drive entrepreneurial motivation (fulfillment, autonomy and control), as well as the entrepreneur's risk-taking and creative abilities.

Chapter 7
Enhancing Policy Evaluation:
The Role of Qualitative Method in Understanding Self-Employment Dynamics

Saadeddine Igamane
https://orcid.org/0000-0002-6675-2278
Sidi Mohamed Ben Abdellah University, Morocco

Abdelhay El Mouatassem
Sidi Mohamed Ben Abdellah University, Morocco

ABSTRACT

This chapter examines the significance of qualitative methodologies in evaluating self-employment policies. The central question is how qualitative approaches enhance policy evaluation in self-employment. The chapter highlights the importance of qualitative studies, such as the biographical approach and life narratives, in offering interpretative perspectives that enrich policy assessments and inform evidence-based decision-making. Qualitative methodologies play a crucial role in providing detailed contextual insights, uncovering complexities, and understanding programmatic effects. By examining contextual influences, implementation strategies, and individual characteristics, qualitative studies help to develop comprehensive theories of change, leading to more responsive and adaptable policies. The chapter emphasizes the need to integrate qualitative and quantitative approaches for a mixed methods policy evaluation. This approach enables policymakers and researchers to develop evidence-based policies that address the diverse needs and challenges of self-employed individuals.

DOI: 10.4018/979-8-3693-5543-5.ch007

INTRODUCTION

Morocco's economy, despite showing signs of growth and development, faces complex socio-economic challenges, particularly in the labor market and self-employment. Data provided by the High Commission for Planning (HCP, 2013-2022) clearly reflect the impacts of climatic, geopolitical, and pandemic factors on the national labor market. The activity rate was 44.3%, with spatial disparities (41.9% in urban areas and 19.8% for women, while for young people aged 18-24, the rate was only 22.8%). The activity rate even declined in 2022 compared to the reference year (2013), dropping from 48.7% to 44.3%. The same observation regarding the worsening situation in the labor market is confirmed when considering the employment rate (which decreased from 44% to 39.1%), the unemployment rate (which increased from 9.6% to 11.8%, with 68.5% being long-term unemployed). The unemployment rate for higher education graduates reached 25.8% in 2022.

Issues such as unemployment, job insecurity, and economic inequalities converge, highlighting the urgent need to understand self-employment dynamics. There is a persistent imbalance between job supply and demand, with significant progress in economic development not fully addressing unemployment, especially among highly qualified graduates struggling to find jobs matching their skills. This skills-job mismatch underscores the need to rethink employment approaches, with direct economic implications (McKee-Ryan & Harvey, 2011).

In response to these ongoing economic challenges, self-employment has become a cornerstone of Morocco's economy. Public policies have shifted focus toward promoting self-employment through initiatives like cooperatives and the self-employed status, aligning with broader economic liberalism and diversification efforts. However, transitioning to self-employment presents its own set of challenges.

Self-entrepreneurship, once the domain of the economic elite, has become an alternative path, now accessible to a broader population, including young people and women often in precarious situations. Support programs for self-employment have driven this change, expanding economic opportunities and fostering empowerment.

Self-employment, often referred to as independent or autonomous work, is categorized into three sub-categories by employment statisticians: independent work without employees, independent work with employees, and unpaid family workers. Independent work without employees includes self-employed individuals, while independent work with employees pertains to employers according to the Economic, Social and Environmental Council (CESE, 2016).

Ensuring full employment, where finding a job is not difficult and any observed unemployment is frictional and short-term, is a key economic policy goal. However, underemployment occurs when a person works below legal standards, earns a low wage, or has lower productivity than their potential, and can also be seen in self-employment.

In Morocco, self-employment should include employers, self-employed workers, cooperative members, or associates, according to the High Commissioner for Planning (HCP, 2022). In 2014, self-employment involved 32.5% of the active population, with a low female participation rate (14.4%) and equal distribution between urban and rural areas. Over 85% of self-employed individuals worked for their own account, with 74.5% lacking a diploma. Besides agriculture and commerce, liberal professions were also represented. Self-employment contributed to an average of 15,000 new jobs annually between 1999 and 2014, mainly among self-employed individuals (82%) and cooperative members (9%). However, national surveys highlighted the precarious nature of these jobs, rarely improving living standards or providing adequate social coverage (CESE, 2016).

The informal sector is also a significant source of employment and self-employment in Morocco. In 2007, informal production units provided 2,216,116 jobs, with employment rates of 53% in commerce, 21.5% in industry, 19% in services, and 9% in construction (HCP, 2006-2007).

This article aims to show how qualitative studies can provide more understanding of public policy impact and relevance, especially in the field of employment long dominated by quantitative studies, particularly by public organizations and engineering offices. In this article, we will discuss the various methodological approaches utilized in assessing the effectiveness of public policies related to employment and self-employment. We will delve into the intricacies of both quantitative and qualitative methods, exploring their applications, advantages, and challenges in evaluating the impacts of these policies. Additionally, we will examine the emerging trend of employing innovative methodologies such as the biographical approach, which offers a holistic understanding of self-employment within broader social, cultural, and economic contexts. Through this exploration, we aim to provide insights into the complexities of policy evaluation and offer guidance for future research endeavors in this field.

1. FROM ECONOMETRIC APPROACHES TO QUALITATIVE STUDIES IN PUBLIC POLICY EVALUATION

This section highlights the complexity and diversity of methodological approaches used to evaluate employment and self-employment policies, emphasizing the importance of combining quantitative and qualitative methods for a comprehensive understanding. It sets the foundation for the study by outlining the specific methodology to be used in evaluating the effectiveness of a self-employment policy, considering contextual elements, and exploring various dimensions of the policy.

It is crucial to examine the relevance of public self-employment policies and understand the unintended effects of the self-employed status on beneficiaries. Additionally, studying the reasons behind the failure of some cooperatives, despite receiving public financial support, is essential for refining future policies. Thus, a mixed-methods approach, combining qualitative and quantitative methods, is necessary to deeply analyze the real impact of these policies on individuals. This balanced approach will capture the subtle nuances of individual experiences while quantifying measurable changes resulting from these policies.

Public employment policies are critical for governments and policymakers. To assess their effectiveness and impacts, researchers use a variety of methods, including econometric and qualitative techniques. These approaches have evolved to provide a deeper and more holistic understanding of the issues at hand.

Traditionally, econometric techniques, along with experimental and quasi-experimental methods, have dominated the analysis of public employment policies. For instance, quantitative data can evaluate the impact of self-employment policies, such as tax incentives or professional training, on the creation and sustainability of individual businesses. However, this approach may not fully capture the contextual factors and social dynamics influencing outcomes.

With the development of econometric and qualitative survey techniques, the evaluation of public policies has become a decision-making aid tool. Beyond policy analysis, we can now speak of analysis for policy (Hill, 2005). It is no longer limited to the analysis of causes and consequences (Dunn, 2011) but extends to understanding an entire process (Thissen & Walker, 2013). Indeed, evaluation can now take place even before the policy is launched, allowing for the choice between several alternatives.

In the field of employment in general and self-employment in particular, evaluation can take place before the actual launch of projects (ex-ante evaluation), during their implementation (formative evaluation), or at the end, within the framework of impact studies, for example.

Recently, the importance of qualitative and multidisciplinary approaches has been recognized for a more comprehensive understanding of public employment policies. Evaluating self-employment policies, for example, requires considering the personal stories and experiences of individuals. Biographical studies and qualitative methods allow researchers to collect testimonies and explore individual trajectories, highlighting the challenges, motivations, and obstacles that shape self-employment decisions. Public action narratives can demystify observed outcomes and contextualize evaluations. For instance, if a self-employment policy does not produce the expected job creation results, policymakers' stories might reveal constraints faced, adjustments made, and lessons learned for future policies.

Recognizing that stakeholders—whether employers, workers, entrepreneurs, or job seekers—are influenced by cultural, social, and psychological factors is crucial in analyzing public employment policies. Cultural norms, attitudes towards risk, social networks, and individual perceptions of success and failure play significant roles in self-employment decisions. Qualitative approaches capture these nuances and better understand the motivations behind individual choices.

In several fields such as economics and education sciences, the use of qualitative methods has been very useful for evaluating the performance of training (Hmioui & Bentalha, 2019), studying the resilience of certain sectors like tourism (Alla, Bentalha, Megzari, & Hmioui, 2022), analyzing social responsibility for enterprises and employees' values (Bentalha B., 2023a) and studying the performance of certain services in enterprises (Bentalha B., 2023b). The case study is also one of the more qualitative techniques used in several fields related to economics, political sciences, and social sciences in general (Lamsiah & Bentalha, 2022).

A self-employment policy, for instance, cannot be effectively evaluated without considering the personal narratives and experiences of the individuals directly affected by it. This is where biographical studies and qualitative approaches become essential. These methods provide valuable tools for gathering personal testimonies and exploring individual career trajectories, offering a deeper understanding of the real-life challenges, motivations, and obstacles that shape self-employment decisions. By delving into these personal experiences, researchers can uncover insights that go beyond mere statistical outcomes and reveal the underlying reasons why a policy might succeed or fail. For example, if a self-employment policy fails to meet its job creation targets, qualitative research can illuminate the constraints faced by policymakers, the adjustments made during the policy's implementation, and the lessons learned for future efforts. Furthermore, in analyzing public employment policies, it is crucial to recognize that stakeholders—whether employers, workers, entrepreneurs, or job seekers—are influenced by a complex interplay of cultural, social, and psychological factors. In the case of self-employment policies, cultural norms, attitudes towards risk, social networks, and individual perceptions of success and failure all play significant roles in the decision to pursue self-employment. Qualitative methods are particularly adept at capturing these nuanced influences, which are often overlooked in quantitative analyses. Additionally, integrating insights from behavioral economics, which focuses on the psychological and non-material aspects of human behavior, offers a more comprehensive framework for understanding public employment policies. Behavioral economics introduces concepts such as cognitive biases, individual preferences, and social norms that challenge traditional assumptions of rational decision-making. Applying these concepts can help researchers develop more nuanced econometric models that account for factors like overconfidence, risk aversion, and societal expectations in self-employment decisions. By incorporating these behavioral insights, researchers can enhance traditional models and provide a richer, more accurate analysis of the factors driving self-employment choices and policy outcomes.

Integrating insights from behavioral economics, which emphasizes non-material and non-monetary aspects of human behavior, provides a more complete perspective in analyzing public employment policies. Concepts from this field can enrich traditional econometric models by considering psychological and cultural variables. For self-employment policies, this could mean understanding how cognitive biases, individual preferences, and social norms influence self-employment choices.

When evaluating the effectiveness of a public self-employment policy, qualitative approaches and behavioral economics insights add a crucial dimension to the analysis. For example, biographical narratives help understand the personal experiences and motivations of individuals affected by self-employment policies. In-depth interviews with entrepreneurs who have benefited from these incentives can reveal why they chose self-employment, the challenges they faced, their successes, and their overall perception of the policy's impact. Biographical narratives provide emotional and contextual depth that quantitative methods cannot capture. They also highlight the most relevant aspects of the policy and identify areas for potential adjustments.

Behavioral economics offers a richer perspective on the motivations and behaviors of individuals regarding self-employment. Using concepts like cognitive biases, time preferences, and social influences can explain why certain policies have different impacts than expected. For instance, despite tax incentives, many individuals might still prefer traditional employment due to a preference for financial security and stability. This deeper understanding of human behavior helps refine policy design to better address individuals' real motivations.

Combining biographical narratives with behavioral economics insights gives researchers a holistic view of the impact of self-employment policies. Analyzing qualitative interviews might reveal entrepreneurs' reluctance due to income uncertainty. Behavioral economics can explain how risk aversion significantly influences their decisions. This could then guide policy adjustments to provide financial safety nets or specific training to help entrepreneurs manage uncertainty.

Integrating qualitative studies and behavioral economics concepts significantly enriches the evaluation of self-employment policies. These approaches account for the complexity of individual motivations, the emotional and psychological factors influencing choices, and guide the development and adjustment of policies to better meet the needs of entrepreneurs and self-employed workers.

Overall, public employment policy analysis benefits greatly from adopting qualitative and multidisciplinary approaches. For areas like self-employment policy, these approaches capture the complexity of factors influencing individual decisions and provide a more complete view of policy impacts. Combining econometric techniques, qualitative methods, and behavioral economics insights allows researchers to deepen their understanding and more effectively inform employment policy decisions.

2. STATE OF THE ART IN METHODOLOGIES EMPLOYED TO EVALUATE EMPLOYMENT AND SELF-EMPLOYMENT

The evaluation of employment and self-employment holds paramount importance in understanding economic and social policies. Researchers have developed a range of approaches and methods to rigorously analyze the impacts of such policies. In this section, we will undertake a thorough examination of the state of the art in employment evaluation studies, with a focus on previous studies specifically addressing self-employment.

We will begin by presenting various studies conducted in the field of employment, where researchers have analyzed the effectiveness of policies aimed at promoting job creation and reducing unemployment. These studies will provide an overview of the different approaches used, ranging from econometric methods to randomized controlled experiments. We will also examine the findings of these studies, highlighting key lessons learned from their application. Next, we will specifically address studies that have focused on self-employment. Policies aimed at supporting entrepreneurs and independent workers have garnered increasing interest in recent years. We will analyze the varied approaches used to evaluate the impact of these policies on the creation and viability of individual businesses. We will explore qualitative methods, behavioral economics-based approaches, and econometric techniques that have been employed to understand the complexity of the self-employment domain. Finally, we will synthesize key insights from these studies regarding evaluation approaches and methods for a better understanding of self-employment policies. Emerging trends, observed successes, and challenges encountered in these studies will help us identify best practices and potential gaps in evaluating the impact of self-employment policies. This comprehensive exploration will guide our understanding of how methods and approaches can be adapted for more accurate and informed evaluation of future policies in the field of self-employment.

2.1. Case Studies in the Evaluation of Public Employment Policies

The review of youth employment programs through various studies conducted globally has highlighted significant attention to quasi-experimental methods to elucidate impacts that are sometimes challenging to discern. Among these studies, several researchers (Kontos, 2008; Fang Cai, 2008) have adopted different approaches, including biographical or quasi-experimental methods, to evaluate the effects of these programs on youth employment. This often involves creating both a treatment group and a control group, thereby enabling the measurement of the policy's causal effect.

However, it is crucial to emphasize that the field of youth employment and employment policies cannot be solely understood through experimental and quasi-experimental methods. The complex nature of these issues requires a more holistic and diversified approach. Indeed, qualitative, multidisciplinary, and contextual approaches play an equally important role in seeking a comprehensive and balanced understanding of policy impacts on youth employment.

Comparison of Employment Policy Impact Evaluation Studies

In their study, Crépon, Duflo, Gurgand, Rathelot, and Zamora (2014) addressed the issue of long-term unemployment among highly educated individuals in France. They evaluated a private intermediation device designed to assist these young unemployed individuals in their job search. The study aimed to determine whether this device promoted stable employment and measured the effect of the program on the beneficiaries' professional futures, specifically the probability of being employed after the program. The evaluation used a quasi-experimental method with randomization, dividing participants into a treatment group (followed by an interim agency) and a control group (not followed). The counterfactual method assumed that both groups would have identical outcomes in the absence of the intervention. The results revealed a displacement effect, where companies preferred to hire the assisted youths over others, favoring the treatment group. The authors recommended an efficiency evaluation considering results versus costs.

Benoteau (2015) focused on unemployed individuals with social and professional access difficulties. The study evaluated multiple parallel devices to determine if these programs favored short-term recruitment and medium-term integration of beneficiaries. The evaluation employed a quasi-experimental method using propensity score matching to create treatment and control groups. The findings indicated variable effectiveness of the devices, depending on the contract's duration, content, and employer type. The study recommended appropriate targeting to avoid windfall effects.

Summary of Impact Evaluation Studies

Several notable studies have contributed to the understanding of employment policy impact evaluations:

Corinna Ghirelli and others (Ghillelli, 2019) evaluated the WELL (Work Experience Laureati et Laureate) program. The final sample comprised 5,816 participants. Using propensity score matching, the study found that participants in WELL were more likely to secure employment.

Jonas Maibom and others (Maibom et al., 2014) examined the effects of PAMT (Active Labor Market Policy Expenditures) on 3,380 participants using randomized controlled trials (RCT). The intensified program reduced employment for non-schooled youths and had no impact on educated youths.

Pierre Cahuc and others (Cahuc et al., 2012) studied the Rebond program, which supports higher education dropouts, involving 2,367 youths. The RCT method revealed that enhanced support positively impacted the number of job interviews and the number and quality of jobs obtained by beneficiaries.

Orazio Attanasio and others (Attanasio et al., 2015) evaluated the Jóvenes program, a training initiative for young urban unemployed individuals aged 18-25. The RCT approach showed a positive and significant effect on the probability of working in the formal sector.

Stijn Broecke (Broecke, 2012) assessed the SIVP (Professional Integration Internship) involving 4,763 youths. The study estimated the probability of undergoing SIVP and its effect on employment probability, finding that SIVP positively impacted employment, especially for high-risk unemployed graduates.

Groh Matthew and others (Groh et al., 2016) investigated a non-technical skills training program for 1,347 female college graduates using the randomized controlled trial (RCT). The results indicated that the training had no significant impact on employment.

These studies highlight the diverse methodologies and outcomes in evaluating employment policies, demonstrating the benefits and limitations of various approaches in understanding the effectiveness of employment interventions across different socio-economic contexts.

Qualitative approaches, such as in-depth interviews, case studies, and participatory observations, capture individual experiences and subtle nuances of the youth employment process. These methods provide rich information on the challenges faced by young people, their expectations, perceptions, and complex interactions with their socio-economic environment. Additionally, multidisciplinary approaches are essential for addressing youth employment from various angles, considering economic, sociological, psychological, and cultural dimensions. A balanced evaluation must account for the diversity of factors influencing the success or failure of these policies.

Lastly, the importance of context cannot be underestimated in evaluating youth employment policies. Economic conditions, local specificities, labor market needs, and socio-political developments play a crucial role in how these policies are perceived and implemented.

Previous studies evaluating public employment policy offer a diverse landscape of methodological approaches used to analyze the effectiveness of various interventions aimed at promoting employment and professional integration. By examining a selection of studies, trends, divergent results, and important lessons for policy development and implementation can be discerned. One widely employed approach is quasi-experimental methodology, which attempts to isolate the causal effect of a policy by comparing a treatment group to a control group. A study conducted by Crépon and others (Crépon et al., 2014) in France, focusing on a private mediation program for highly educated young unemployed individuals, revealed that the program promoted stable employment. While this approach yields significant results by creating a relevant comparison between groups, it may also face issues such as displacement effects or musical chairs effects in the labor market.

However, the diversity of methodological approaches stands out. A study by Benoteau (2015) in France used propensity score matching to evaluate the effect of hiring under subsidized contracts on the career trajectory of individuals facing difficulties in accessing employment. This method emphasizes creating a control group comparable to the treatment group in terms of observable characteristics. This study found that the effectiveness of the programs varied depending on the duration and content of the contracts, as well as the type of employer. This highlights the importance of appropriately targeting policies to avoid windfall effects.

When examining the results of studies focused on self-employment, a range of methodological approaches is observed. For example, a study by Corinna Ghirelli and others (Ghirelli et al., 2019) in Italy used matching to demonstrate that the WELL program increased the probability of employment for participants. In contrast, a study by Jonas Maibom et al. (2014) in Denmark revealed that the intensified program had no significant impact on the employment of non-educated youth and negatively impacted educated youth.

Internationally, a study by Orazio Attanasio and others (Attanasio et al., 2015) in Colombia used a method based on randomized experiments to show that the Jóvenes program had a positive and significant effect on the probability of working in the formal sector. Similarly, a study by Broecke in Tunisia (Broecke, 2012) on the SIVP on the SIVP program demonstrated that it had a positive effect on the employment probability of youth, especially for high-risk unemployed graduates.

However, null results are also observed, as illustrated by the study conducted by Matthew Groh and others (Groh et al., 2016) in Jordan, where non-technical skills training had no significant impact on the employment of college graduates.

Thus, previous studies on the evaluation of public employment policy offer a range of methodological approaches to measure the effectiveness of different interventions. While some methods may provide significant results, other studies highlight mixed or null results. This variability underscores the importance of carefully considering methods and adapting policies based on the specific context to ensure optimal outcomes.

In analyzing employment-related public policies, a multidisciplinary approach is essential for understanding the multiple dimensions of the potential effects of a given policy or program. The impacts of a policy are not always limited to direct effects and can extend to areas initially not targeted, such as pollution levels, living standards, the well-being of targeted and non-targeted individuals, as well as the legal framework. These impacts, although important, are often challenging to quantify precisely. Evaluation relies on comparing the state of non-existence of the policy with the outcome resulting from its implementation.

This consideration of various dimensions and non-immediately evident impacts is particularly relevant when examining policies related to self-employment. Indeed, the implications of self-employment extend beyond simply creating new businesses or job opportunities. They also touch on aspects such as individual autonomy, labor market flexibility, and even the economic stability of a given region.

To illustrate these concepts, we will examine some previous studies on self-employment, highlighting the different methodological approaches used and the results obtained. This comparison will allow us to better understand how various approaches can influence conclusions and draw lessons for improved evaluation of self-employment public policies.

2.2 Case Studies in the Evaluation of Public Self-Employment Policies

Studies on self-employment have primarily focused on entrepreneurship, particularly within the informal sector, which remains a significant source of employment opportunities even in developed countries like China and England. There is now a recommendation to integrate this sector into public policies by replacing initiatives aimed at its eradication with formalization strategies. Qualitative and biographical approaches are strongly encouraged as they allow for a better understanding of individual trajectories and on-the-ground realities.

However, studies on public policies in this field seem to be less abundant. While the informal sector has long been seen as a problem to be eliminated, programs for its formalization are emerging to reduce unemployment. The status of self-entrepreneur has been established in several nations, such as France and Morocco, to formalize this sector. At the international level, research on self-employment is less frequent compared to studies on wage and professional integration programs for youth and women. The informal sector is not limited to developing countries, as it is also a reality in more advanced nations like China and England.

The following paragraphs provide a summary of key studies evaluating self-employment policies, focusing on identified problems, research questions, evaluation methods, and key findings.

Kontos (2008) examines the situation of women in the labor market, with a particular focus on self-employment. The primary research question investigates how support policies for self-employment influence women's experiences. This study employs a biographical approach, utilizing life narratives to gain deeper insights into the lived experiences of women entrepreneurs. The results underscore the importance of complementing quantitative analyses with qualitative ones. Specifically, the study finds that women's entrepreneurship is deeply embedded in their biographical processes, highlighting the necessity for policy evaluations to consider these personal and contextual factors.

Williams (2005) addresses the role of undeclared work in England, considering it as a potential source of employment and wealth creation. The research explores how perceptions of undeclared work influence public policy approaches. Using a qualitative approach based on 861 face-to-face interviews, the study reveals that undeclared work contributes significantly to added value and self-employment, with participants engaging in such activities often taking calculated risks. The findings suggest that public policies should not aim to eradicate the undeclared sector but rather find ways to formalize it, acknowledging its substantial role in creating self-employment opportunities.

Fang Cai and Meiyan Wang (2008) investigate income disparities and other factors among self-employed farmers across various regions in China. The study seeks to determine whether there is a significant income difference between rural and urban areas, considering the recognized technological advancements in urban regions. Utilizing a counterfactual method, the study finds no significant income disparities between rural and urban self-employed farmers. This result challenges the common assumption that urban areas inherently offer higher income opportunities due to technological advancements, suggesting that income gaps between these regions are not as pronounced as previously thought.

These studies highlight the multifaceted nature of self-employment and the importance of using diverse methodological approaches to evaluate self-employment policies. Kontos (2008) emphasizes the need for qualitative insights to understand the biographical embedding of women entrepreneurs, while Williams (2005) advocates for recognizing the value of undeclared work in creating self-employment opportunities. Fang Cai and Meiyan Wang (2008) challenge assumptions about urban-rural income disparities, suggesting that self-employment policies should be tailored to the nuanced realities of different regions. Together, these studies underscore the necessity of integrating both qualitative and quantitative methods to develop more effective and contextually relevant self-employment policies.

These studies demonstrate that evaluating public policies on self-employment requires a multidisciplinary approach, combining quantitative and qualitative methods. Life narratives, examination of undeclared work, and the use of counterfactual methods provide different and complementary perspectives. These studies highlight the importance of considering the diversity of individual experiences, recognizing the value of the undeclared sector, and questioning pre-existing assumptions for more informed evaluations of self-employment policies.

3. CRITICAL REVIEW AND METHODOLOGICAL INSIGHTS FROM THE STATE OF THE ART

The analysis of previous studies on the evaluation of self-employment policies highlights a wide range of methodological approaches that have been employed to measure the effectiveness of these government initiatives. This variety of approaches, although seemingly heterogeneous, is crucial in understanding and improving public policies by providing nuanced insights and substantial lessons to guide the future design and implementation of these policies.

Self-employment policies aim to stimulate the creation of individual businesses and promote employment. However, rigorous evaluation of their effectiveness requires thoughtful methodological approaches. Among the frequently used approaches are quasi-experimental methods that establish a relevant comparison between a treatment group (benefiting from the policy) and a control group (not affected by the policy). This method allows for isolating the causal effect of the policy under study. However, inherent limitations such as displacement effects and musical chairs effects must be considered to avoid simplistic interpretations of the results.

Another approach, propensity score matching, aims to balance the characteristics of treatment and control groups to minimize selection biases. This method provides a better comparison by making the groups more comparable but requires robust data and may not capture all relevant factors.

Additionally, evaluations of self-employment have also turned to qualitative approaches, including observation, life stories, and semi-structured interviews. These methods allow for exploring individual experiences and capturing subtle impacts that may escape quantitative evaluation. However, the subjective interpretation of these qualitative data can introduce biases and requires thorough analysis to derive robust conclusions. An innovative perspective that has emerged in these studies is the biographical approach. This approach recognizes that self-employment is embedded in social networks, economic and legal contexts, as well as individual life trajectories. It offers a holistic understanding of the factors influencing self-employment and allows for grasping complex interactions with other policies. Nevertheless, its implementation can be complex, requiring in-depth qualitative data and analysis that takes into account the multitude of variables at play.

Thus, the diversity of methodological approaches in evaluating self-employment policies reflects not only a plurality of techniques but also the breadth of issues to consider. Each approach brings its own advantages and limitations, and their combination can provide a more comprehensive and balanced view of policy effects.

These methodological insights are of paramount importance for the design and implementation of future public policies. By recognizing the strengths and weaknesses of each method, researchers and policymakers can adopt more appropriate and informed approaches, thereby aiming to maximize the positive impact of self-employment policies on job creation, economic growth, and social well-being.

3.1 Challenges of Quantitative Approaches

Quasi-experimental methods have become common tools in employment evaluation studies. These methods allow for a meaningful comparison between a treatment group, affected by the policy under study, and a control group unaffected by it, thus isolating the causal effect of the policy. However, they face significant challenges. One challenge is displacement effects, where the policy may lead to unintended consequences in other areas, shifting individuals rather than creating new job opportunities. Another challenge is musical chairs effects, where individuals simply change status without generating sustainable new job opportunities, potentially distorting the assessment of a policy's real impact. Relying solely on quasi-experimental methods could therefore underestimate or overestimate the true effects of the policy by neglecting these potential side effects.

In response to these limitations, another commonly used methodological approach is propensity score matching. This method aims to balance observable characteristics between treatment and control groups, making them more comparable. It involves selecting individuals in the control group whose characteristics are similar to those in the treatment group, mitigating some potential selection biases and enhancing the internal validity of the results. However, propensity score matching has its own limitations. Its success depends largely on the quality and availability of data. Additionally, it may not account for all unobservable factors that could influence the results, compromising the external validity of the conclusions and their generalizability to other contexts.

The use of quasi-experimental methods and propensity score matching in employment evaluation studies demonstrates an ongoing effort to achieve reliable and accurate results. Each of these approaches has its own strengths and weaknesses, and it's crucial to consider them in a broader context. Combined judiciously and used with a critical eye, these methods can provide valuable insights into the effectiveness of employment policies and contribute to better decision-making in public policy.

In the realm of self-employment evaluation, a notable trend is the emphasis on qualitative approaches as essential instruments for a thorough understanding of this complex phenomenon. Methods such as observation, life narratives, and semi-structured interviews prove invaluable for exploring the multiple facets of self-employment and highlighting impacts that cannot be immediately quantified. By focusing on individual experiences and personal narratives, these qualitative methods fill the gaps left by traditional quantitative analyses and offer a more holistic perspective.

Direct observation allows researchers to engage with the field of self-employment, observing the dynamics and social interactions that shape this sector. It can help capture unreported or informal elements often overlooked in purely quantitative analyses. Similarly, life narratives provide rich and contextualized information about the experiences of individuals engaged in self-employment. These narratives can reveal motivations, challenges, and personal aspirations that guide career choices and trajectories.

Semi-structured interviews add a human dimension to the evaluation of self-employment policies. They give researchers the opportunity to ask open-ended questions and delve deeply into the perceptions, attitudes, and experiences of the individuals involved. This qualitative approach enables the collection of nuanced and detailed information about self-employment strategies, encountered challenges, achieved successes, and unmet needs.

However, it is important to acknowledge that qualitative methods are not without limitations. Sample size and representativeness can influence the validity of conclusions drawn from these methods. The results of a qualitative inquiry may not be generalizable to the entire target population. Moreover, these methods are subject to interpretation biases and subjectivities inherent in qualitative analysis.

Despite these challenges, qualitative approaches remain indispensable for shedding light on the understanding of self-employment. They provide contextual information and data rich in nuances, essential for unveiling the complex and multidimensional aspects of this phenomenon. By skillfully combining qualitative and quantitative data, it is possible to draw more robust and balanced conclusions. These qualitative methods thus make a crucial contribution to the evaluation of self-employment policies by bringing data to life and highlighting the human experiences that shape this field.

3.2 Specific Challenges of the Biographical Approach

A pioneering approach emerging in self-employment evaluation is the biographical approach. This method fundamentally acknowledges that self-employment is deeply rooted in a complex set of social networks, interpersonal relationships, economic and legal contexts, as well as individual biographical processes. The uniqueness of this approach lies in its ability to broaden the traditional evaluation perspective by embracing the entirety of the self-employment phenomenon and placing it in a broader context.

The biographical approach contextualizes self-employment by considering not only economic aspects but also the social, cultural, and personal influences that shape career choices and trajectories. It explores how these different dimensions interact and how self-employment fits into a life continuum. By integrating these elements, it offers a more comprehensive understanding of the motivations, challenges, and successes of individuals engaged in self-employment.

However, implementing the biographical approach may pose significant challenges. Collecting in-depth qualitative data requires time, resources, and a commitment to a thorough understanding of individual experiences. Moreover, navigating a multidimensional landscape requires a flexible and adaptable methodology to capture the complexity of interactions among different factors.

CONCLUSION

The analysis of previous studies in the field of evaluating employment and self-employment public policies reveals a wealth of methodological approaches used to measure the effectiveness of these interventions. These studies have highlighted the complexity of evaluating public policies in these areas, as well as the importance of considering a multitude of factors and dimensions to understand their real impacts.

Quantitative approaches, such as quasi-experimental methods and propensity score matching, have provided solid frameworks for assessing the causal effects of employment policies. However, they do not always capture the diversity of individual experiences and may sometimes overlook unobservable factors. Therefore, the integration of qualitative approaches has been recognized as crucial for understanding the complexity of the self-employment phenomenon.

Econometric techniques and qualitative methods have been developed to serve this field of intervention, which is beginning to attract researchers from various disciplines. However, experimental and quasi-experimental techniques remain the most widely used. It is only recently that we have seen writings highlighting the importance of qualitative and multidisciplinary approaches in understanding the impacts and effects of public employment policies in general. All stakeholders living in a society are framed by laws, customs, and unique ways of thinking and acting, which inevitably influence the implementation and execution of public policies themselves. The inherent cultural variable must be considered in explanatory models.

Economists' recourse to the insights of behavioral economics (Tversky & Kahneman, 1974; Kahneman, 1979) allows them to combine different types of analyses. Indeed, behavioral economics aims to integrate non-material and non-monetary aspects to understand and explain phenomena, enabling econometric models to go beyond purely economic variables inspired by traditional theories of homo economicus and rational choice.

Qualitative methods, such as observation, life narratives, and semi-structured interviews, allow for capturing subtle and non-quantifiable aspects of the impacts of self-employment policies. These approaches offer in-depth information on individual experiences, thus complementing traditional quantitative analyses.

Furthermore, the biographical approach proves to be innovative in self-employment evaluation. It emphasizes the broader context in which self-employment takes place, considering the social, cultural, economic, and personal influences that shape career choices and trajectories. This approach provides a more holistic perspective and helps better understand how self-employment interacts with other policies and socio-economic factors.

Drawing inspiration from these insights, future research on self-employment can engage in a mixed-method approach that combines qualitative and quantitative methods to gain a comprehensive understanding of the effects and challenges of self-employment policy.

However, for more understanding of the issue developed in this article, there is a need for more references on the application of qualitative methods for assessing employment public policy, which would help strengthen the theoretical foundation of the research. One limitation of qualitative methods is the inherent subjectivity, which can influence the interpretation of data and findings. Additionally, qualitative research often involves small, non-random samples, making it difficult to generalize the results to a larger population. Conducting qualitative fieldwork also requires specific skills and abilities from researchers, including strong interpersonal skills, the ability to build trust with participants, and proficiency in qualitative data analysis techniques.

In summary, previous studies on the evaluation of self-employment policies present a varied range of methodological approaches, from quasi-experimental methods to qualitative and biographical approaches. Each of these approaches brings unique perspectives, with their specific advantages and limitations. This diversity underscores the importance of combining these methods to obtain a holistic and nuanced understanding of the effectiveness of self-employment policies.

The cultural aspect, which has been addressed in the texts, plays a crucial role in the evaluation of self-employment policies. The attitudes, beliefs, and behaviors of stakeholders are deeply influenced by cultural factors, which must be considered for accurate contextual evaluation. A multidisciplinary approach, combining quantitative and qualitative methods, is therefore essential to inform the design and effectiveness of self-employment policies.

These studies also highlight the relevance of the informal sector as a significant contributor to self-employment and emphasize the need to reconsider pre-existing perceptions to maximize the benefits of these policies. By adopting innovative methodologies such as the biographical approach, it is possible to navigate the complex dimensions of self-employment and guide policy decisions toward more effective, sustainable, and tailored outcomes to individual and societal realities.

The examination of quantitative and qualitative techniques in the assessment of public policies related to employment and self-employment highlights the necessity of a thorough and well-rounded approach to comprehending the effects of policies. Although propensity score matching and quasi-experimental techniques provide strong frameworks for quantifying causal effects, qualitative approaches frequently struggle to capture the complex, subjective experiences of individuals. Qualitative techniques, such as semi-structured interviews and life narratives, offer insightful perspectives on these individual experiences by revealing the intricate relationship between socioeconomic variables and self-employment laws. This research is made even more insightful by the incorporation of behavioral economics, which shows how social networks, cultural norms, and cognitive biases affect policy decisions. Future studies in this area should aim to apply qualitative approaches farther while addressing issues like subjectivity and the difficulties in extrapolating results from small, non-random samples. In order to properly address these issues, researchers must also acquire advanced abilities for doing qualitative fieldwork. Future research can provide a more comprehensive and nuanced knowledge of the effects of self-employment regulations by combining qualitative and quantitative methodologies, ultimately aiding in the development of more inclusive and effective policy solutions. Adopting a mixed-method approach will help to improve both the theoretical and practical knowledge base and the policy-making process.

REFERENCES

Alla, L., Bentalha, B., Megzari, G., & Hmioui, A. (2022). Les effets du COVID-19 sur les chaines logistiques: Une analyse qualitative. Reconfigurations du tourisme en contexte de la crise du Covid-19: Quelles démarches pour quelles résiliences au Maroc et ailleurs?, 1(1), 130-153.

Attanasio, O., Kugler, A., & Meghir, C. (2015). Subsidizing vocational training for disadvantaged youth in Colombia: Evidence from a randomized experiment (NBER Working Paper No. 21390). National Bureau of Economic Research.

Benoteau, I. (2015). Quels effets du recrutement en contrat aidé sur la trajectoire professionnelle ? Document d'études de la DARES, 192.

Benoteau, I. (2015). The effects of subsidized contracts on the career trajectories of employees facing difficulties in accessing employment: Evidence from propensity score matching. *Economie & Statistique*, 477(1), 167–198. 10.3406/estat.2015.10589

Bentalha, B. (2023a). Influence des actions sociales de la RSE sur les valeurs perçues des employés: Étude qualitative des œuvres sociales de l'entreprise tajmouati. Repères et Perspectives Economiques, 7(2).

Bentalha, B. (2023b). A Framework for Exploring the Global Performance of Service Supply Chain Management: A Qualitative Study. Government Impact on Sustainable and Responsible Supply Chain Management, 106-129.

Broecke, S. (2012). Decomposing Wage Inequality in Tunisia. African Development Bank (AfDB).

Cahuc, P., Carcillo, S., & Zylberberg, A. (2014). *Labor Economics*. MIT Press.

CESE. (2016). L'auto-entreprise, levier de développement et d'intégration du secteur informel. Auto - Saisine n°27.

Crépon, B., Duflo, E., Gurgand, M., Rathelot, R., & Zamora, P. (2014). Do labor market policies have displacement effects: Evidence from a clustered randomized experiment. *The Quarterly Journal of Economics*, 128(2), 531–580. 10.1093/qje/qjt001

Dunn, W. N. (2011). *Public Policy Analysis. Pearson* (5th ed.). Graduate School of Public and International Affairs.

Fang Cai, M. W. (2008). A Counterfactual Analysis on Unlimited Surplus Labor in Rural China. *China & World Economy*.

Ghirelli, C., Havari, E., Santangelo, G., & Scettri, M. (2019). Evidence from an Italian region. *International Journal of Manpower*, 40(3), 500–524. 10.1108/IJM-02-2018-0062

Ghirelli, C., Havari, E., Santangelo, G., & Scettri, M. (2019). Evidence from an Italian region. *International Journal of Manpower*, 40(3), 500–524. 10.1108/IJM-02-2018-0062

Groh, M., Krishnan, N., McKenzie, D., & Vishwanath, T. (2016). *Do wage subsidies provide a stepping-stone to employment for recent college graduates? Evidence from a randomized experiment in Jordan.* World Bank. 10.1162/REST_a_00584

HCP. (2014). Enquête nationale sur le secteur informel 2013 - 2014: Rapport de synthèse.

HCP. (2022). La situation du marché du travail en 2021. Rabat: Haut-Commissariat au Plan. Récupéré sur https://www.hcp.ma/La-situation-du-marche-du-travail-en-2021_a2810.html

HCP. (2006-2007). Enquête nationale sur le secteur informel - Rapport de synthèse. Rabat: HCP.

Hill, M. J. (2005). *The Public Policy Process*. Pearson Longman.

Hmioui, A., Alla, L., & Bentalha, B. (2019). Évaluation des formations en ligne: Etude qualitative des perceptions des utilisateurs des MOOC au Maroc. In ICT-Days' 19.

Kahneman, D. T. (1979). Prospect Theory: An analysis of decision under risk. Econometrica, 47, pp. p. 263-291.

Kontos, M. (2008). The biographical embeddedness of women's self-employment. Motivations, strategies and policies. In Apitzsch, U. (Ed.), *K. M.* Self-Employment Activities of Women and Minorities. 10.1007/978-3-531-90816-8_5

Kontos, M. (Ed.). (2008). *Self-employment activities of women and minorities: Their success or failure in relation to social citizenship policies*. VS Verlag für Sozialwissenschaften., 10.1007/978-3-531-90532-7

Lamsiah, A., & Bentalha, B. (2022). Economic Diplomacy and Economic Intelligence Amalgamation: The Moroccan-Nigerian Economic Exchange as a Case Study. *Alternatives Managériales Economiques*, 4(4), 582–603.

Maibom, J. (2014). *The Effects of Debt Relief on Labor Market Outcom.*

McKee-Ryan, F. M., & Harvey, J. (2011). "I have a job, but...": A review of underemployment. *Journal of Management*, 37(4), 962–996. 10.1177/0149206311398134

McKee-Ryan, F. M., & Harvey, R. J. (2011). I have a job, but . . .: A review of underemployment. *Journal of Management*, 37(4), 962–996. 10.1177/0149206311398134

Thissen, W. A., & Walker, W. E. (2013). *Public Policy Analysis: New Developments. Springer.*

Williams, C. C. (2005). The undeclared sector, self-employment and public policy. *International Journal of Entrepreneurial Behaviour & Research*, 11(4), 244–257. 10.1108/13552550510603289

Chapter 8
Evolving Experiences in Management Education Through a Longitudinal Qualitative Study

Soufiane Elmouhib
Sidi Mohamed Ben Abdellah University, Morocco & HECF Business School, Morocco

Mourad Es-Salmani
Sidi Mohamed Ben Abdellah University, Morocco & HECF Business School, Morocco

Abdelkarim Moussa
Sidi Mohamed Ben Abdellah University, Morocco

ABSTRACT

This study employs longitudinal qualitative research methods to explore the dynamic learning experiences of students in a newly introduced management education program. Grounded in theoretical perspectives on time, lived experiences, and developmental trajectories, the chapter underscores the importance of capturing the fluidity of educational journeys over an extended period. An in-depth case study was conducted at USMBA University, following a cohort of five students through their first year in the Excellence Program in Management Sciences. Data were collected through semi-structured interviews at beginning and end of the academic year. Iterative qualitative coding and analysis techniques were applied to identify evolving patterns, critical incidents, and transformations in students' perspectives, skill development, and experiences. The findings reveal the complex interplay between

DOI: 10.4018/979-8-3693-5543-5.ch008

institutional factors and personal factors shaping individual learning trajectories.

INTRODUCTION

While educational programs are designed to provide learners with the knowledge, skills, and competences required to pursue successful careers, it is important to consider the fluidity and the changes of such experiences over time. In this sense, longitudinal qualitative research emerges as a powerful methodological approach to delve into the complexities and nuances of learning trajectories, providing insights that can inform pedagogical practices, curricular design, and support systems within management education.

The paper presents the rationale for adopting longitudinal qualitative research in the context of management education. It unfolds the conceptual basis that articulates the importance of the recognition of subjective and objective levels of time in the constitution of learning processes and outcomes, Section 1.1; the notion of lived experiences as a process of unfolding, Section 1.2; and the conceptualization of trajectories as developmental paths, Section 1.3. The paper discusses then the applications of longitudinal qualitative research across different disciplines, Section 2.1, to indicate their potential in bringing rich insights. It further discusses the critical interfaces of institutional frameworks and human experiences, Section 2.2, by recognizing the mutual and simultaneous influence that shapes processes of learning and development. In addition, the paper indicates that the development of research partnerships, Section 2.3, is crucial for enhancing the quality, validity, and usefulness of longitudinal qualitative research conducted in the field of management education.

Methodological issues that the paper addresses relate to strategies for commitment to long-term projects, Section 3.1, the need for mutual reflexivity, Section 3.2, strategies for integrating theoretical framework articulation for temporal dynamics, Section 3.3, and the concurrent application of iterative data collection strategies and multi-strand data analysis techniques, Section 3.4.

While existing research has explored various aspects of management education, there remains a significant gap in our understanding of how students' learning experiences evolve over time within newly introduced programs. Much of the current literature focuses on snapshots of student experiences or program outcomes, failing to capture the dynamic nature of learning journeys (Gangl, 2002; Verina et al., 2021). Additionally, the interplay between institutional factors and personal experiences in shaping these journeys is often overlooked. This study aims to address these gaps by employing a longitudinal qualitative approach to explore the fluid and evolving nature of student experiences in a new management education program. By doing so, it seeks to provide a more nuanced and comprehensive understanding of the learning process, which can inform both theory and practice in management education.

The empirical part of this paper deals with a longitudinal case study at USMBA University concerning a cohort of students enrolled in the Excellence Program in Management Sciences for the academic year 2023-2024. This study combines interviews that were carried out at different moments in time: one just before the beginning of the program and another one at the end of the academic year. The overarching research questions guiding this study are:

1. How do the learning experiences and perspectives of management students evolve over time within the Excellence program?
2. What is the interplay between institutional factors (curriculum design, teaching methods, support systems) and students' personal experiences in shaping their learning journeys?
3. How can longitudinal qualitative research contribute to the theoretical understanding of student learning experiences and inform the practical implementation of effective management education programs?

By addressing these research questions, the paper seeks to make significant contributions to both the theoretical discourse on student learning and development, as well as the practical implementation of inclusive and transformative management education programs that foster personal and professional growth for diverse student populations.

1. CONCEPTUALIZING TIME IN LONGITUDINAL QUALITATIVE RESEARCH

Understanding the role of time is crucial in longitudinal qualitative research, particularly within the field of management education. This section explores how time can be conceptualized to enhance our comprehension of learning processes and outcomes. By examining both the subjective and objective dimensions of time, the unfolding of lived experiences, and the understanding of developmental trajectories, researchers can gain a holistic view of how time influences education.

1.1. Subjective and Objective Dimensions of Time

Understanding the dynamics of time in management education requires that both subjective and objective dimensions are further explored. Time is a multidimensional phenomenon, which impacts the process and outcome of learning in multiple ways. Dimensions of time have been considered for a more comprehensive understanding of how time affects the learning process. The objective dimension of time involves the measurable and quantifiable aspects of time that are observable and recordable. Within management education, structured academic time includes course timelines, assignment deadline dates, and exam schedules. They offer the curriculum organization elements for learning objectives, considering that they have to be accomplished within a specific period (Johansson, 2023).

For instance, a management education program could spread over a few semesters, each semester containing several weeks, and each week comprising several hours assigned to lectures, seminars, and practical sessions. Such a design facilitates planning both for how the course content will be delivered and when the students can be measured. The planning of the objective dimension also includes such considerations as pacing, the timing of feedback, and the time at which evaluation will take place through exams and projects. As Johansson (2023) points out, time perception and time management are key factors in educational leadership, in view of the implications of digitalization and other reforms to be implemented (Johansson, 2023).

In this regard, the principal considers time to be a resource that can be managed and apportioned in order to avail events and experiences for the teachers, such as teacher training or integration workshops for new teaching models. In this manner, this consideration is commensurate with the structured nature of programmed activities, where enough time is allotted for activities that encourage teachers to learn and adopt new digital tools as part of their teaching methods. This provides a way of achieving an educational aim in the required time while being vigilant for improvement to both teaching methods and student learning outcomes on an ongoing basis (Johansson, 2023).

Contrary to the objective aspect of measuring time, the subjective dimension of time concerns individual, psychological perception of time by the personages (Bergson, 2002). This involves how students and educators perceive the flow of time, which may happen quite differently from one another, depending on one's involvement, level of stress, work volume, or family situation. One of the most important reasons that a contemporary management educator would be interested in subjective time is that it is directly related to devising a supportive learning environment that takes account of differences among individuals (Bergson, 2014).

The view on what constitutes time might be significantly various for different students due to their workload, interest in the subject, and learning speed. For example, a very interested student will believe that three hours of a lecture are a very short time, while a less interested one may think that it is extremely long (Schatzki, 2013). Equally, deadlines can be perceived as both surmountable and overwhelming if the students have good time management and a number of commitments, respectively (Heidegger et al., 2008). Time may also be subjective for the educator and influenced by such aspects as the teaching workload, preparation time, and administrative responsibilities. In instances where the educator feels overwhelmed by their tasks, time may be felt as constraining, in contrast to when one is well-organized and supported, where they may feel time being flexible and manageable (Kemmis et al., 2014).

To effectively capture the fluidity of learning experiences over time, it is essential to integrate both dimensions: subjective and objective. Through this integrated approach, an explanation of the ways in which time influences the processes and results of learning can be achieved at a more comprehensive level (Kemmis et al., 2014). For instance, although the objective timelines give structure, considering the subjective experience of students will inform changes in methods of instruction, such as flexible deadlines or varied pacing in content delivery to consider varying learning styles (Schatzki, 2013). Educators can make use of knowledge in both the objective and subjective concepts of time to fashion out a class environment that is structured and, at the same time, able to accommodate the needs of individual students.

By understanding that students may actually feel and experience time differently (Heidegger et al., 2008), educators are able to adjust their practices in bringing about engagement and supporting various learning processes for students (Kemmis et al., 2014). This integrated approach will produce more inclusive and effective educational experiences that recognize the complexities of time in learning (Bergson, 2014).

Longitudinal, qualitative research in management education is especially well suited to the investigation of such dimensions. It is through the in-depth interviewing of students, educators, and even reflective journals, in focus groups, over a longer period of time, that researchers can access the continuously changing perception and experience of time. This methodology allows the identification of patterns and changes in how time is experienced and managed throughout the academic journey.

1.2. Lived Experiences Unfolding Over a Period

It will be important for the study, in light of the notion of "lived experiences unfolding over a period," to understand how learning and development are taking place over time in management education. This approach to research will therefore be centered on an understanding of experiences as lived and interpreted dynamically in educational settings by individuals over time (Farrell, 2020).

Lived experiences refer to the subjective, personal accounts of an individual as one moves through an educational journey. Such experiences are not fixed but rather dynamic, changing, growing, and transforming as students go on to encounter different learning activities, challenges, and accomplishments. It is through this unfolding experience that the researcher gets to understand more about learning processes and personal development, which would, at times, be ignored in cross-sectional studies.

Although the period is long, experiences are captured through repeated observations and interactions with participants, such that researchers are able to trace changes and identify patterns in the changes. The approach ensures that researchers gain a rich, nuanced understanding of how students' perceptions, attitudes, and behaviors change over time (Farrell, 2020). An illustration is how initial apprehension from a student regarding a difficult subject at the beginning slowly changes into confidence and mastery of the subject as the student advances in coursework and receives feedback.

Furthermore, studying lived experiences longitudinally helps to highlight the contextual factors that influence learning: the impact of social interactions, institutional practices, and external events on students' educational trajectories. By analyzing such factors, researchers come to understand in a better way the fluidity of learning experiences and how they are shaped by broader social and institutional contexts (Farrell, 2020).

1.3. Understanding Developmental Trajectories

In the area of longitudinal qualitative research in management education, the bases on which developmental trajectories are recognized and captured have shown the subtle changes in learners throughout the learning process. In this respect, developmental trajectory can be defined as the course or sequence of events of developmental changes that a student undergoes in the course of their education (Grosemans & De Cuyper, 2021). The very concept is indispensable to understanding in what way students develop with regard to their knowledge, skills, and attitudes through education.

Developmental trajectories offer a framework for understanding the continuous, and often non-linear, nature of learning. As opposed to snapshot studies, which register a single moment in time, it is within longitudinal research that the dynamics of these trajectories come to light, thereby unveiling the multifold and fluctuating nature of the students' educational experiences. Such an approach enables researchers to capture moments of epiphany or struggle that are critical in developing the learner.

This study allows a view of the development trajectories and insight into how educational interventions add up, the impact of different pedagogical strategies, and the roles of different contextual factors in making up learning outcomes (Akkermans et al., 2013). For example, the trajectory might depict how early exposure to collaborative projects enhanced a student's teamwork skills or how consistent feedback from instructors led to a deeper understanding of complex concepts.

Understanding these trajectories also indicates the individual differences among learners, where the path of each learner is shaped by their background, prior knowledge, personal motivation, and support they are given (Grosemans & De Cuyper, 2021). Qualitative longitudinal research, in this way, brings out these individual stories in ways that allow a deeper and more personalized understanding of learning processes. This can further help in more relevant and effective educational practices to ensure that the diverse needs of students are addressed.

Further, understanding the long-term effects of management education is realized through the tracking of developmental trajectories. It thus presents a way of assessing the influence of learning experiences on career choices, professional growth, and lifelong learning habits, for example, leadership. Verbruggen et al. (2015) brought the example of a student who has developed effective leadership skills through this developmental trajectory and goes into major leadership positions at the professional level; this is a great example of the sustainable effects of learning experiences (Verbruggen et al., 2015).

2. LONGITUDINAL QUALITATIVE RESEARCH ACROSS DISCIPLINES

Longitudinal qualitative research (LQR) has emerged as a powerful approach for exploring complex, dynamic phenomena across various academic disciplines. By studying participants' experiences as they unfold over extended periods, LQR captures the nuances, transitions, and evolutions that shape individual and collective trajectories (Balmer et al., 2021). This section delves into the application of LQR in diverse fields, including education, sociology, and business, showcasing its versatility and potential for generating rich insights (Bernardi, 2022; Cooke et al., 2012; Taylor, 2020). Furthermore, it examines the crucial intersections between institutional frameworks and human experiences, emphasizing the reciprocal influence that shapes learning and development processes (Treviño-Eberhard & Kaufmann-Kuchta, 2022; Hanft & Knust, 2009). Finally, the importance of fostering research collaborations in LQR is highlighted, as these partnerships enhance the robustness, validity, and impact of findings in management education and beyond (Bourner & Brook, 2019; Matsuo & Nagata, 2020; Brook & Pedler, 2020).

2.1. Examples From Various Academic Fields

Longitudinal qualitative research has been welcomed in a variety of academic disciplines, where each leverages its unique features to examine complex, evolving phenomena over time. These examples illustrate the versatility and depth of insight that longitudinal studies can offer to researchers, highlighting their impact across diverse learning and research contexts.

In education, longitudinal qualitative (LQR) research is used to examine the lived experience of participants as those experiences unfold through time. For example, LQR can investigate how medical students integrate their personal and professional identities during clinical clerkships (Balmer et al., 2021). Following students for an extended period allows one to capture the stops, starts, detours, transitions, and reversals that characterize the learning journey. This approach deepens our understanding of the dynamic processes guiding student development and educational outcomes (Balmer et al., 2021).

LQR has been used by education researchers to study human development in the context of formal education. To do so, Saldana in 2003 proposed the use of framing and descriptive questions that are sensitive to more linear streams of events, followed by analytic and interpretive questions about how time may be at work in the phenomenon of interest (Zacharias et al., 2002). This framework may help investigators remain open to thinking about time as both fixed and fluid and, in doing so, may better illuminate the complexities of the students' lived experiences.

In sociology, QLR has increasingly been acknowledged as a potentially productive venue for investigating family changes as they take place (Bernardi, 2021). It provides a way for the researcher to "grasp how people make meaning about the twists of their lives and their contexts as they unfold" (Bernardi, 2021). In the potential for uncovering the complexity of changes in the making and unmaking of family relationships across the life course (Bernardi, 2022), many QLR studies, for example, include analyses of how family or social groupings move across generations. That work presents how social, economic, and cultural factors affect the time of their lives, illuminating social change and continuity processes in communities (Brannen, 2006). This is of crucial importance to studies that focus on issues such as social mobility, intergenerational reproduction of values, and long-term policy change effects (Bernardi, 2021; Millar & Ridge, 2013).

In business and management, LQR is one useful approach in providing a deep understanding of processes that change over time. The way it has, for example, can be found in organizational change processes, leadership, and careers (Cooke et al., 2012). Thus, researchers in leadership development may conduct longitudinal studies of emerging leaders over a few years to study how their leadership style and strategy evolve as they face different challenges and opportunities (Taylor, 2020). Similarly, when longitudinal studies are done to track the use of new business strategies or technologies in companies, it has been possible to present the subtle processes and key elements that contribute to a positive response to organizational change initiatives (Taylor, 2020). Data gathered at a number of time points permit exploration of the complexities and evolutions that are characteristic of organizational phenomena in LQR (Cooke et al., 2012; Taylor, 2020).

As evidenced by these examples from various academic fields, LQR harnesses the power to capture fluidity and complexity in human experiences. Through a study of phenomena across time, the identification of patterns, critical turning points, and an understanding of the processes shaping individual and collective trajectories can be elucidated. This approach is critical for the generation of robust, contextually rich insights that inform theory, practice, and policy in different disciplines.

2.2. Intersections of Institutional Frameworks and Human Experiences

Understanding the interplay between the institutional frameworks and the human experience is one of the major keys in the realm of longitudinal qualitative research to be able to understand how processes of learning and development take place over time (Treviño-Eberhard & Kaufmann-Kuchta, 2022). This institutional framework, implying policies, organizational cultures, and educational structures, delimits the environment in which a person navigates his or her journey of personal and professional growth. Such frameworks are, at the same time, presenting opportunities and constraints that influence human experiences in manifold and often unpredictable ways (Knust & Hanft, 2009).

One of the most important areas is that of the influence of institutional norms and values on the learning trajectory of individuals. For example, in continuing higher education, the emphasized pedagogic approaches, including case studies or experiential learning, may be of significant influence on adult learners in acquiring knowledge and developing critical thinking skills over time. These educational practices, embedded in institutional policies, shape the identities of adult learners as professionals in their fields. The longitudinal approach allows researchers to capture the dynamic nature of the relationships between these factors (Dollhausen et al., 2013). Following individuals through extended periods, researchers can observe how institutional policy shifts or changes in organizational culture within institutions alter personal experiences and learning outcomes over a long period of time. Such a longitudinal study, for example, could show how a change in policy that revolves around continuing higher education accreditation or new leadership in a university influences the participation and academic performance of adult learners over several years.

At the same time, human experiences provide a distinctive, subjective view of institutional structures. Personal stories and lived experiences create deep and subtle ways to learn the ways in which individuals perceive and respond to the institutional settings in which they live (Knust & Hanft, 2009). In these many accounts, it is not uncommon to hear the tension or mismatch between institutional demand and personal aspiration as a signal toward opportunities to change policy and innovation in the enterprise of continuing higher education.

In addition, longitudinal qualitative research reveals the reciprocal influence between individuals and institutions. Institutional framework's structure individual experience, while individuals are also active contributors and co-creators of these frameworks. This two-way relationship emphasizes the need to understand macro-level institutional processes in relation to micro-level human experiences to grasp the intricacies of learning and development in contexts of continuing higher education.

2.3. Fostering Research Collaborations

In light of the growing importance in the changing landscape of management education, collaborative research has become paramount to capturing the dynamic and multifaceted nature of learning experiences over time. With an emphasis on extended temporal engagement and deep exploration, participatory action research (PAR) provides a unique platform for interdisciplinary and cross-institutional collaborations (McTaggart, 1991). These collaborations not only enrich the research process but also contribute to a more holistic understanding of educational phenomena.

Collaborative efforts in PAR can span from partnerships between academic institutions to those with industry practitioners. For example, universities may collaborate to track the learning trajectories of students across diverse educational contexts, thus offering comparative insights that no single institution could possibly provide. Such partnerships could then lead to standardized data collection instruments, shared methodologies, and innovative analytical techniques that boost the robustness and validity of research findings (Bourner & Brook, 2019).

Also, the involvement of industry partners in PAR studies can bridge theoretical knowledge and practical application. By integrating practitioner perspectives, the researcher will appreciate more about how management theories and educational practices are applied and developed in actual settings (Teehankee, 2018). Such collaborations also open the way for varied participant groups to be accessed, ensuring the research is broad in experiences and scope.

A number of factors are important for the success of a research collaboration. Trust and mutual respect are the basis of a relationship and, thus, open communication toward a common vision (Bourner & Brook, 2019). Agreed roles, responsibilities, and equitable resourcing of the various inputs and outputs help in managing expectations and ensuring the research progresses smoothly. At the same time, a number of regular meetings and updates help in maintaining momentum and address emerging challenges. Additionally, collaborative PAR can highly contribute to building capacity in management education. Junior researchers and doctoral students involved in such projects benefit from the mentorship and the early engagement in high-impact research afforded to them (Udas, 1998). Such a collaborative environment promotes learning and the professional development of the conduct of the research in management education.

Longitudinal qualitative research is key in understanding the long-term impact of management education initiatives in the process of fostering research collaborations. This is because the research can help identify how longitudinal experiences influence the career trajectory of participants and the broader impact on management and organization practices. Collaborative longitudinal research can be implemented at more than one institution, including multiple stakeholders, thereby enabling comparisons across contexts, resulting in the identification of best practice principles within management education. This can aid the formulation of strong theoretical frameworks and recommendations based on the findings, which might be useful in informing policies and practices of management education.

3. METHODOLOGICAL CONSIDERATIONS

In longitudinal qualitative research within the management education context, employing iterative data collection and multi-strand data analysis is essential for understanding the evolving nature of learning and support over time. This methodological approach allows researchers to continuously refine their understanding of the phenomena under investigation, adapting data collection and analysis to emerging insights and changing contexts. By cycling through data collection and analysis, researchers can adjust their focus based on previous findings, ensuring a dynamic and flexible research design. Multi-strand data analysis, involving various analytical perspectives, enables a comprehensive understanding of complex phenomena such as student support roles and experiences. Together, these methods enhance the validity and depth of the research, providing a robust framework for studying the fluidity of learning and support in a longitudinal context.

3.1. Committing to Long-Term Research Projects: Didactical Design Research

the Didactical Design Research (DDR) is committed to long-term research projects running for many years. Developing teaching and learning settings under an intertwined and iterative sequence with studying the learning trajectories that the students get involved with throughout the process in an empirical manner commits the researcher to a resourceful design (Gravemeijer & Cobb, 2006). This design should be strategic and flexible enough to accommodate and evolve with the participants' experiences. The initial plan would thus identify research objectives and learning contents, methodological approaches to be taken, and how to deal with empirical findings from each cycle by opening and being open to flexibility (Prediger & Schnell, 2014). Among other provisions, the design must include stages of data collection that are well spread out to be able to capture significant developments in participants' learning trajectories or learning journeys.

Adequate resources in terms of finances and the research team are important for the sustenance of DDR over time. The people involved in the research need financial support to access research funds from different funding agencies to facilitate data collection and analysis, offer support to the participants, and be on the ground to meet the participants. One key is to build a team with this diversity from the beginning because it is harder to introduce it at a later stage (Prediger & Schnell, 2014). Close collaboration through developing trust-based relationships with participants is critical. Building and maintaining relationships based on trust by regular communication, information sharing on progress, and participant feedback builds the kind of rapport that would keep participants involved in the study throughout the process. Laying transparently on the table issues of ethical concern, such as involving participants through informed consent and issues of confidentiality, has to be done.

Data from longitudinal DDR projects grow over time, and hence the critical need for effective data management. DDR projects have to work with systematic routines on secure data storage, coherent coding schemes, and careful documentation of processes of collection and analysis (Prediger & Schnell, 2014). Data management supports comparison and synthesis of longitudinal data across different phases.

Finally, researchers need to anticipate the challenges unique to longitudinal studies, including attrition of participants, changes in context, and changes in the research questions themselves. Some strategies in DDR research that might help in this regard include maintaining a bigger sample at the outset and using a mixed method of data collection and analysis (Cobb et al., 2003).

3.2. Encouraging Mutual Reflexivity

Encouraging mutual reflexivity is a significant methodological consideration in longitudinal qualitative research, especially within the medical education context. According to Balmer et al. (2021), mutual reflexivity refers to the recognition of dynamic lived experiences between researchers and participants as the experiences unfold through time. Not only does it allow quality and depth improvement in research, but it also sets the stage for a more nuanced and dynamic interpretation of the data (Balmer et al., 2021).

First, mutual reflexivity acknowledges that both researchers and participants possess their unique perspectives, biases, and experiences, which they bring to the research table (Reid et al., 2018). Balmer et al. (2021) state that by having an atmosphere where such perspectives can be openly exchanged and examined, researchers stand a chance of getting richer, more comprehensive insights into the phenomena being studied. That kind of collaborative approach facilitates the surfacing of hidden assumptions, shakes the foundation of existing stories, and uncovers the convoluted dynamics between individual experiences and the larger educational context.

To give an instance, Balmer et al. (2021) suggest that encouraging mutual reflexivity includes recursive data generation. This method involves inviting participants to reach back into their data to imagine forward. This can be realized through regular, structured interviews, where participants can vocalize their thoughts, feelings, and interpretations of the experiences obtained in learning. On the other hand, researchers can share their preliminary analyses and interpretations to attract participants to give feedback and suggest alternative explanations that possibly describe the data in a better way (Balmer et al., 2021). An integrative process of this nature serves to improve and increase the understanding of the data, hence ensuring that findings are well grounded in the lived realities of the participants.

Another important feature of mutual reflexivity concerns the role of reflective journals or diaries. As seen in the works of Monrouxe and Sweeney (2013) or Ottrey et al. (2021), participants can be motivated to take regular records of their thoughts and experiences, which could be used as data by the researcher (Ottrey et al., 2021, 2021). Equally, as Balmer et al. (2021) stipulates, researchers should maintain reflective journals, indicating the progressive thoughts, questions, and insights that the researcher gets during the study. Comparing and contrasting the reflections reveals how the research processes itself influenced findings and can also identify the potential biases and blind spots that can occur.

The other feature of mutual reflexivity is that it should be characterized by transparency, accountability, and ethical scrupulousness as stated by Reid et al. (2018) and Baker et al. (2016). In effect, the researcher should be transparent about their positionality and its likely impact on research (Baker et al., 2016; Reid et al., 2018). This includes transparency about the researchers' theoretical frameworks, methodological choices, and potential conflicts of interest. By doing so, the researchers can be able to build trust with the participants and thus create a collaborative and respectful research environment among the two (Balmer et al., 2021).

The last aspect of mutual reflexivity concerns the publication of research findings. This means that, as Balmer et al. (2021) says, researchers should include participants in both the interpretation of and presentation of findings. That is, participants need to be given a voice and be heard in the most effective manner. This can be conducted through papers that are co-authored, collaborative conference presentations, or through feedback presentations that can enable participants to review and comment on research findings.

3.3. Integrating Theoretical Frameworks for Temporal Dynamics

In carrying out longitudinal qualitative research within the field of management education, the alignment of theoretical frameworks that accommodate temporal dynamics is very important. The said frameworks give a lens to enable researchers to look into how learning experiences and educational outcomes change with time in a structured manner. The addition of theories that consider temporal elements can give longitudinal researchers insight into the processes and factors that guide managerial learners' growth.

The conceptual foundation of most research on the returns to education is the life course perspective (Elder et al., 2003; Ulrich Mayer, 2004). It highlights the importance of timing and context in the life of an individual, factors that can shape educational trajectories. The life course perspective enables researchers to investigate how significant events, transitions, and the wider sociocultural environment interact with and shape the consequences of education

For example, the point in time at which one enters a management program during an economic recession or a booming economy can give rise to various learning experiences, career paths, and finally, different returns to education (Blossfeld & Hofmeister, 2008; Gangl, 2002). Aspects of studying the influences of education taking place at different life stages, such as entry into the labor market, mid-career (Blossfeld & Hofmeister, 2008), and later in life, where health and health behavior as non-monetary returns to education become more important, are underlined by the life course perspective.

As Ding and Curtis (2021) establish, narrative inquiry is a strong approach for capturing temporal dynamics in longitudinal research. It seeks personal stories and accounts that capture changes that have taken place in the experiences of persons over time. The analysis of these narratives can show patterns, turning points, or significant incidents that shape the students' identity and adjustment to university life (Ding & Curtis, 2021). Ding and Curtis (2021) show that narrative inquiry can help to highlight the development of identity among first-year university students and highlight the complexity of transition to university and the particular challenges of international students.

Ding and Curtis (2020): A narrative inquiry design was employed for their study, which attempted to fathom the experiences of one mainland Chinese student as she studied in the university in Hong Kong (Ding & Curtis, 2021). Using in-depth interviews, journal entries, and email communication, rich qualitative longitudinal data were gathered, from which it was possible to trace the identity change in Christy during her first year at university. Patterns and turning points, critical incidents, and the like that have been tagged to have an influence on the identity formation process of Christy are identified initial excitement, struggles, emotional breakdown, and adaptation. Ding and Curtis (2020) brought to light the complex interplay of environmental factors, cognitive processes, and the development of identity in the context of the university transition. These results also call for the use of narrative research to deepen the temporal dynamics of identity formation and the challenges of first-year university students, especially those from culturally diverse backgrounds (Ding & Curtis, 2021).

Incorporating temporal structuring from organizational studies can further contribute to the understanding of how time is perceived and organized within educational settings. Temporal structuring focuses on how institutional practices and routines establish temporal rhythms that shape individuals' experiences (Orlikowski & Yates, 2002). In the mobile learning context, it is important for students to keep pace with academic semesters and deadlines, as well as the timing of assessments, to help make cognitive load bearable and keep students as engaged as possible with learning outcomes.

From the work of Zhampeissova et al. (2020), one can deduce that a high level of thematic and temporal structuring of the educational process in mobile learning environments results in optimum educational content and scheduled learning activities that improve learning outcomes and minimize the cognitive load of a student. The authors underlined that the distribution of the cognitive load in an educational course has to be optimized in order to achieve efficiency in students' mental resources and aid them in memorizing the studied materials (Zhampeissova et al., 2020). Moreover, the introduction of tools that would ensure student attention and active student involvement in the learning process will further support the appropriate management of the cognitive load in mobile learning settings (Zhampeissova et al., 2020). Such thematic social interaction between learners may also add value to the learning experience and facilitate the achievement of educational objectives.

By integrating these theoretical frameworks, researchers can then put together a thorough and nuanced understanding of the temporal dynamics at play in management education. This approach not only enriches the analysis of longitudinal qualitative data but also provides practical insights for educators and policymakers to design and implement more effective educational interventions that are responsive to the evolving needs of learners over time.

3.4. Iterative Data Collection and Multi-Strand Data Analysis

In longitudinal qualitative research in the management education context, iterative collection and multi-strand analysis are vital in grasping the fluidity of learning and support over time (Kellogg, 2024). This approach to the methodology enables researchers to enhance their understanding of the phenomena being investigated continuously, ensuring the responsiveness of data collection and analysis to emergent insights and shifting contexts (Thomson & Holland, 2003).

Iterative data collection cycles through the process of data collection, analysis, and then using the analysis to inform the subsequent collection of data (Patton, 2002). This approach is of particular benefit in longitudinal research because it allows researchers to adjust the approach and the focus of data collection and analysis based on the findings from earlier cycles and the context at the time of data collection (Kellogg, 2024). For instance, from the exploration of the experiences of support for college students, the initial interview might discover an unforeseen theme or unanticipated challenge, then the researcher is facilitated to expound on that theme or challenge by probing into it in the subsequent interviews or even different sources of data from observations, focus groups, etc. (Kellogg, 2024).

This iterative process will keep the research design dynamic and flexible so that when new directions present themselves during the study, there is space and capability to follow up on these (Kellogg, 2024). It also provides more in-depth engagement with the participants, as researchers can return to them with more focused and relevant questions, which will enhance the richness and depth of the data gathered.

Multi-strand data analysis is the application of varied analytical perspectives or frameworks in interpreting the data collected (Kellogg, 2024). The approach means that complex phenomena, such as support roles and experiences of college students, should be understood from different lenses of analysis (Lincoln & Guba, 1986). By using numerous strands of analysis, researchers can draw distinct dimensions of the data to provide an all-rounded approach to the research questions (Kellogg, 2024).

For example, in a college student study, one strand of data analysis could include theme coding searching through the data for common patterns and themes in the types of support college students provide. These include tangible help, informational assistance, and emotional support (Kellogg, 2024; Krause, 1995). Another strand could include narrative analysis of these support journeys to further understand the details of individual students (Kellogg, 2024). A third strand may use an asset-based perspective to illuminate the support strengths and contributions of students, countering the current deficit-focused literature (Yosso, 2005).

The combination of different strands of analysis works to triangulate findings in the research, which increases the validity and trustworthiness of the study (Lincoln & Guba, 1986). It also allows researchers to account for the multidimensionality of the experiences of student support indeed, they are considering not only what support the students provide but how students go about providing that support and the assets that they bring to their networks (Kellogg, 2024).

The unique power of iterative data collection and multi-strand data analysis together in a single study is the strong methodological design of a longitudinal qualitative study about college student support. By its very nature, iterative data collection allows for a simultaneous process of analyzing collected data. Therefore, the outcome of one element of analysis aids in directing the following data collection, in a constant feedback loop that bolsters the research along the way (Thomson & Holland, 2003). Finally, the two are integrated to ensure that the study's design can remain responsive to the shifting nature of the students' developing support roles and the broader educational and social context.

4. EMPIRICAL STUDY: LONGITUDINAL CASE STUDY OF MANAGEMENT STUDENTS' LEARNING EXPERIENCES

The preceding sections have laid a conceptual foundation for understanding the importance of longitudinal qualitative research within Management Education. Building upon these conceptual underpinnings, the empirical study presented in this Section aims to exemplify how longitudinal qualitative research can be employed to gain a nuanced understanding of students' learning experiences within a newly introduced management education program. By following a cohort of students over their first academic year, this study seeks to capture the evolution of their perspectives, challenges, and personal growth trajectories.

4.1. Research Context and Setting

This longitudinal qualitative study was conducted at the Faculty of Law, Economics and Social Sciences at USMBA University. The research focused on a cohort of five first-year students enrolled in the inaugural promotion of the Excellence program in Management Sciences during the 2023-2024 academic year.

4.1.1. Description of the Management Education Program

The Excellence program in Management Sciences is a newly introduced three-year undergraduate program at USMBA University, designed to provide students with a comprehensive and rigorous education in various aspects of management. The program aims to develop well-rounded professionals with strong theoretical knowledge and practical skills, preparing them for successful careers in diverse business environments.

The curriculum of the Excellence program is structured around core management disciplines, including organizational behavior, marketing, finance, accounting, operations management, and strategic management. Additionally, the program places a strong emphasis on the development of both hard and soft skills essential for professional success. Hard skills, such as data analysis, financial modeling, and project management, are integrated into the coursework, while soft skills like communication, teamwork, leadership, and critical thinking are honed through various pedagogical approaches.

The program incorporates a variety of teaching methods and pedagogical approaches, including lectures, case studies, group projects, simulations, and experiential learning opportunities. Students are encouraged to actively participate in class discussions, engage in teamwork, and apply their knowledge to real-world business scenarios.

One unique feature of the Excellence program is its emphasis on experiential learning through internships and practical training. This practical component aims to bridge the gap between academic knowledge and industry requirements, enhancing the students' employability and preparedness for the labor market.

4.1.2. Rationale for Selecting the Student Cohort

The decision to focus this longitudinal study on the inaugural cohort of the Excellence program in Management Sciences at USMBA University was driven by several compelling factors:

- **First Promotion:** Being the first cohort to experience this newly implemented program, these students could provide valuable insights into the initial implementation and evolution of the program over time.
- **Unique Opportunity:** As the Excellence program was introduced in the 2023-2024 academic year, this cohort represented a unique opportunity to capture the students' learning experiences from the very beginning of their journey.
- **Small Cohort Size:** With only five students in the inaugural cohort, a longitudinal qualitative study could delve deeply into their individual experiences, allowing for rich and nuanced data collection.
- **Program Innovation:** The Excellence program introduced several innovative elements, such as experiential learning and a strong emphasis on practical skills development. Studying this cohort could shed light on how these innovations shaped the students' learning experiences over time.

- **Institutional Interest:** As a newly introduced program, USMBA University had a keen interest in understanding the experiences of the first cohort to inform potential program improvements and future iterations.

By selecting this cohort of five first-year students from the Excellence program's inaugural promotion, the longitudinal study aimed to provide valuable insights into the implementation and evolution of this innovative management education program, contributing to the broader understanding of students' learning experiences in higher education.

4.2. Participants and Data Collection Methods

4.2.1. Participant Selection and Characteristics

The participants in this longitudinal qualitative study were the cohort of five first-year students enrolled in the inaugural promotion of the Excellence program in Management Sciences at USMBA University during the 2023-2024 academic year. This cohort was purposefully selected due to the unique opportunity it presented to capture the students' learning experiences from the inception of this newly introduced program.

The participants formed a relatively homogeneous group in terms of their age, ranging from 18 to 21 years old, which is typical for first-year undergraduate students at the faculty. The cohort consisted of both male (3 students) and female (2 students) students, providing a balanced representation of gender diversity. Additionally, the participants came from diverse socioeconomic backgrounds, ranging from lower-middle-class to upper-class families.

In terms of prior academic performance, the participants had varying levels of achievement, with some having excelled in high school and others with more moderate academic records. Their educational and professional aspirations also differed, with some aiming for further graduate studies, while others intended to pursue entrepreneurial ventures or corporate careers after completing the program.

Two participants had limited work experience through part-time jobs or internships, while the remaining three had no prior professional experience beyond academic settings. Culturally, the cohort was relatively homogeneous, with all participants coming from the local region and sharing a similar linguistic and cultural background.

4.2.2. Interviews, Observations, and Document Analysis Over Multiple Time Points

To capture the dynamic and evolving nature of the students' learning experiences, data collection in this longitudinal study employed multiple methods across two distinct time points during the 2023-2024 academic year.

- **Interviews:** Semi-structured, in-depth interviews were conducted with each participant at two different stages:
    - Initial Interview (Beginning of the Academic Year): These face-to-face interviews, lasting approximately 60-90 minutes, took place during the first few weeks of the program. The interviews were guided by a semi-structured protocol that explored the participants' initial expectations, motivations, and experiences as they embarked on their journey in the Excellence program. Key areas of inquiry included their reasons for selecting the program, their perceptions of the program's structure and curriculum, and their anticipated challenges and opportunities.
    - Follow-up Interview (End of the First Academic Year): Conducted towards the end of the first academic year, just before the final exams, these follow-up interviews aimed to capture the participants' reflections on their learning experiences throughout the year. The interviews, lasting around 90-120 minutes, followed a semi-structured format and focused on exploring any transformations or shifts in the participants' perspectives, their overall assessment of the program, and the strategies they employed to navigate challenges and opportunities. Participants were encouraged to share specific examples and critical incidents that shaped their learning journeys.

All interviews were audio-recorded and transcribed verbatim for subsequent analysis.

By combining these multiple data collection methods and employing a longitudinal approach, the study aimed to capture the richness and complexity of the participants' learning experiences as they unfolded over time, enabling a nuanced exploration of the fluidity and transformative nature of their educational journeys. While this study primarily used interviews, other qualitative methods like netnography can also provide rich insights into participant experiences and perceptions (Hmioui et al., 2020).

4.3. Data Analysis Approach

4.3.1. Systematic Interpretation of Qualitative Data: Employing Thematic Analysis

While the initial coding process utilized NVivo software to facilitate organization and identification of key themes, a more rigorous analytical approach was subsequently applied to ensure depth and reliability in our findings. Specifically, we employed Thematic Analysis as described by Braun and Clarke (2006), a method renowned for its flexibility and effectiveness in identifying, analyzing, and reporting patterns within qualitative data (Braun & Clarke, 2006).

According to Braun and Clarke's, a six-phase process is needed to ensure reliability in our findings:

- Familiarization: We familiarized ourselves with the data through repeated readings of interview transcripts at both time points, beginning and end of the academic year. In the process, we have undergone active reading, noting down initial ideas and potential patterns.
- Initial Coding Generation: This involved systematic coding of interesting features across the dataset. This means finding relevant phrases or sentences or paragraphs labeling them as corresponding to our questions. For example, some codes were generated, such as 'initial academic challenges', 'evolving study strategies', and 'perceived program value'.
- Searching for themes: We grouped codes together under potential themes, collecting all data associated with each potential theme. For example, various codes relating to academic adaptation were grouped together under the overarching heading of 'Academic resilience and growth.'
- Reviewing Themes: We checked whether the themes worked in relation to the coded extracts and the entire dataset. This entailed creating a thematic 'map' of the analysis, which provided an overview of coherence within themes and clear distinctness between themes. For example, we reflectively considered whether the theme 'academic resilience and growth' should be a distinct category or integrated into 'personal development'.
- Defining and Naming Themes: This step enabled a further specification of the nature of each theme and the bigger picture that our analysis told. We provided clear definitions of what each theme referred to. For example, 'academic resilience and growth' was defined as 'The process by which students overcome the initial academic challenges and in due course develop effective learning strategies.'

- Production of the Report: In this last phase of analysis, we chose vivid extract examples, which were compelling, and returned the analysis to our research questions and literature. This made sure that our findings were duly underpinned by data but also contributed toward the greater theoretical understanding.

By combining multiple data collection methods and employing a longitudinal approach, the study aimed to capture the richness and complexity of the participants' learning experiences as they unfolded over time. This approach enabled a nuanced exploration of the fluidity and transformative nature of their educational journeys. While this study primarily used interviews, other qualitative methods, such as netnography, can also provide rich insights into participant experiences and perceptions (Bentalha & Hmioui, 2022).

4.3.2. Qualitative Coding and Analysis Techniques

The qualitative data analysis process employed a rigorous coding procedure facilitated by NVivo software. The coding scheme was meticulously developed to align with the research objectives and capture the salient themes emerging from the interview data. The following primary codes and their respective child codes were utilized in the study:

Figure 1. Overview of research codes and sub-codes

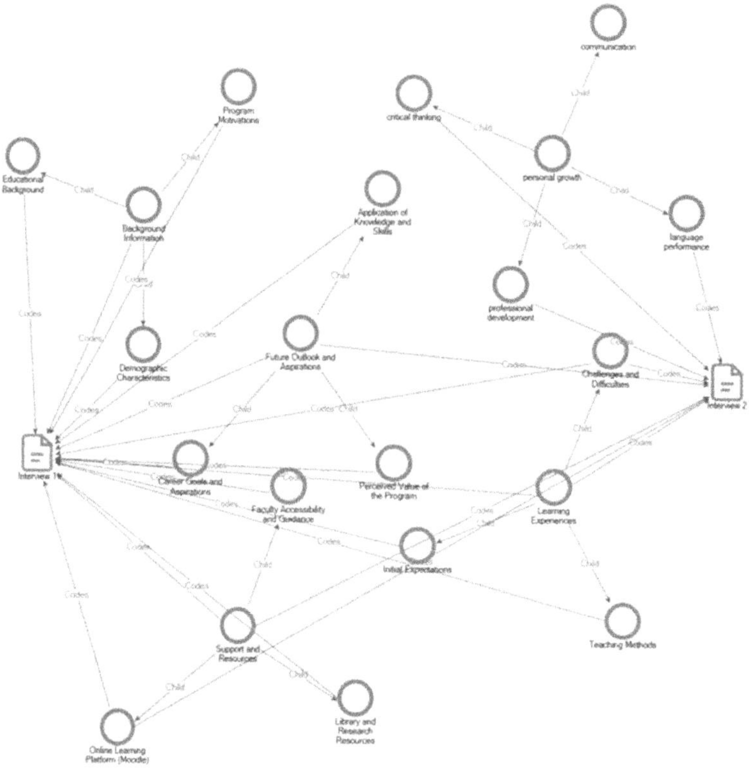

Source: Authors

- **Background Information**

  Demographic Characteristics: This code encompassed details about the participants' age, gender, nationality, and other relevant demographic information (Venegas-Muggli, 2020).

  Educational Background: This code captured the participants' prior educational experiences, including their academic qualifications, previous institutions attended, and any relevant specializations or majors (Xing et al., 2021).

  Program Motivations: This code explored the participants' motivations and reasons for enrolling in the new excellence program, such as personal aspirations, career goals, or recommendations from others (Rahman et al., 2022).

- **Learning Experiences**

Initial Expectations: This code captured the participants' initial expectations and perceptions about the new excellence program, including their anticipated learning outcomes, teaching approaches, and overall academic experiences (Campos et al., 2022).

Teaching Methods: This code focused on the participants' experiences with various teaching methods employed in the program, such as lectures, case studies, group projects, or experiential learning activities (Puranik, 2020).

Challenges and Difficulties: This code identified the challenges and difficulties encountered by the participants during their learning journey, such as time management issues, academic workload, or personal obstacles (Tunku Ahmad, 2020).

Personal Growth: This code explored the participants' perceptions of their personal growth and development throughout the program, including aspects such as professional development, language performance, critical thinking, and communication skills (Weigold et al., 2021).

- **Support and Resources**

Faculty Accessibility and Guidance: This code explored the participants' perceptions of the accessibility and guidance provided by faculty members, including their availability for consultations, responsiveness to queries, and overall support in the learning process (Bong & Chen, 2024).

Online Learning Platform (Moodle): This code examined the participants' experiences with the online learning platform (e.g., Moodle), including its usability, functionality, and effectiveness in facilitating their learning (Liu et al., 2020).

Library and Research Resources: This code assessed the participants' perspectives on the availability and quality of library resources and research materials, as well as their ability to access and utilize these resources effectively (Shoaib et al., 2022).

- **Future Outlook and Aspirations**

Career Goals and Aspirations: This code delved into the participants' career aspirations and goals, exploring how the new excellence program aligned with their desired professional trajectories (Healy et al., 2022).

Application of Knowledge and Skills: This code focused on the participants' perceptions of how the knowledge and skills acquired during the program could be applied in their future careers or personal endeavors (Elkhder et al., 2022).

Perceived Value of the Program: This code captured the participants' overall assessment of the value and significance of the new excellence program, including their perceptions of its impact on their personal and professional development (Del Río-Rama et al., 2021).

The coding process involved a systematic and iterative approach, where the interview transcripts were thoroughly reviewed, and relevant segments were assigned appropriate codes and child codes. This meticulous process facilitated the identification of recurring patterns, themes, and insights, providing a comprehensive understanding of the students' learning experiences in the new excellence program.

4.3.3. Strategies for Tracking Temporal Dimensions and Developmental Trajectories

Specific strategies were implemented as this study aimed at capturing temporal dimensions and developmental trajectories of the learning experiences of the students. The analysis was therefore focused on identifying changes, transitions, and evolving perspectives across these two interview time points: the beginning and the end of the academic year.

A major strategy was the comparison of coded data from the first and second interview of each participant. This was important in identifying shifts in their expectations, challenges, perceptions, and aspirations over time. The analysis also looked into what possible factors or events might have caused these changes be it in specific learning experiences, support mechanisms, or personal growth (Creswell, 2007).

In addition, the analysis explored the developmental paths of the learning experiences of students, developing their perceptions and skills, and how they gained development throughout the academic year. It included the identification of patterns, critical incidents, or pivotal moments that formed their learning journey and determined the directions for the future (Creswell, 2007).

These strategies aim to provide a comprehensive understanding of the dynamic and evolving nature of student learning experiences, capturing the fluidity and complexities of personal and academic development over the years within the new program of excellence.

4.4. Key Findings and Discussion

4.4.1. Evolution of Students' Learning Experiences and Perspectives

The longitudinal design of the study made it possible to capture the changes and shifts in students' experiences and perceptions of learning over the academic year in the first year of the Excellence License. Based on the analysis of the beginning and end-of-year interviews, several key patterns and shifts were identified:

- **Adaptation to Academic Rigor and Workload**: Students were excited about the beginning of the program, the interactive teaching, and practical applications. However, they were aware of the difficulties associated with strong academic requirements and a large number of students in the class. During the year, successful adaptation to the workload and pace was noticed for some students, while others still lagged behind, particularly students with non-traditional profiles or those with predetermined competency shortages, such as in quantitative and computational abilities.
- **Development of Practical and Transferable Skills**: Overall, the students significantly developed their transferable competencies, including communication, teamwork, critical and analytical abilities, and problem-solving skills. Cooperative projects, case studies, and presentations proved to be of utmost value in developing these competencies. By the end of the year, students were confident that they could apply this knowledge and these skills further in their future careers.
- **Language Skills and Understanding of Cultures**: Students appreciated the program's strong emphasis on French and English language courses, and many of them reported noticeable growth in their language skills, recognizing these as critical factors for their success in the global business world. Additionally, due to the diverse backgrounds of the students, they emphasized the need for cultural awareness and the ability to work in multicultural environments.
- **Personal and Professional Development**: Beyond academic achievements, the students demonstrated significant personal and professional development over the year. They reported increased confidence, resilience, personal identity, and a strengthened sense of purpose. Some also clarified their career interests, aligning them with the knowledge and competencies gained in the program.
- **Challenges and Support Systems**: Even though the program provides multiple forms of support, such as access to faculty, online platforms, and library

resources, some students felt that specific assistance and mentorship were needed, especially for those facing unique challenges or from unconventional backgrounds. Large class sizes were often cited as a barrier to receiving the proper attention and guidance.

- **Perceptions of Program Value and Future Outlook**: By the end of the year, almost all students reported that the Excellence License program had been very valuable to them personally and professionally. However, students facing major challenges and those with low confidence in certain areas were uncertain about their future careers and further education intentions.

Figure 2. Word cloud

Source: Authors

Based on the word cloud, it appears to be depicting the key themes and topics that emerged from the longitudinal study conducted on the experiences of students in the Excellence program at USMBA University.

The prominent words and phrases provide insights into various aspects of the study:

- "Excellence program" and "management" suggest that the research focused on examining the learning experiences of students enrolled in this specific program related to management education.
- Words like "university," "resources," "class," "professors," "family," and "online platform" indicate that the study explored the institutional factors, support systems, and learning environments provided by the university for the Excellence program.
- Terms such as "interview," "consent," "background," "part," and "case" imply that the study involved interviews or case studies as part of the qualitative data collection process, potentially exploring the participants' backgrounds and experiences.

- The words “skills,” “learning,” “future,” “career,” and “outlook” suggest that the research investigated the development of skills, learning outcomes, and future career aspirations of the students in relation to the Excellence program.
- Words like “experiences,” “initial,” “support,” “closing,” and “project” could indicate that the study tracked the students' experiences from the beginning of the program, the support they received, and potentially their experiences with a final project or capstone.
- The presence of words such as “confidentiality,” “determined,” and “consent” implies that the study followed ethical research practices, ensuring confidentiality and obtaining consent from participants.
- Words like “challenging,” “disciplines,” “computational,” “introduction,” and “coursework” point to the academic rigor and interdisciplinary nature of the Excellence program, potentially exploring the challenges students faced in various courses or disciplines.
- The word “develop” suggests that the study may have examined the personal and professional development of students throughout the program.

Overall, the word cloud provides a visual representation of the key concepts and themes explored in the longitudinal study, reflecting the focus on the Excellence program, management education, student experiences, institutional factors, skill development, career aspirations, and the methodological aspects of the research.

The longitudinal analysis revealed a dynamic and evolving journey for the students, highlighting their adaptability, resilience, and personal growth alongside the challenges they encountered. The insights gained from this study can inform strategies to enhance the program's effectiveness and support systems, ultimately fostering a more inclusive and transformative learning experience for future cohorts.

4.4.2. Interplay of Institutional Factors and Personal Experiences

Through the longitudinal data analysis, it was established that there was a constant, vibrant interplay among the aforementioned institutional and personal factors that characterized the Excellence program. This interplay played a significant role in determining the students' learning journeys, their perceptions, the challenges they faced, and how their developmental trajectories unfolded over the year. Additionally, the interplay between institutional factors and personal experiences is not unique to education; it is also observed in other fields, such as supply chain management (Bentalha, 2023) . This highlights the importance of considering both macro and micro-level factors in understanding complex phenomena.

- **Institutional Factors:**

Curriculum Design and Academic Rigor: The broad-based curriculum, covering a wide array of management disciplines and placing significant emphasis on practical applications, was highly appreciated by the students. Nevertheless, some students found the course academically intense and the pace challenging, especially non-traditional students or those with deficits in important skills.

Teaching Methods and Learning Environment: Students appreciated the use of interactive teaching and learning methods, such as case studies, group projects, and experiential learning. However, although students valued how large class sizes facilitated group work, these same sizes impeded active class participation, individual attention, and effective class discussions.

Faculty Support and Accessibility: Students appreciated the approachability and interest of the faculty. The faculty was available for student consultations and provided proper guidance. Nevertheless, some students felt that personalized mentorship and more directed attention were needed, especially for those who presented unique challenges or came from diverse backgrounds.

Resources and Support Systems: The program had various resources, such as the Moodle platform, library access, and language courses. These resources helped enhance learning and skill development. Nevertheless, more effective support systems were needed to address students' personal needs and backgrounds.

- **Personal Factors and Experiences:**

Educational and Family Backgrounds: Diverse educational and family backgrounds among students contributed to varying levels of preparedness, different motivations, and different expectations upon joining the Excellence License program. Non-traditional students and those from rural backgrounds faced additional challenges in quickly adapting to academic requirements and typical university life.

Socioeconomic Circumstances: Juggling personal and financial obligations with academic demands was a concern for some students, and such pressures could have compromised their ability to fully commit to the program.

Language Proficiency and Cultural Awareness: Students' proficiency in French and English, along with their cultural awareness and ability to work in a multicultural environment, were critical to their learning experiences and personal development in the program.

Prior Knowledge and Skill Sets: Prior knowledge and skills, especially in quantitative analysis, computational abilities, and communication, influenced students' ability to manage the program's curriculum and remain engaged in specific courses or coursework.

The interaction between these institutional factors and personal experiences suggests the need for a more comprehensive and individualized approach to student support to enhance student success. Although the program offers ample resources and developmental opportunities—both personal and professional—focusing on the specific needs and characteristics of the students could add significant value to their learning experiences and development paths.

4.4.3. Discussion of Key Findings

Several meaningful patterns and findings that are worthwhile discussing came from the longitudinal analysis of the students' experiences in the Excellence program.

First, it brings out how student perceptions and skills will have changed over the year, thus making the program transformational. According to Vermunt, 2005, it is in these transformations that one gets indicators of an effective learning environment. The fact that clear portable skill improvements were noted, especially in communication and critical thinking skills, agrees with the need for the development of these competencies during higher education, as put forward by Tuononen et al. (2022).

The struggles of non-traditional/skill-deficit students indicate that pedagogies must be more inclusive. This makes intrinsic sense, since Hammond and Jackson were urging in 2015 for culturally responsive education that prioritizes the diverse needs of all learners in the learning environment. In consideration of this, even though the Excellence program does offer many rich learning experiences, some things could be done better to meet the demands of students who come from diverse backgrounds.

Very strong in our analysis is the interplay of institutional factors and personal experiences setting the stage for learning. According to Fellenz, effective management education has to consider both organizational structures and individual learner characteristics (Fellenz, 2006). Our results show that on the one hand, institutional factors—like curriculum design and teaching methods—strongly affect learning experiences; and that on the other hand, there are personal factors equally relevant, like prior knowledge and socio-economic conditions.

The appreciation of interactive ways of teaching and practical applications the students showed therefore justifies the argument put forth by Bong and Chen, 2024, for incorporating diverse pedagogies into management education. Large class sizes promote interactive learning at the expense of personalized attention, as Liu et al., 2020, argued in their study over resource allocation decisions in higher education.

Finally, the language skills and cultural awareness that students develop showcase global competence in management education as of growing importance. This aligns with the findings of Zhampeissova et al. (2020), who stressed the urgency to internationalize business curricula.

These findings collectively provide invaluable insights into the complex dynamics of student learning experiences in management education. They not only identify the strengths of the program of Excellence and areas where improvements are needed but also provide a basis for making evidence-based enhancements in the curriculum and support systems.

4.4.4. Contributions to Theory and Practice in Management Education

This longitudinal study therefore makes important contributions to the theoretical understanding of the student learning experience but also to the practical implementation of effective programs for management education. Exploring the dynamic interplay between institutional factors and personal experiences, the research provides insights to inform future developments in the field.

- **Development and Student Learning Theories**: The longitudinal approach of the study and its focus on students' shifting perspectives and experiences corresponds well to the existing theories and the frameworks for understanding student development, personal growth, and transformative processes of learning (Vermunt, 2005). The presented findings emphasize the complexities and nonlinearity of learning journeys, which question the traditional linear models and strengthen the necessity of using holistic approaches.
- **Inclusive Pedagogy and Culturally Responsive Education**: Examination of different backgrounds and experiences of students in the research raises the question of the importance of pedagogy inclusive of every perspective and culturally responsible education. It does underline, therefore, the importance of dealing with the problems students from nontraditional and underrepresented groups face, thus implying an inclusive and supportive learning environment (Hammond & Jackson, 2015; Vermunt, 2005).
- **Program Evaluation and Curriculum Development**: Insights gained from the research on the dynamic interplay between institutional factors and personal experiences provide a full understanding of the strengths and areas that need improvement within the Excellence program (Creswell, 2007). This knowledge will be useful for designing program evaluation frameworks and guiding efforts in curriculum development to better conform to student needs and to provide effective learning experiences.

- **Transferable Skills Development**: Building transferable skills in communication, teamwork, problem solving, and critical thinking relevant to higher education students is an aspect of the findings that contributes to the larger conversation around the role of higher education in fostering appropriate preparation for careers in the future and for participation in society (Tuononen et al., 2022).
- **Tailored Support Systems and Mentorship**: The research identifies that the support systems should be tailored and the mentorship provided should be personalized to meet the needs and backgrounds of different students. Using this information, institutions can design specific activities like peer mentoring programs, academic coaching, or workshops to enhance student success and engagement (Fellenz, 2006).
- **The importance of creating inclusive learning environments**, embracing diversity, and fostering cultural awareness. This can be achieved through faculty training, curriculum revision, and the inculcation of culturally responsive pedagogies that will validate and value the students' varied experiences and perceptions (Hammond & Jackson, 2015).
- **Faculty Development and Instructional Strategies:** The study recommends faculty development in interactive teaching and experiential learning activities. More collaborative learning, more case studies, and more real-world applications have the potential of increasing student engagement and transferring more knowledge into practical contexts (Bong & Chen, 2024).
- **Resource Allocation and Institutional Support:** In this context, the lessons from this research can guide institutions to provide the resources for effective student success. This could range from investment in smaller-sized classrooms, more library resources, on-line learning platforms, or academic and skill development support centers (Liu et al., 2020).

This study can provide valuable insights for both theory and practice on the next generation of management education, thus ensuring inclusive, transformative, and effective learning experiences for diverse student populations.

4.4.4. Reflections on the Longitudinal Qualitative Approach and Study Limitations

Being a longitudinal study, it helps to explore developmental trajectories and is crucial in identifying pivotal or turning points in the difference of the students' perceptions and skill acquisition. This temporal dimension provides an exceptional view of learners' transformational journey, bringing out the complexities and non-linearity of the learning process (Derrington, 2019).

Moreover, the nature of the study is qualitative, thus soliciting data collection with rich, contextualized information that allows a nuanced understanding of students' lived experiences. The open-ended, semi-structured interviews with the students effectively helped in bringing forth their stories, challenges, and aspirations in their own words, revealing in-depth and diverse perspectives.

However, it is crucial to admit that the following are some of the limitations of the study and the longitudinal qualitative approach:

- **Sample Size and Generalization:** The study focused on a particular group of students in the Excellence program at USMBA University. The findings might be interesting; however, the relatively small sample size is a problem if they are to be generalized to other programs and educational contexts. For this reason, it would be interesting to work with a more considerable and variegated sample to improve result transferability.
- **Participant Attrition:** Longitudinal studies generally experience participant dropout because of various reasons. This may lead to biases or incomplete data. Although the current study did not experience a significant level of this, future studies should consider a reduction strategy to minimize dropout rates and thus ensure the continuity of data collection.
- **Researcher Subjectivity:** Qualitative research is, by definition, implicated by the perspectives, backgrounds, and interpretations of the researcher. Measures were taken to ensure rigor and transparency, including the use of consistent coding techniques and peer review, but the potential for researcher bias cannot be entirely eliminated.
- **Temporal Constraints**: The study captured the experiences of the students within a single academic year. Whereas this provided a good insight, it is equally important to consider learning as a lifelong process, and the perspectives and experiences of students may continue to change and develop beyond the period covered by this study.
- **Methodological Limitations:** Though interviews formed the leading collection tool, inclusion of more qualitative methods like focus groups or even observational methods would have enhanced the data and facilitated more insight into the learning experiences.

While these limitations are acknowledged, it's important to contextualize them within the broader landscape of longitudinal qualitative research in education. As Derrington (2019) notes, such studies often grapple with issues of generalizability and sample size (Derrington, 2019). However, the rich, contextualized data they provide offers unique insights that quantitative methods might miss. Future research could address these limitations by employing mixed-methods approaches or by conducting comparative studies across multiple institutions. Additionally, extending the study duration beyond a single academic year could provide more comprehensive insights into the long-term impacts of the Excellence program on students' professional development and career trajectories.

Notwithstanding these, this study, with its longitudinal qualitative design, has significantly added to our understanding of the dynamic interplay of institutional factors and personal experiences in management education. The insights gained may be used for future program development, instructional strategies, and support systems that aim to create inclusive and transformational learning experiences for diverse student populations.

CONCLUSION

This paper has convincingly argued that longitudinal qualitative research is a robust approach for capturing the fluidity of learning experiences in management education (Balmer et al., 2021; Gravemeijer & Cobb, 2006). Through an exploration of the conceptual underpinnings that frame the importance of time, lived experiences, and developmental trajectories (Bergson, 2014; Heidegger et al., 2008; Kemmis et al., 2014; Schatzki, 2013), the paper has constructed a theoretical underpinning in understanding the dynamic nature of learning processes.

In longitudinal qualitative research within diverse disciplinary areas (Balmer et al., 2021; Bernardi, 2021; Cooke et al., 2012), the value and flexibility of this methodology have been brought to the fore. That was also a paper stressing the examination of intersections between institutional frames and human experiences (Knust & Hanft, 2009; Treviño-Eberhard & Kaufmann-Kuchta, 2022), acknowledging mutual influences through which learning and development processes are shaped. Further, it was highlighted that research collaboration is important as a way of increasing the strength, validity, and potential of impact of findings in management education and beyond.

The power of this approach to be realized in practice has been evidenced by the empirical longitudinal case study carried out at USMBA University (Section 4). The study followed a cohort of students enrolled in the inaugural promotion of the Excellence program in Management Sciences through a whole academic year and revealed the evolution of their learning experiences (Section 4.4.1), the interplay between the institutional factors and personal experience (Section 4.4.2), as well as the development of their perspective and skills.

Key insights will be provided into the processes by which students adapt to academic rigour and workload, develop practical and transferable skills, enhance their proficiency in languages and raise their level of cultural awareness, and the challenges and support systems involved (Section 4.4.1). Moreover, the research has observed institutional factors like curriculum design, teaching approach, and support systems and the relationship of these with the personal experiences and background of the students (Section 4.4.2), which has brought out the need for inclusive and customized approaches to ensure the success of the students. This longitudinal qualitative study has made contributions to the theoretical understanding of the student learning experience and practical implementation of effective management education (Section 4.4.3).

Though longitudinal qualitative research of this paper has shown value, yet it has its own share of limitations on counts of sample and generalizability, potential subject dropout, subjectivity on the part of the researcher, time limitations, and methodological issues (Section 4.4.4). These limitations point to the future work that must be undertaken to extend and refine the longitudinal qualitative method use in management education.

Future studies may move along these lines:

- Expanding the time-span of longitudinal studies to encapsulate the entire academic life of students, from the time of joining the institution till their graduation and even further to their professional life to understand the long-term effects of the management education programs (suggested in Conclusion section).
- The development of comparative longitudinal studies over diverse institutions, programs, or cultural environment may prove to be of immense significance to understand the effects of diverse educational environments on the learning experiences and patterns of students (suggested in Conclusion section).
- Applying mixed-method research, combining longitudinal qualitative studies with quantitative measures of student performance, engagement, and outcomes, may contribute to offering a more holistic view of whether management education initiatives are effective (Cobb et al., 2003; suggested in Conclusion section).

Such a methodological stance harnesses the fluidity of the learning experience over time and thus offers a rich tapestry of insights that can be used in the formation and crystallization of pedagogical practices and institutional policies geared toward increasing the overall quality and inclusivity of management education programs.

REFERENCES

Akkermans, J., Brenninkmeijer, V., Huibers, M., & Blonk, R. W. B. (2013). Competencies for the Contemporary Career : Development and Preliminary Validation of the Career Competencies Questionnaire. *Journal of Career Development*, 40(3), 245–267. 10.1177/0894845312467501

Baker, L., Phelan, S., Snelgrove, R., Varpio, L., Maggi, J., & Ng, S. (2016). Recognizing and Responding to Ethically Important Moments in Qualitative Research. *Journal of Graduate Medical Education*, 8(4), 607–608. 10.4300/JGME-D-16-00384.127777677

Balmer, D. F., Varpio, L., Bennett, D., & Teunissen, P. W. (2021). Longitudinal qualitative research in medical education : Time to conceptualise time. *Medical Education*, 55(11), 1253–1260. 10.1111/medu.1454233847408

Bentalha, B. (2023). *A Framework for Exploring the Global Performance of Service Supply Chain Management : A Qualitative Study* (p. 106-129). 10.4018/978-1-6684-9062-4.ch007

Bentalha, B., & Hmioui, A. (2022). *Smart Service Supply Chain and Just Walk Out Technology : A Netnographic Approach* (p. 223-236). 10.1007/978-3-030-94191-8_18

Bergson, H. (2014). *Time and free will : An essay on the immediate data of consciousness* (Routledge, T.Francis Group, Trans.). F. L. Pogson. 10.4324/9781315830254

Bernardi, L. (2021). Qualitative longitudinal research in family sociology. In Schneider, N. F., & Kreyenfeld, M. (Eds.), *Research Handbook on the Sociology of the Family*. Edward Elgar Publishing., 10.4337/9781788975544.00015

Blossfeld, H.-P., & Hofmeister, H. A. (Eds.). (2008). *Globalization, uncertainty and women's careers : An international comparison* (Paperback ed). Edward Elgar.

Bong, W. K., & Chen, W. (2024). Increasing faculty's competence in digital accessibility for inclusive education : A systematic literature review. *International Journal of Inclusive Education*, 28(2), 197–213. 10.1080/13603116.2021.1937344

Bourner, T., & Brook, C. (2019). Comparing and contrasting action research and action learning. In Mertler, C. A. (Ed.), *The Wiley Handbook of Action Research in Education* (pp. 185–206). Wiley-Blackwell. 10.1002/9781119399490.ch9

Brannen, J. (2006). Cultures of Intergenerational Transmission in Four-Generation Families. *The Sociological Review*, 54(1), 133–154. 10.1111/j.1467-954X.2006.00605.x

Braun, V., & Clarke, V. (2006). Using thematic analysis in psychology. *Qualitative Research in Psychology*, 3(2), 77–101. 10.1191/1478088706qp063oa

Campos, M., Peixoto, F., Bártolo-Ribeiro, R., & Almeida, L. S. (2022). Adapting as I Go : An Analysis of the Relationship between Academic Expectations, Self-Efficacy, and Adaptation to Higher Education. *Education Sciences*, 12(10), 10. Advance online publication. 10.3390/educsci12100658

Cobb, P., Confrey, J., Disessa, A., Lehrer, R., & Schauble, L. (2003). Design Experiments in Educational Research. *Educational Researcher*, 32(1), 9–13. 10.3102/0013189X032001009

Cooke, A., Smith, D., & Booth, A. (2012). Beyond PICO : The SPIDER tool for qualitative evidence synthesis. *Qualitative Health Research*, 22(10), 1435–1443. 10.1177/1049732312452938822829486

Creswell, J. W. (2007). *Qualitative inquiry and research design : Choosing among five approaches, 2nd ed* (p. xvii, 395). Sage Publications, Inc.

Del Río-Rama, M. de la C., Álvarez-García, J., Mun, N. K., & Durán-Sánchez, A. (2021). Influence of the Quality Perceived of Service of a Higher Education Center on the Loyalty of Students. *Frontiers in Psychology*, 12, 671407. Advance online publication. 10.3389/fpsyg.2021.67140734163410

Derrington, M. L. (2019). *Qualitative Longitudinal Methods : Researching Implementation and Change*. SAGE Publications, Inc., 10.4135/9781071814277

Ding, F., & Curtis, F. (2021). 'I feel lost and somehow messy' : A narrative inquiry into the identity struggle of a first-year university student. *Higher Education Research & Development*, 40(6), 1146–1160. 10.1080/07294360.2020.1804333

Dollhausen, K., Lattke, S., Scheliga, F., Andrä, W., Spexard-Heinemann, A., Geffers, J., & Banscherus, U. (2013). *Opening Higher Education to Adults. Final Report.*

Elder, G. H., Johnson, M. K., & Crosnoe, R. (2003). The Emergence and Development of Life Course Theory. In Mortimer, J. T., & Shanahan, M. J. (Eds.), *Handbook of the Life Course* (pp. 3–19). Springer US., 10.1007/978-0-306-48247-2_1

Elkhder, E. E., Al-Zubair, B. T., Mohammed, E.-R. H., & Onia, S. I. (2022). Application of Knowledge Management in The Sudanese Universities : An Analytical Study. *Indonesian Journal of Education and Social Studies*, 1(1), 1. Advance online publication. 10.33650/ijess.v1i1.3411

Farrell, E. (2020). Researching Lived Experience in Education : Misunderstood or Missed Opportunity? *International Journal of Qualitative Methods*, 19, 160940692094206. 10.1177/1609406920942066

Fellenz, M. R. (2006). Toward Fairness in Assessing Student Groupwork : A Protocol for Peer Evaluation of Individual Contributions. *Journal of Management Education*, 30(4), 570–591. 10.1177/1052562906286713

Gangl, M. (2002). Changing Labour Markets and Early Career Outcomes : Labour Market Entry in Europe Over the Past Decade. *Work, Employment and Society*, 16(1), 67–90. 10.1177/09500170222119254

Gravemeijer, K., & Cobb, P. (2006). Design research from a learning design perspective. *Educational design research.*

Grosemans, I., & De Cuyper, N. (2021). Career competencies in the transition from higher education to the labor market : Examining developmental trajectories. *Journal of Vocational Behavior*, 128, 103602. 10.1016/j.jvb.2021.103602

Hammond, Z., & Jackson, Y. (2015). *Culturally responsive teaching and the brain : Promoting authentic engagement and rigor among culturally and linguistically diverse students.* Corwin, a SAGE company.

Healy, M., Hammer, S., & McIlveen, P. (2022). Mapping graduate employability and career development in higher education research : A citation network analysis. *Studies in Higher Education*, 47(4), 799–811. 10.1080/03075079.2020.1804851

Heidegger, M., Macquarrie, J., & Robinson, E. S. (2008). *Being and time.* Harper-Perennial/Modern Thought.

Hmioui, A., Bentalha, B., & Alla, L. (2020). *La netnographie dans les recherches marketing : La communauté virtuelle comme consom'acteur vecteur d'efficacité marketing. 4*, 631-652. 10.48374/IMIST.PRSM/ame-v2i4.23559

Johansson, E. (2023). Reconceptualizing Time to Understand How to Lead Digitalization in Education. *Forskning Og Forandring*, 6(1), 3–21. 10.23865/fof.v6.5344

Kellogg, B. T. (2024). An Asset-Based Approach to Understanding College Students as Sources of Support. *Journal of College Student Retention*, 25(4), 820–845. 10.1177/15210251211025212

Kemmis, S., Wilkinson, J., Edwards-Groves, C., Hardy, I., Grootenboer, P., & Bristol, L. (2014). *Changing Practices, Changing Education.* Springer Singapore., 10.1007/978-981-4560-47-4

Knust, M., & Hanft, A. (2009). *Continuing Higher Education and Lifelong Learning*. Springer Netherlands., 10.1007/978-1-4020-9676-1

Krause, N. (1995). Negative interaction and satisfaction with social support among older adults. *The Journals of Gerontology. Series B, Psychological Sciences and Social Sciences*, 50(2), 59–P73. 10.1093/geronb/50B.2.P597757834

Lincoln, Y. S., & Guba, E. G. (1986). But is it rigorous? Trustworthiness and authenticity in naturalistic evaluation. *New Directions for Program Evaluation*, 1986(30), 73–84. 10.1002/ev.1427

Liu, Z.-Y., Lomovtseva, N., & Korobeynikova, E. (2020). Online Learning Platforms : Reconstructing Modern Higher Education. *International Journal of Emerging Technologies in Learning*, 15(13), 4–21. 10.3991/ijet.v15i13.14645

McTaggart, R. (1991). Principles for Participatory Action Research. *Adult Education Quarterly*, 41(3), 168–187. 10.1177/0001848191041003003

Millar, J., & Ridge, T. (2013). Lone mothers and paid work : The 'family-work project'. *International Review of Sociology*, 23(3), 564–577. 10.1080/03906701.2013.856161

Orlikowski, W. J., & Yates, J. (2002). It's About Time : Temporal Structuring in Organizations. *Organization Science*, 13(6), 684–700. 10.1287/orsc.13.6.684.501

Ottrey, E., Rees, C. E., Kemp, C., Brock, T. P., Leech, M., Lyons, K., Monrouxe, L. V., Morphet, J., & Palermo, C. (2021). Exploring health care graduates' conceptualisations of preparedness for practice : A longitudinal qualitative research study. *Medical Education*, 55(9), 1078–1090. 10.1111/medu.1447533617656

Patton, M. Q., & Patton, M. Q. (2002). *Qualitative research and evaluation methods* (3rd ed.). Sage Publications.

Prediger, S., & Schnell, S. (2014). *Investigating the Dynamics of Stochastic Learning Processes : A Didactical Research Perspective*. Its Methodological and Theoretical Framework, Illustrated for the Case of the Short Term–Long Term Distinction., 10.1007/978-94-007-7155-0_29

Puranik, S. (2020). *Innovative teaching methods in higher education*. BSSS Journal of Education., 10.51767/je0907

Rahman, H., Hasibuan, A. F., Syah, D. H., Sagala, G. H., & Prayogo, R. R. (2022). Intrapreneurship : As the outcome of entrepreneurship education among business students. *Cogent Education*, 9(1), 2149004. 10.1080/2331186X.2022.2149004

Reid, A.-M., Brown, J. M., Smith, J. M., Cope, A. C., & Jamieson, S. (2018). Ethical dilemmas and reflexivity in qualitative research. *Perspectives on Medical Education*, 7(2), 69–75. 10.1007/S40037-018-0412-229536374

Schatzki, T. R. (2013). *The timespace of human activity : On performance, society, and history as indeterminate teleological events* (First paperback edition). Lexington Books.

Shoaib, M., Ali, N., Anwar, B., & Abdullah, F. (2022). Library services and facilities in higher education institutions during coronavirus disease (COVID-19) in Pakistan. *Journal of Information Science*. Advance online publication. 10.1177/01655515221141035

Taylor, C. J. (2020). *A qualitative longitudinal evaluative study of individual acceptance of leader concepts.*

Teehankee, B. (2018). Critical Realist Action Research and Humanistic Management Education. *Humanistic Management Journal*, 3(1), 71–90. 10.1007/s41463-018-0033-2

Thomson, R., & Holland, J. (2003). Hindsight, foresight and insight : The challenges of longitudinal qualitative research. *International Journal of Social Research Methodology*, 6(3), 233–244. 10.1080/1364557032000091833

Treviño-Eberhard, D., & Kaufmann-Kuchta, K. (2022). Does It Make a Difference? Relations of Institutional Frameworks and the Regional Provision of Continuing Higher Education in England and Spain. *Education Sciences*, 12(2), 132. 10.3390/educsci12020132

Tunku Ahmad, T. B. (2020). Learning from Others : Their Experiences, Challenges and Difficulties. *IIUM Journal of Educational Studies*, 7(2), 1–2. 10.31436/ijes.v7i2.294

Tuononen, T., Hyytinen, H., Kleemola, K., Hailikari, T., Männikkö, I., & Toom, A. (2022). Systematic Review of Learning Generic Skills in Higher Education—Enhancing and Impeding Factors. *Frontiers in Education*, 7, 885917. Advance online publication. 10.3389/feduc.2022.885917

Udas, K. (1998). Participatory action research as critical pedagogy. *Systemic Practice and Action Research*, 11(6), 599–628. 10.1023/A:1022184120067

Ulrich Mayer, K. (2004). Whose Lives? How History, Societies, and Institutions Define and Shape Life Courses. *Research in Human Development*, 1(3), 161–187. 10.1207/s15427617rhd0103_3

Venegas-Muggli, J. I. (2020). Higher education dropout of non-traditional mature freshmen : The role of sociodemographic characteristics. *Studies in Continuing Education*, 42(3), 316–332. 10.1080/0158037X.2019.1652157

Verbruggen, M., Van Emmerik, H., Van Gils, A., Meng, C., & De Grip, A. (2015). Does early-career underemployment impact future career success? A path dependency perspective. *Journal of Vocational Behavior*, 90, 101–110. 10.1016/j.jvb.2015.08.002

Verina, N., Titko, J., & Shina, I. (2021). Digital transformation outcomes in higher education : Pilot study in Latvia. *International Journal of Learning and Change*, 13(4/5), 459. 10.1504/IJLC.2021.116661

Vermunt, J. (2005). Relations between student learning patterns and personal and contextual factors and academic performance. *Higher Education*, 49(3), 205–234. 10.1007/s10734-004-6664-2

Weigold, I. K., Weigold, A., Ling, S., & Jang, M. (2021). College as a Growth Opportunity : Assessing Personal Growth Initiative and Self-determination Theory. *Journal of Happiness Studies*, 22(5), 2143–2163. 10.1007/s10902-020-00312-x

Xing, Y., Hu, Y., & Zhou, J.-Z. (2021). Higher education and family background : Which really matters to individual's socioeconomic status development in China. *International Journal of Educational Development*, 81, 102334. 10.1016/j.ijedudev.2020.102334

Yosso, T. J. (2005). Whose culture has capital? A critical race theory discussion of community cultural wealth. *Race Ethnicity and Education, 8*(1), 69-91. 10.1080/1361332052000341006

Zacharias, M. E., Saldaña, J., & Saldana, J. (2002). Longitudinal Qualitative Research : Analyzing Change through Time. *Canadian Journal of Education / Revue canadienne de l'éducation, 27*(2/3), 341. 10.2307/1602237

Zhampeissova, K., Gura, A., Vanina, E., & Egorova, Z. (2020). Academic Performance and Cognitive Load in Mobile Learning. *International Journal of Interactive Mobile Technologies*, 14(21), 78. 10.3991/ijim.v14i21.18439

Chapter 9
Investigating the Determinants of ESG Disclosure Practices in Morocco:
A Qualitative Analysis

Ouissal El Aziz
Hassan II University of Casablanca, Morocco

Abdelkarim Asdiou
Hassan II University of Casablanca, Morocco

ABSTRACT

Non-financial disclosure refers to the publication of indicators that do not result from the financial statements. In Morocco, several companies publish their ESG reports annually, but the quality and quantity of information presented varies from one company to another. This chapter per the authors seeks to understand the factors likely to impact the scale and scope of ESG reporting done by Moroccan companies. This work began with a theoretical exploration of the determinants of non-financial disclosure, which enabled the authors to draw up a theoretical model incorporating the main determinants proposed in the literature. Subsequently, an exploratory qualitative approach will be implemented in order to contextualise the theoretical model identified. In terms of results, the contextualised model contains eight explanatory variables, the size of the company, the industry sector, the governance structure, profitability, CSR performance, stakeholders pressure, the cost of producing reports, and the desire to acquire a competitive advantage.

DOI: 10.4018/979-8-3693-5543-5.ch009

INTRODUCTION

Social and environmental concerns are increasingly part of the business strategies of many companies in different sectors (Dentchev, 2004; Hillman & Keim, 2001). In recent years, however, it has become clear that the integration of social policies and procedures as such is not enough. It needs to be accompanied by communication with stakeholders to highlight the efforts made by the company as part of its responsible commitment.

Communication between the company and its various stakeholders consists in presenting in a document the information on social, environmental and governance policies and actions. More precisely, it covers different pillars of the company's responsible commitment. ESG report, non-financial report, CSR report and several other terms are used to express the report in which ESG or non financial information is published (Rezaee, 2016). In view of the growing importance of this practice, companies are increasingly committed to implementing sustainable business models based on sustainability issues (Morioka et al., 2017), as well as implementing disclosure strategies (Viehs et al., 2014).

In Morocco, several strategic orientations have been implemented in recent years. In particular, to meet new international requirements and cope with various social pressures and changes. To this end, in 2019 the AMMC amended Circular n°03/19 on financial transactions and disclosures. These changes include the transition from the voluntary status of non financial disclosure to a legal obligation for publicly-traded companies. As a result, the companies concerned are increasingly committed to publishing ESG information, the quality and quantity of which varies from one company to another.

This practice, has rarely been studied in Morocco. Research has generally focused on the interaction between CSR and organizational and financial performance, or on its relationship with HRM (Cherkaoui, 2020). Moreover, little research has focused on the determinants of non financial disclosure in Moroccan companies. The Moroccan field presents little data on the subject, and ESG disclosure practices remain an unexplored area in Morocco.

We considered it essential to carry out an exploratory study because Morocco is in an emerging phase in terms of non financial disclosure. For its part, the literature does not present enough data, so we are in a phase of understanding and exploration. The qualitative approach is therefore the one best suited to the scientific needs of our research. Conducting a qualitative exploratory study will ensure a thorough understanding of the phenomenon. This will enable us to contextualise the theoretical contributions to the Moroccan context in order to better define the motives for this communication, and to prepare the ground for any possible quantitative study. As qualitative research aims to provide an answer to the question of why, rather than how (Bentalha et al., 2020).

Based on all these considerations, an exploratory qualitative study was conducted to investigate the determinants of non financial reporting done by Moroccan companies. Identifying the determining factors of the publication of ESG reports will enable us to establish a contextualized theoretical model of the determinants of extra-financial communication specific to Moroccan companies.

THEORICAL FRAMEWORK

Non-Financial Information Definition

The concept of non-financial communication has gradually broadened the spectrum of information in the fields of accounting and financial reporting. By introducing a wide range of information, particularly on social, environmental and governance issues. Although there is growing interest in this practice, the concept is particularly vague. It is characterised by a considerable diversity of points of view. This means that there is no unanimous appreciation of the concept of non-financial disclosure (Eccles & Krzus, 2010; Haller et al., 2017). However, we can propose a few definitions proposed earlier, ***Table 1***;

Table 1. Non financial communication defintions

(Cohen et al., 2000)	Non-financial information consists of data that is not derived from the financial statements due to the fact that financial information is primarily derived from the financial statements.
(Robb & Zarzeski, 2001)	Non-financial communication consists of qualitative information presented in companies' annual reports that is not included in the summary financial statements.
(Abdolmohammadi et al., 2006)	Non-financial information provides non-financial indicators of the company's performance.
(Bassen & Kovács, 2008)	Non-financial information is considered to be qualitative information describing the company's strategies, results and structures.
(Amir et al., 2003)	Non-financial information relates to items that are not provided in the financial reports.

Source: Researchers constructs 2024

Definitions Analysis

In terms of definition, non-financial disclosure is characterised by the absence of a common definition. However, it is possible to identify a number of characteristics cited by several authors that can define the morphology of the term. Firstly, the source and place of publication of this information, which is generally outside the financial statements.

Thus, another specificity related to non-financial disclosure compared to the financial one. The latter reports on past performance achieved while non-financial disclosure makes it possible to present an idea on the future performance of the company. Therefore, this information is intended for all stakeholders and not only shareholders.

Theories and Determining Factors

Theories

In previous studies of non-financial communication, a combination of theories has been used to clarify the issues involved. As a result, we are faced with the inexistence of a universally applicable approach that can provide explanations for all the issues in this area (Omran & Ramdhony, 2015). The facets of non-financial communication are so numerous that a single theoretical approach alone cannot justify them. It is therefore wiser to opt for a multiple theoretical framework comprising different theories.

Legitimacy theory: Legitimacy theory is widely used in non-financial disclosure studies (Dentchev, 2004; Khan et al., 2013).

Firstly, legitimacy is linked to the appropriateness of the actions taken by the company on the basis of a generalised perception in a social context made up of values, beliefs and norms (Perrow, 1970). Legitimacy theory has its origins in the concept of the 'social contract', according to which the company's activities are determined by the limits set by society (Gray et al., 1996). This will enable the company to ensure its survival through the support of its stakeholders, as long as its activities are beneficial to society.

This theory states that companies are required to operate in a socially acceptable way, in order to have access to resources and to maintain their existence (Guthrie & Parker, 1989). The entity can only be truly active if society accepts it as a responsible actor. As a result, society's relationship with the company is governed by a social contract. The latter stipulates that the company can carry out its activity and benefit from resources, but only if it develops its social contributions.

Legitimacy makes it possible to understand the extent to which the company's governance structure can have an impact on the company's responsiveness to its various stakeholders. It is often used to justify the motivations behind CSR communication (C. Chen et al., 2008). Because this type of communication is able to limit threats to legitimacy (Deegan, 2002). Non-financial communication can be used by managers as a tool to enhance the legitimacy of a company operating in an emerging context (Dentchev, 2004).

In short, non-financial disclosure can be used to gain greater legitimacy by presenting more information about a company's responsible commitment. By providing more information on CSR, the company will be judged positively within society.

Stakeholder theory: A company's responsibility is not limited to its duty to its shareholders. It is as wide-ranging as that, due to the diversity of internal and external stakeholders. A stakeholder group can be represented by any group or individual influenced by the company's activity, or who may have an impact on it (Freeman, 2010). All of these stakeholders provide essential resources for the functioning of the company (Donaldson & Preston, 1995).

Stakeholder theory highlights all the actors to whom the company has responsibilities (Boatright, 2014). A strong relationship of trust between the company and its stakeholders is established by taking account of the latter's expectations. We talk more specifically about satisfying the needs of different stakeholders, which is generally manifested in the sustainable and efficient management of resources. CSR performance and disclosure can therefore be used to highlight a company's responsible behaviour.

Non-financial disclosure is a good governance practice because it helps to promote and maintain a healthy relationship with stakeholders. It allows the company to resolve conflicts resulting from the demands of different stakeholders (Adams, 2004).

Institutional theory: In general, the concept of institution refers to the socio-economic values, principles and practices commonly accepted by society (Judge et al., 2010).

From an economic perspective, institutional theory can be indirectly associated with the concept of 'instrumentality', it can be also linked to economic efficiency (Zattoni & Cuomo, 2008). This means that the various actors in society aim to maximise their personal interests by competing for resources. On the other hand, the sociological approach of the same theory highlights the importance of social legitimacy. As well as competing for economic resources, members of society also strive for social acceptance in order to have a better chance of existing (Zattoni & Cuomo, 2008).

From the point of view of Scott (2005), institutional theory provides a view of the way in which social structures, norms, beliefs and principles can frame corporate behaviour. Hence the importance of non-financial disclosure, which offers the company more stability, legitimacy and, consequently, resources. This gives the company more opportunities to survive (Suchman, 1995).

As a result, companies that properly present their non-financial information may consider lower agency costs (Berrone & Gomez-Mejia, 2009), better financial performance and more flexible access to resources (Bansal, 2005).

Institutional theory has been used in previous studies (Jackson & Apostolakou, 2010; Muthuri & Gilbert, 2011), to clarify the impact of country specificities on corporate behaviour with regards to non-financial communication. It argues that corporate survival requires social acceptance (Ashforth & Gibbs, 1990).

Agency theory: Under the neo-classical theory, the firm is no longer considered as a single actor, but can be represented by a 'knot of contracts' between different actors (shareholders, suppliers, creditors, etc.). This is the famous agency theory, which aims to establish the relationship between the functioning of the firm and the behaviour of its partners. It highlights the conflicts of interest that may exist between the company's various stakeholders (Jensen & Meckling, 1976).

It is more common to discuss the divergence of interests that exist between shareholders and directors, the shareholders seek the maximization of dividends while the latter aim to increase the company's wealth and size. For this reason, and in order to limit the behaviour of managers which tends towards their own interests, shareholders must pay them an indemnity called «agency costs» (Rezaee & Tuo, 2017). More specifically, these are control costs, structuring costs and opportunity costs. Disclosure of information that goes beyond summary statements helps to reduce these costs by making the behaviour of directors more visible (Rezaee & Tuo, 2017).

Signalling theory: Signaling theory makes it possible to highlight how communication can impact the behavior of actors. This theory complements the agency theory by considering that leaders can use information sharing when they achieve good results, in order to mention the quality of their management (Hassanein & Hussainey, 2015).

The information asymmetries that exist between managers and stakeholders, including shareholders, can therefore be explained through signalling theory. This theory provides an explanation of how the publication of information can reduce these asymmetries. It involves a process of sending information ('Signals') from the party providing more information to the other party. In return, the receiver interprets and reacts to the signal according to the nature of the information shared with them (Rezaee & Tuo, 2017). Generally speaking, the company itself can use disclosure of different aspects to improve its image and reputation on the market, as a means of 'signalling' to investors and stakeholders the company's commitment to social and environmental issues. The theories analyzed in this section can be summarized as follows, ***Table 2***.

Table 2. Summary of mobilized theories

Theories	The basic idea behind the theory	Application to non-financial disclosure
Agency theory	Agency theory describes the governance and regulatory mechanisms that can be used to control the behavior of managers, so that they act in the shareholder's interest and not just in their own.	Disclosure is seen as a mechanism for reducing information asymmetries between managers and shareholders.
Legitimacy theory	According to legitimacy theory, a company's survival depends on its respect for the norms, values and beliefs established by society.	Non-financial disclosure is likely to reinforce the legitimacy and social acceptance of the company, by highlighting its responsible behavior.
Stakeholder theory	Stakeholder theory is based on the necessity of taking into account the expectations and needs of all the company's stakeholders, not just its shareholders.	Non financial disclosure strengthens a company's relationship with its stakeholders. Providing them with more information reduces the conflicts that arise from information asymmetries, satisfies their disclosure needs and makes the company more transparent in terms of its CSR commitment.
Institutional theory	Institutional theory stipulates that the company's institutional environment is likely to have an impact on its decisions	Information sharing can help the company to face up to institutional pressures for transparency and responsible commitment.
Signal theory	Signal theory examines the use of signals by companies to transmit information that can resolve situations of information asymmetry or impact receiver behavior.	The company can use non financial disclosure as a means of "signalling" to investors and stakeholders the company's responsible commitment. In this way, managers can signal the quality of its management.

Source: Reaserchers constructs 2014

The Determinants of Non-Financial Disclosure

Company size: Previous research (Brammer & Pavelin, 2004; Gallo & Jones-Christensen, 2011) has largely confirmed the significant impact of company size on the adoption and scope of non-financial reporting. Larger organisations have a higher social and environmental impact and greater social visibility (Cowen et al., 1987), which may encourage them to present more information in order to justify and legitimize their actions (Hackston & Milne, 1996; Jensen & Meckling, 1976).

By having more stakeholders, large companies are exposed to increased pressure from their stakeholders (Fortanier et al., 2011). Although these companies have a high visibility in various spheres of the public, their responsibilities and expectations are also more significant (Fernando & Pandey, 2012). The stakeholders of a large company are more numerous and more diverse, which puts more pressure on the entity to present appropriate ESG Repoting (Al-Tuwaijri et al., 2004; Jamali, 2010).

Industry sector: The literature suggests that the company's business sector has a significant impact on the disclosure of social and environmental aspects, insofar as the petroleum, chemical and mining sectors require more disclosure on worker health and safety measures and environmental management (Clarkson et al., 2011; Line et al., 2002). Whereas service sectors in general, including the financial sector, give priority to social issues, including labour practices (Kolk, 2002; Line et al., 2002).

In general, the degree and nature of the impact of a company's activity affects the scope and structure of its communication. Organisations operating in sectors with a higher impact on the environment, such as the transport and energy sectors, are likely to disclose more environmental information than others (Hackston & Milne, 1996).

Governance structure: Several researchers teach that the effectiveness of the governance mechanism leads to a reduction in agency costs related to the separation between ownership and management. Disclosure of non-financial information can be considered as a governance mechanism aimed at reducing information asymmetries (Wise & Ali, 2009). And so in case of asymmetry of information within the company, it is required to disclose more in order to reduce this asymmetry. Effective corporate governance is based on the fact that the disclosure of non-financial information is a managerial necessity to better defend the interests of stakeholders, measure the quality of management of leaders and above all have more transparency and credibility.

Company Age: Older businesses have a better understanding of their communities, and they aspire to more citizenship by providing more information about corporate social responsibility (Dissanayake et al., 2016). They recognize the importance of non-financial communication in maintaining good customer relationships and attracting new customers.

The company's responsible commitment: Theoretically, an entity decides to publish information only when it believes that the information is favourable to it; otherwise, if the information is considered unfavourable, the entity will not disseminate it (Dye, 2001). The more committed a company is to social, environmental and governance issues, the more information it will have to publish. We can therefore conclude that there is a link between a company's societal or environmental performance and the non-financial information it discloses.

The variation in the amount of disclosure from one company to another depends on whether the company has good or bad news to communicate. In other words, entities with a high performance on several aspects have more information to disclose. As a result, low corporate commitment may explain the weakness of the information disclosed.

Corporate profitability: Based on the contributions of agency theory, with good profitability, managers are committed to provide more detailed information to the company's shareholders in order to maintain or develop a good image and guarantee their stability (Campbell, 2007; Gamerschlag et al., 2011). On the other hand, when profitability is low, they tend to withhold information and make less known, to conceal results and lost profits.

Regulatory framework: A country's legal and regulatory framework has an important role to play in influencing the extent of ESG disclosure practices (Campbell, 2007; Roe, 2003). Several studies have been conducted at this level, confirming that laws and regulations impact the social performance of the company (Amaeshi et al., 2006).

Figure 1. Theorical model

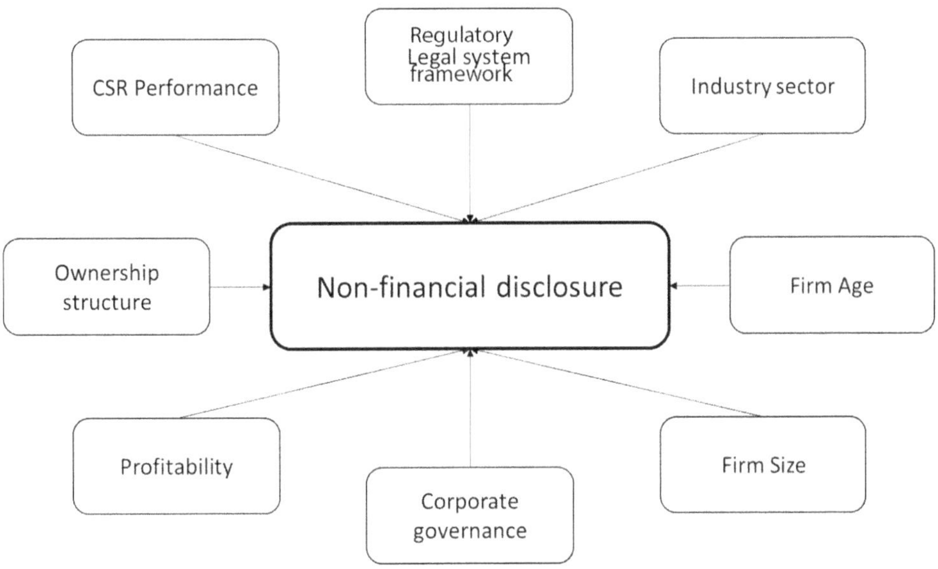

Source: Researchers constructs 2024

RESEARCH METHODOLOGY

We aim to describe the elements that contribute to determine the nature and scope of non financial information presented by Moroccan companies. To achieve this, we first conducted a theoretical exploration of the determinants previously proposed. A literature review revealed a *Theorical model* (*Figure 1*) of the determinants of non financial disclosure, including the main explanatory variables mobilized. This theoretical synthesis allowed us to highlight certain considerations to carry out our study, including the contextualization of constructs.

The subject of CSR information and its determinants has generally received relatively limited attention in the Moroccan context. Starting with Aamoum & Soufiane (2021), who examined the determinants of intangible capital disclosure. Meanwhile, Asdiou & Mokhtari (2019) investigated the determinants and the real motivation of the use of CSR indicators by Moroccan companies. We therefore set out to enrich the Moroccan literature, alongside the work of Farrat & Hajji (2022), and Aziz & Asdiou (2024) in terms of analyzing the determinants of Extra financial disclosure.

It therefore seemed wiser to conduct an exploratory qualitative study. Since this approach allows us to clarify our research subject. Conducting a qualitative study ensures the understanding of the modalities and reasons for events in concrete situations (Wacheux, 1996). It should be highlighted that qualitative research is a process that repeats itself (Bentalha, 2023).

Choice of the Research Approach

As an approach to knowledge production, exploration enables new interactions to be established between theoretical concepts, by setting up new concepts within a given body of theory (Charreire & Huault, 2001). The main explanatory factor for our choice lies in the nature of our research object and the context in which it will be conducted. We are looking for a more appropriate understanding of the phenomenon of non financial disclosure, while highlighting the characteristics of the Moroccan context.

CSR practices are generally perceived as an emerging field in Morocco. The perception of these practices differs from one company to another (Cherkaoui, 2016). Several internal and external factors are likely to impact Moroccan companies' decisions and actions in terms of CSR communication. This highlights the need for contextualization, through qualitative study, in order to explore the particularities of the context under study. As well as Qualitative research aims to explore a situation that has been little or never studied by researchers (Hmioui et al., 2019).

Thanks to a qualitative approach, the theoretical model developed above can be finalized and contextualized in the Moroccan context. This suits our research object, as we seek to understand the phenomenon of non-financial disclosure in a specific context.

Several determinants of non-financial disclosure have been proposed in the literature, so we need to explore those that have an effect on the Moroccan context. In this light, we thought it yould be wise to opt for an exploratory qualitative study in order to contextualize the theoretical model derived from the literature.

We can position ourselves in a hybrid exploration (Allard-Poesi & Maréchal, 2014), an iterative research path. It refers to the back-and-forth between theories and the research field.

Our exploratory study focuses on two lines of analysis: Firstly, understanding the concept and its various pillars as a preparatory step for the second theme, which concerns the determinants of Moroccan corporate CSR disclosure.

Data Collection

Data Collection Tool

As part of a qualitative study, four data collection tools can be distinguished: documentary study, interviews, observation and questionnaires (Ketele & Roegiers, 1996). The choice at this level is strongly linked to the characteristics of the research, in particular, the research question and subject (Flick, 2011) For our research question, the Moroccan context provides us with limited empirical data, so it makes more sense to opt for semi-structured interviews.

Exploratory interviews enable us to highlight aspects related to our research topic, thus complementing our theoretical work (Blanchet & Gotman, 1992). The interview represents a qualitative study, providing an insight into situations and judgments. It involves a limited number of subjects, and is characterized by a specific framework for conducting interviews with one or more individuals (Pin, 2023).

This data collection tool provides the researcher with a better understanding of the elements collected by raising additional thoughts and reflections, which perfectly reflects the nature of exploratory study (Seidman, 2013). Thus, for a subject that is not well covered in a given context, the most appropriate data collection tool is interviews.

Because not enough information is available for quantitative analysis. The latter requires the prior arrangement of the variables to be considered (Giordano, 2003). The interviews were conducted individually with each participant. They are described as semi-structured, since they are organized around an interview guide with predefined themes (Blanchet & Gotman, 1992).

Population/Sample

Our study was based on a sample of non-financial disclosure professionals from all sectors. More specifically, we're talking about CSR managers, CSR experts and CSR disclosure consultants. These are profiles familiar with the implementation of responsible practices, which are related to corporate social responsibility and its communication.

Unlike quantitative approaches, in which individuals in the sample are randomly selected, the aim is not to have as many subjects as possible to guarantee the representativeness of our sample. In qualitative studies, the choice of participants is not random; rather, the researcher seeks to obtain a variety of perspectives on the subject studied from a small sample of participants (Giordano, 2003). The exploratory study thus provides an opportunity to report on participants' experiences in relation to a given theme (Rezaul, 2018).

Therefore, the choice of these profiles is mainly related to their attributes acquired through direct and/or indirect contacts with companies publishing ESG reports. In this respect, the profiles selected represent those best placed to inform us about our research question. This diverse group of participants is able to provide us with a varied perspective on the determinants of non-financial disclosure.

We used the principle of saturation to define our sample size (Giordano, 2003), which means that data collection should be stopped when new information can no longer be gathered. This saturation point is reached when the production of new knowledge seems to be getting less and less. The characteristics of our sample are presented in detail in Appendix 2, *Table 8*.

Development of the Interview Guide

The interview guide refers to a preparatory instrument that helped us organize the operationalization of our research question (Giordano, 2003). This instrument acts as a reference point, guiding the discussion towards the research topic. We relied on an interview guide when carrying out our survey, to ensure a certain homogeneity and consistency when conducting our interviews (Giordano, 2003).

The speakers were asked to comment on the determining factors of the non-financial disclosure of Moroccan companies. This was done in accordance with the interview guide, in particular the order of questions and the points to be addressed, ranging from the general to the specific.

The interview guide proposed for this study contains two main parts, presented as follows in Appendix 1, *Table 7*.

The first phase is a descriptive step, the aim of which is to evoke the various basic notions and concepts related to the research topic through general questions. As proposed by Gavard-Perret et al. (2008), the aim is to ensure a "funnel-like" progressiveness. Each participant's perception of the state of non financial disclosure in Morocco will allude to the research problem.

The second part is devoted to the explanation of the phenomenon studied, pushing the reflection towards an explanatory practical aspect. The aim is to provide a synthesis of the determinants of non-financial disclosure in Morocco, based on the theoretical model developed. The significance of each determinant in the Moroccan context is discussed.

Data Analysis

To analyze the interview data, we opted for content analysis. It is a technique for processing and analyzing content, based on a systematic and objective description of messages, in order to deduce information about the conditions of production of this knowledge (Bardin, 1977). Content analysis represents a set of sophisticated and constantly evolving methods that apply to a wide variety of "discourses" based on deduction and inference.

It is an interpretive effort that oscillates between two axes: on the one hand, the determination of objectivity, and on the other, the fertility of subjectivity (Bardin, 1977). It takes place in three stages: pre-analysis, exploitation of data, and processing and interpretation of results. The method of content analysis seemed to us the most appropriate for processing the information collected. In fact, it facilitates the understanding of the elements collected and the identification of those that most relevant.

Thematic analysis, as the name suggests, is based primarily on predefined themes. This method of analysis consists in segmenting the discourse into basic units in order to assemble them into coherent and exhaustive classes (L, 2016; Pras et al., 2003). The themes, previously established and included in the interview guide, were mobilized to classify the collected discourse. The major themes identified were broken down into several sub-themes for analysis. We used Nvivo software to transcribe, organize and process the speeches, in order to highlight the dispersion and frequency with which key research concepts were cited.

The themes and sub-themes of the thematic analysis are as follows ; *Table 3.*

Table 3. Content analysis themes and sub-themes

Themes	Sub-themes
Theme 1: Non financial disclosure	Definition of non financial disclosure
	Typology of non financial disclosure
	Publication channels
	Non-financial disclosure in Morocco

continued on following page

Table 3. Continued

Themes	Sub-themes
Theme 2: Determinants and explanatory factors	Company size
	Company age
	Legal system
	CSR performance
	Ownership sector
	Profitability
	Corporat governance
	Industry sector

Source: Researchers constructs 2024

RESULTS AND DISCUSSION

Descriptive Analysis

Data analysis was carried out at the same time as data collection (Alla et al., 2022). We began processing the collected data with a content description, using the Nvivo tool. This enabled us to explore meanings, themes, motifs or symbols in the text, giving priority to the quality of information over its quantity. We proceeded to describe the textual characteristics of the collected speeches.

Figure 2. Word cloud

Source: Generated by Nvivo

The Word cloud (*Figure 2*) figure represents the 100 most frequent words in our interviews. In other words, the terms most frequently used by our interviewees.

In the figure, the terms ESG, disclosure and reporting are at the center of the presentation. This placement confirms the relevance of the interviews conducted, since these terms represent our dependent variable. With regard to the independent variables, most of the elements proposed in the theoretical model appear in the figure. In particular, size, whose importance is evident through the location and term size in the figure. Profitability, CSR strategy and sector of activity are all well placed, and are therefore likely to have been confirmed in the theoretical model.

Table 4. Matrix analysis

The participants	Company size	Industry sector	Company Age		CSR Performance	Coporate governance	Ownership structure
Participant 1	10	5	7	3	9	1	2
Participant 2	12	3	8	6	7	5	1
Participant 3	9	6	3	9	10	4	3
Participant 4	6	5	5	2	2	7	4
Participant 5	10	2	2	1	9	6	2
Participant 6	5	3	1	4	3	3	1
Participant 7	7	4	7	1	8	2	1
Total	59	28	33	26	48	28	14
Percentage	25%	11%	14%	11%	20%	12%	6%

Source: Generated by Nvivo

The Matrix analysis, in *Table 4*, allows us to take the textual analysis a step further, in terms of the frequency of occurrence of our variables. The results at this level confirm the observations made earlier. Starting with company size, which can be considered to have the highest frequency, followed by CSR performance and profitability.

THEMATIC CONTENT ANALYSIS

Theme 1: Non-Financial Disclosure: Complex Concepts

Sub-Themes 1 and 2: Definition of Non-Financial Disclosure and Typology of Information Presented on the Subject

In an attempt to define the notion of non financial disclosure and the various elements related to it, the participants proposed a general presentation of the concept according to each participant's personal perception. This enabled us to frame the discussion by defining the terms to be used at the outset. Various definitions of extra-financial communication have been proposed, with a significant level of mastery and familiarity with the field of research. While we can't come up with a precise and common definition, we can cite a number of elements that characterize the concept of non financial disclosure, as shown in ***Table 5.***

Table 5. Summary of the characteristics of non-financial disclosure

Where the information is published/ source of information	**Outside the financial statements**
Utility	Inform on the company's future performance
Category of stakeholders concerned	All the company's stakeholders
Type of information	Non-financial / non-monetary indicators
Information typology / Information spectrum	CSR / ESG / Sustainable development

Source: Researchers constructs 2024

Sub-Theme 3: Publication Channels

The majority of participants favor the adoption of a single distribution channel, the ESG report, enabling greater homogeneity and comparability between the information presented by the different actors. In addition, the annual publication of a report promotes the company's commitment to the publication of non-financial information.

Another remark was thus made at this level, that of the use of several publication channels in parallel with the ESG report, including social networks. This is in order to ensure the publication of information in several periods of the year.

Sub-Theme 4: Non-Financial Disclosure in Morocco

Participants were unanimous in stating that non financial disclosure has evolved significantly since the introduction of the new regulations in 2019. Hence the importance of the regulatory framework in this area.

Another point that has been highlighted is the wide diversity of ESG reporting by Moroccan companies. The level of ESG reporting done by Moroccan companies varies considerably, dividing them into two categories: those with good ESG reports and those just seeking to meet their legal obligations. This point opened the discussion to the second phase, which concerns the determinants of ESG communication. In fact, these determining factors may provide explanations for the current level of extra-financial communication in Morocco.

Theme 2: Determinants and Explanatory Factors

The determinants identified in the literature were proposed to the participants, in order to obtain their perception of the significance of their impact on the Moroccan context. Respondents were free to suggest other factors that were not mentioned. The summary of responses is as follows in ***Table 6*** ;

Table 6. Summary of interview responses

Sub-themes	Participant 1	Participant 2	Participant 3	Participant 4	Participant 5	Participant 6	Participant 7
Company size	Significant positive impact	Significant positive impact	Significant positive impact	Significant positive impact	Significant positive impact	No significant impact	Significant positive impact
Industry sector	Impact on the report structure	Significant impact	Less significant impact	Significant impact	Significant impact	Impact on the nature of the information provided	Significant impact
Company age	Significant impact	Impact on the quality of published information	Significant impact	Significant impact	No significant impact	Significant impact	Impact on the extent of commitment
Profitability	No significant impact	Significant positive impact	Significant positive impact	Significant impact	No significant impact	Significant positive impact	Significant positive impact
CSR performance	Significant positive impact	Significant positive impact	Significant positive impact	Significant positive impact	Significant positive impact	Significant positive impact	Significant positive impact
Corporate governance	Significant positive impact	Significant positive impact	Significant positive impact	No significant impact	Significant positive impact	No significant impact	Significant positive impact
Ownership structure	No significant impact	No significant impact	No significant impact	Significant impact	No significant impact	No significant impact	No significant impact
Other elements to be added		Report production costs	Stakeholder pressure		Looking for a unique competitive advantage		Stakeholder pressure

Source: Generated by Nvivo

The statements made regarding the effectiveness of the variables identified in the literature were characterized by a degree of unanimity. CSR performance is the exception, with all participants unanimously agreeing on its impact on the company's non financial disclosure.

However, most respondents acknowledge the importance of the impact of the sector as a factor likely to affect the type of CSR information presented by companies. Except that this impact cannot determine the quality or degree of commitment of the company, its effect lies in the structure of the report.

Company's size, for its part, is a factor likely to have an impact on the CSR communication of Moroccan companies. All participants confirmed the importance of its impact as a determinant and thus explanatory factor of the quality and quantity of information presented by companies at this level.

In general, all the variables included in the theoretical model previously identified can have an impact on the Moroccan context. Three explanatory variables were excluded from our theoretical model: the regulatory framework, the age of the company and the ownership structure.

The regulatory framework has not been eliminated, it has been integrated into the variable related to stakeholder pressure. This element was strongly mentioned in the interviews conducted. Stakeholders have a major impact on the company, which can affect its actions and decisions. Thus, one of the aspects of the company's responsible commitment lies in meeting stakeholder expectations.

Furthermore, the age variable seems to have no impact in the Moroccan context. The interviewees preferred to be content with the size variable, which in turn determines extra-financial communication. This is linked to the visibility and activity of the company, which are impacted by size. As a result, age is not an explanatory variable for a company's commitment, since it must adopt responsible behavior and integrate sustainable dimensions into its strategy, regardless of its age or the number of years it has been in existence.

The speeches collected reveal other factors that may explain the non financial disclosure of Moroccan companies. These factors have not been identified in the literature. Firstly, pressure from stakeholders, which has a major impact on the company's decisions. The company's performance in the various areas can be linked to the pressure exerted by stakeholders. Stakeholder pressure is reflected in the fulfilment of expectations.

The results thus reveal the effect of the cost of producing reports on their quality. This element may refer to the company's availability of funds for the production and preparation of ESG reports, which are relatively costly.

One of the participants focused on competitive advantage and its relationship with extra-financial communication. He argued that a company's desire to acquire a unique competitive advantage can be translated into better extra-financial communication. This will enable the company to differentiate itself from its competitors.

Discussion of the Findings

Comparing the results of the interviews with the theoretical model established on the basis of the literature enabled us to form the following model. This model incorporates all the theoretical contributions in terms of determinants, thus integrating the specificities of the Moroccan context in this area. The exploratory approach we implemented enabled us to contextualise the initial theoretical model, in order to arrive at a Contextualised model, in *Figure 3*, to the characteristics of Moroccan companies.

In this study, we have attempted to examine theoretical advances in non-financial reporting, and to sift through the factors likely to impact the optimal adoption of non-financial reporting by Moroccan companies.

Based on a theoretical exploration, we were able to frame the characteristics of the Moroccan context in this area. The size of the company and its CSR performance were identified from the theory and confirmed in our interviews. In line with studies carried out in other contexts, in particular (Abdul Rahman & Alsayegh, 2021; Aggarwal & Singh, 2019; Kuzey & Uyar, 2017).

A company's commitment to non financial disclosure is positively correlated with its size. Company size can represent a fairly powerful determinant with a significant impact on ESG disclosure practices, in different contexts and with different characteristics. These findings can be explained by the increased pressure to which larger entities are exposed, with growing activity and increasingly significant impact. Larger structures are more visible to the various actors and are more concerned about their reputations. On the other hand, a company's CSR performance or responsible commitment to the various aspects of CSR can significantly determine the quantity and quality of ESG information presented by companies (Abdul Rahman & Alsayegh, 2021; Gao et al., 2016).

This correlation seems more logical, insofar as companies with higher performance in this area are more committed to non financiak disclosure, in particular to legitimize their activities and highlight their responsible commitment (Hummel & Schlick, 2016).

Age represents the number of years of existence or activity, despite the significance of this determiant in some contexts (Chakroun et al., 2017; Kumar et al., 2021). However, it does not have the same effect in Morocco. The majority of our participants were unable to express an opinion on the viability of this element as a determinant. The same was true of the ownership structure, which had no effect on the ESG disclosure carried out by the Moroccan companies. This is due to the capital structure of Moroccan companies, which is characterized by low foreign ownership.

For the other determinants, participants were not totally unanimous about the significance of their impact. However, we can keep them at the level of our theoretical model by the majority of participants.

The specificity of Morocco in relation to other contexts lies in a number of determinants raised by the exploratory study and not included in our theoretical model. Firstly, the cost of producing reports, a factor that can hinder the production of ESG reports and sometimes impact the quality of the information presented in them. The fact that producing ESG reports according to international standards is relatively expensive, which may lead the company to draw up its ESG reports in-house and settle for the minimum required to fulfill its legal obligations.

Through ESG disclosure, managers can seek to gain a unique competitive advantage. Strong competition in a given sector may push a company to develop its responsible commitment in order to differentiate itself. Competitive advantage can be sought with stakeholders, whose impact is major. Hence the need to take stock of the pressure exerted by stakeholders and the potential impact they can have on the company, in order to develop its non financial disclosure.

As a result, pressure from all stakeholders is having a significant impact on the extra-financial communication of Moroccan companies. Starting with the market regulator, who is likely to impact on the behavior of companies as a major stakeholder. Suppliers, subcontractors and customers are also stakeholders who are increasingly aware of the need to integrate non-financial criteria into their choice of partner. Since responsible commitment is not limited to the behavior of the actor itself, it also concerns its partners.

Figure 3. Contextualised model

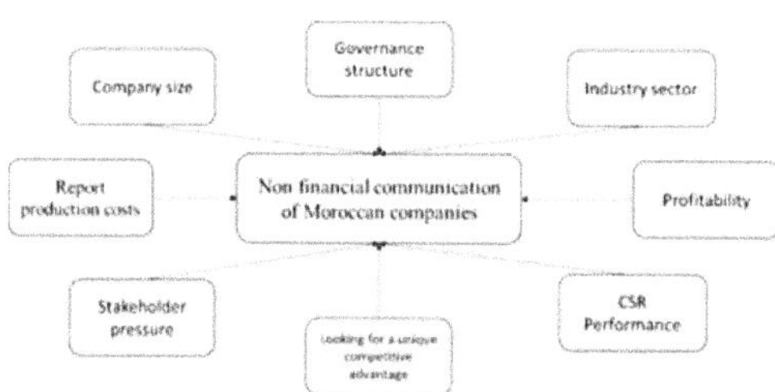

Source: Researchers constructs 2024

CONCLUSION

This work contributes to the exploration of the determinants of ESG reporting by Moroccan companies. This was achieved through an exploratory qualitative study. As CSR practices represent a relatively new field for Morocco, contextualization is essential.

To this end, we have attempted to draw up a theoretical model that incorporates the contributions made in the literature. The combination of several theories enabled us to address the main factors likely to impact corporate behavior in terms of ESG communication. By mobilizing agency theory, stakeholder theory, legitimacy theory, signal theory and institutional theory, we have developed an initial theoretical model that highlights non financial disclosure with its various determinants.

The aim was to identify a model of the determinants of extra-financial communication that takes into account the specific features of the Moroccan context. To this end, a qualitative study was carried out. This enabled us to contextualize the initial theoretical model in the Moroccan context.

The results show that variables such as age, size, industry sector, profitability, governance structure and CSR performance have an impact on ESG reporting in Morocco, as has been argued in the literature. Moroccan companies have the particularity of seeking a unique competitive advantage through ESG disclosure, and of dealing with pressure from stakeholders, which are determinants of the quality of ESG reports published. The cost of producing ESG reports also represents an obstacle for Moroccan companies. The contextualized model, which takes all these considerations into account, is better able to capture the determinants of ESG reporting in Morocco.

The research field represents one of the main contributions of the present research. Extra-financial communication refers to a recent practice in Morocco, thus presenting an emerging field of research. It provides an understanding of the determining factors of non financial communication in Morocco, by developing a model that takes into account the particularities of the Moroccan context. We have thus been able to propose a model of the determinants of ESG disclosure specific to the Moroccan context. This can be used as a reference by managers to measure the importance of the non financial communication they carry out, as well as the determinants that can guide their Reporting.

The contextualization work carried out paves the way for a confirmatory study to validate and test the contextualized model, which can represent the starting point for future research.

The theoretical framework for extra-financial communication is highly diverse. Numerous theories have been mobilized, and several determinants have been proposed. This diversity has not been taken into account in the present study. We relied on a limited number of theories and determinants, given the inability of a qualitative approach to support a large enough number of variables. The methodological limitations of our work are essentially linked to sample size and composition. Our sample consisted of nine interviews, which seems limited and not very diversified.

Non-financial disclosure in its various forms has become a recurrent practice, with corporate behavior in this area constantly evolving in line with environmental, social and societal requirements. Nevertheless, non-financial disclosure and, more specifically, each company's ESG report reflects its internal characteristics as well as its competitive, social and regulatory environment. That's why this exploratory study was a necessity in order to better assess them in a specific context.

REFERENCES

Aamoum, H., & Soufiane, N. (2021). L'étude de la divulgation volontaire d'informations à propos du capital immatériel à la lumière de la théorie de l'agence et la théorie du signal : Quels déterminants pour quels objectifs? The study of the voluntary disclosure of information about intangible capital in the light of agency theory and signal theory: Which determinants for which objectives? 2, 2658-8455. 10.5281/zenodo.5531367

Abdolmohammadi, M., Simnett, R., Thibodeau, J. C., & Wright, A. M. (2006). Sell-Side Analysts' Reports and the Current External Reporting Model. *Accounting Horizons*, 20(4), 375–389. 10.2308/acch.2006.20.4.375

Abdul Rahman, R., & Alsayegh, M. F. (2021). Determinants of corporate environment, social and governance (ESG) reporting among Asian firms. *Journal of Risk and Financial Management*, 14(4), 167. 10.3390/jrfm14040167

Adams, C. A. (2004). The ethical, social and environmental reporting-performance portrayal gap. *Accounting, Auditing & Accountability Journal*, 17(5), 731–757. 10.1108/09513570410567791

Aggarwal, P., & Singh, A. K. (2019). CSR and sustainability reporting practices in India : An in-depth content analysis of top-listed companies. *Social Responsibility Journal*, 15(8), 1033–1053. 10.1108/SRJ-03-2018-0078

Al-Tuwaijri, S. A., Christensen, T. E., & Hughes, K. E.II. (2004). The relations among environmental disclosure, environmental performance, and economic performance : A simultaneous equations approach. *Accounting, Organizations and Society*, 29(5), 447–471. 10.1016/S0361-3682(03)00032-1

Alla, L., Jaouhari, E. L., & Bentalha, B. (2022). Repenser l'entrepreneuriat au sein du secteur informel au Maroc à l'ère du post Covid-19 : Pour approche prospective de la formalisation intégrative. *Alternatives Managériales et Economiques*, 4(1), 180–202.

Allard-Poesi, F., & Maréchal, G. (2014). Construction de l'objet de la recherche. In T. R.-A. et al (Éd.), Méthodes de recherche en management (p. 48-76). Dunod. https://hal.science/hal-01123768

Amaeshi, K., Adi, A. B. C., Ogbechie, C., & Amao, O. O. (2006). Corporate Social Responsibility in Nigeria : Western Mimicry or Indigenous Influences? (SSRN Scholarly Paper 896500). 10.2139/ssrn.896500

Amir, E., Lev, B., & Sougiannis, T. (2003). Do financial analysts get intangibles? *European Accounting Review*, 12(4), 635–659. Advance online publication. 10.1080/0963818032000141879

Asdiou, A., & Mokhtari, B. (2019). The Indicators of the Corporate Social Responsibility : A Search for Efficiency or a Concern for Legitimacy Towards Stakeholders? A Proposed Explanatory Model. [IJSSMET]. *International Journal of Service Science, Management, Engineering, and Technology*, 10(1), 62–76. 10.4018/IJSSMET.2019010105

Ashforth, B. E., & Gibbs, B. W. (1990). The Double-Edge of Organizational Legitimation. *Organization Science*, 1(2), 177–194. 10.1287/orsc.1.2.177

Aziz, O. E., & Asdiou, A. (2024). Les déterminants de la communication extra-financière : Cas des entreprises Marocaines cotées à la bourse de valeurs mobilières de Casablanca. Alternatives Managériales Economiques, 6(1), Article 1.

Bansal, P. (2005). Evolving sustainably : A longitudinal study of corporate sustainable development. https://onlinelibrary.wiley.com/doi/abs/10.1002/smj.441

Bardin. (1977). L'Analyse du contenu, https://www.persee.fr/doc/colan_0336-1500_1977_num_35_1_4449

Bassen, A., & Kovács, A. M. (2008). Environmental, Social and Governance Key Performance Indicators from a Capital Market Perspective.

Bentalha, B. (2023). Influence des actions sociales de la RSE sur les valeurs perçues des employés : Étude qualitative des œuvres sociales de l'entreprise Taajmouati. Repères et Perspectives Economiques, 7(2), Article 2. 10.34874/IMIST.PRSM/RPE/43586

Bentalha, B., Hmioui, A., & Alla, L. (2020). La performance des entreprises de services : Un cadrage théorique d'un concept évolutif. *Alternatives Managériales et Economiques*, 2, 58–78.

Berrone, P., & Gomez-Mejia, L. R. (2009). Environmental performance and executive compensation : An integrated agency-institutional perspective. *Academy of Management Journal*, 52(1), 103–126. 10.5465/amj.2009.36461950

Blanchet, A., & Gotman, A. (1992). *L'enquête et ses méthodes : L'entretien*. Nathan.

Boatright, J. R. (2014). Ethics and the conduct of business (7. ed., Pearson new internat. ed). Pearson.

Brammer, S., & Pavelin, S. (2004). Voluntary social and environmental disclosures by large UK companies. *Business Ethics (Oxford, England)*, 23(2/3), 86–99. 10.1111/j.1467-8608.2004.00356.x

Campbell, J. L. (2007). Why would corporations behave in socially responsible ways? An institutional theory of corporate social responsibility. *Academy of Management Review*, 32(3), 946–967. 10.5465/amr.2007.25275684

Chakroun, R., Matoussi, H., & Mbirki, S. (2017). Determinants of CSR disclosure of Tunisian listed banks : A multi-support analysis. *Social Responsibility Journal*, 13(3), 552–584. 10.1108/SRJ-04-2016-0055

Charreire, S., & Huault, I. (2001). Le constructivisme dans la pratique de re- cherche : Une évaluation à partir de seize thèses de doctorat.

Chen, C., J., Patten, D. M., & Roberts, R. W. (. (2008). Corporate Charitable Contributions : A Corporate Social Performance or Legitimacy Strategy? *Journal of Business Ethics*, 82(1), 131–144. 10.1007/s10551-007-9567-1

Cherkaoui, A. (2016). Pratiques RSE des PME au Maroc : Une analyse perceptuelle auprès des dirigeants casablancais. Question(s) de management, 14(3), 13-26. 10.3917/qdm.163.0013

Cherkaoui, A. (2020). Du reporting de la performance globale à la stratégie de reporting ESG : Pratiques des entreprises faisant appel public à l'épargne au Maroc. *Journal Of Social Science and Organization Management*, 1(1), 1. Advance online publication. 10.48434/IMIST.PRSM/jossom-v1i1.22348

Clarkson, P. M., Overell, M. B., & Chapple, L. (2011). Environmental Reporting and its Relation to Corporate Environmental Performance. *Abacus*, 47(1), 27–60. 10.1111/j.1467-6281.2011.00330.x

Cohen, J. R., Krishnamoorthy, G., & Wright, A. M. (2000). Evidence on the Effect of Financial and Nonfinancial Trends on Analytical Review. *Auditing*, 19(1), 27–48. 10.2308/aud.2000.19.1.27

Cowen, S. S., Ferreri, L. B., & Parker, L. D. (1987). The impact of corporate characteristics on social responsibility disclosure : A typology and frequency-based analysis. *Accounting, Organizations and Society*, 12(2), 111–122. 10.1016/0361-3682(87)90001-8

Cuervo-Cazurra, A., & Aguilera, R. (2004). *The Worldwide Diffusion of Codes of Good Governance*. 10.1093/acprof:oso/9780199269761.003.0015

Deegan, C. (2002). The Legitimising Effect of Social and Environmental Disclosures – A Theoretical Foundation. *Accounting, Auditing & Accountability Journal*, 15, 282–311. 10.1108/09513570210435852

Dentchev, N. (2004). Corporate Social Performance as a Business Strategy. *Journal of Business Ethics*, 55(4), 395–410. 10.1007/s10551-004-1348-5

DiMaggio, P. J., & Powell, W. W. (1983). The Iron Cage Revisited : Institutional Isomorphism and Collective Rationality in Organizational Fields. *American Sociological Review*, 48(2), 147. 10.2307/2095101

Dissanayake, D., Tilt, C., & Xydias-Lobo, M. (2016). Sustainability reporting by publicly listed companies in Sri Lanka. *Journal of Cleaner Production*, 129, 169–182. 10.1016/j.jclepro.2016.04.086

Donaldson, T., & Preston, L. E. (1995). The Stakeholder Theory of the Corporation : Concepts, Evidence, and Implications | Academy of Management Review. https://journals.aom.org/doi/abs/10.5465/amr.1995.9503271992

Dye, R. A. (2001). An evaluation of « essays on disclosure » and the disclosure literature in accounting. *Journal of Accounting and Economics*, 32(1-3), 181–235. 10.1016/S0165-4101(01)00024-6

Eccles, R. G., & Krzus, M. P. (2010). *One Report : Integrated Reporting for a Sustainable Strategy*. John Wiley & Sons.

Farrat, O., & Hajji, Z. (2022). Contribution à l'analyse des déterminants de la divulgation des informations RSE sur les sites web au Maroc : Cas des organismes financiers cotés dans la bourse de Casablanca. 3(1).

Fernando, A., & Pandey, I. (2012). Corporate social responsibility reporting : A survey of listed Sri Lankan companies. *Journal for International Business and Entrepreneurship Development*, 6(2), 172–187. 10.1504/JIBED.2012.048569

Flick, U. (2011). Designing Qualitative Research. *Sage (Atlanta, Ga.).*

Fortanier, F., Kolk, A., & Pinkse, J. (2011). Harmonization in CSR reporting : MNEs and global CSR standards. *MIR. Management International Review*, 51(5), 665–696. 10.1007/s11575-011-0089-9

Freeman, R. E. (2010). *Strategic Management : A Stakeholder Approach*. Cambridge University Press. 10.1017/CBO9781139192675

Gallo, P., & Jones-Christensen, L. (2011). Firm Size Matters : An Empirical Investigation of Organizational Size and Ownership on Sustainability-Related Behaviors. Business & Society -. *Business & Society*, 50(2), 315–349. 10.1177/0007650311398784

Gamerschlag, R., Möller, K., & Verbeeten, F. (2011). Determinants of voluntary CSR disclosure : Empirical evidence from Germany. *Review of Managerial Science*, 5(2-3), 233–262. 10.1007/s11846-010-0052-3

Gao, F., Dong, Y., Ni, C., & Fu, R. (2016). Determinants and economic consequences of non-financial disclosure quality. *European Accounting Review*, 25(2), 287–317. 10.1080/09638180.2015.1013049

Gavard-Perret, M.-L., Gotteland, D., Haon, C., & Jolibert, A. (2008). Méthodologie de la recherche.

Giordano, Y. (2003). Conduire un projet de recherche. Une perspective qualitative (p. 318). Editions Management et Société. https://shs.hal.science/halshs-00440011

Gray, R. H., Owen, D. L., & Adams, C. A. (1996). Accounting and Accountability : Changes and Challenges in Corporate Social and Environmental Reporting. https://durham-repository.worktribe.com/output/1123501/accounting-and-accountability-changes-and-challenges-in-corporate-social-and-environmental-reporting

Guthrie, J., & Parker, L. D. (1989). Corporate Social Reporting : A Rebuttal of Legitimacy Theory. *Accounting and Business Research*, 19(76), 343–352. 10.1080/00014788.1989.9728863

Hackston, D., & Milne, M. J. (1996). Some determinants of social and environmental disclosures in New Zealand companies. *Accounting, Auditing & Accountability Journal*, 9(1), 77–108. 10.1108/09513579610109987

Haller, A., Link, M., & Groß, T. (2017). The Term 'Non-financial Information' – A Semantic Analysis of a Key Feature of Current and Future Corporate Reporting. *Accounting in Europe*, 14(3), 407–429. 10.1080/17449480.2017.1374548

Hassanein, A., & Hussainey, K. (2015). Is forward-looking financial disclosure really informative? Evidence from UK narrative statements. *International Review of Financial Analysis*, 41(C), 52–61. 10.1016/j.irfa.2015.05.025

Hillman, A. J., & Keim, G. D. (2001). Shareholder Value, Stakeholder Management, and Social Issues : What's the Bottom Line? *Strategic Management Journal*, 22(2), 125–139. 10.1002/1097-0266(200101)22:2<125::AID-SMJ150>3.0.CO;2-H

Hmioui, A., Alla, L., & Bentalha, B. (2019, avril). Évaluation des formations en ligne : Etude qualitative des perceptions des utilisateurs des MOOC au Maroc. ICT-Days'19. https://hal.science/hal-02943698

Hummel, K., & Schlick, C. (2016). The relationship between sustainability performance and sustainability disclosure – Reconciling voluntary disclosure theory and legitimacy theory. *Journal of Accounting and Public Policy*, 35(5), 455–476. 10.1016/j.jaccpubpol.2016.06.001

Jackson, G., & Apostolakou, A. (2010). Corporate Social Responsibility in Western Europe : An Institutional Mirror or Substitute? *Journal of Business Ethics*, 94(3), 371–394. 10.1007/s10551-009-0269-8

Jamali, D. (2010). The CSR of MNC Subsidiaries in Developing Countries : Global, Local, Substantive or Diluted? *Journal of Business Ethics*, 93(S2), 181–200. 10.1007/s10551-010-0560-8

Jensen, M. C., & Meckling, W. H. (1976). Theory of the firm : Managerial behavior, agency costs and ownership structure. *Journal of Financial Economics*, 3(4), 305–360. 10.1016/0304-405X(76)90026-X

Judge, T. A., & Klinger, R. (2008). Job satisfaction : Subjective well-being at work. In *The science of subjective well-being* (pp. 393–413). The Guilford Press.

Judge, T. A., Piccolo, R. F., Podsakoff, N. P., Shaw, J. C., & Rich, B. L. (2010). The relationship between pay and job satisfaction : A meta-analysis of the literature. *Journal of Vocational Behavior*, 77(2), 157–167. 10.1016/j.jvb.2010.04.002

Ketele, J.-M. D., & Roegiers, X. (1996). *Méthodologie du recueil d'informations : Fondements des méthodes d'observations, de questionnaires, d'interviews et d'études de documents*. De Boeck Université.

Khan, A., Muttakin, M., & Siddiqui, J. (2013). Corporate Governance and Corporate Social Responsibility Disclosures : Evidence from an Emerging Economy. *Journal of Business Ethics*, 114(2), 207–223. 10.1007/s10551-012-1336-0

Kolk, A. (2002). The Effectiveness of Self-regulation. European Management Journal. https://www.academia.edu/33220513/The_Effectiveness_of_Self_regulation

Kumar, K., Kumari, R., Poonia, A., & Kumar, R. (2021). Factors influencing corporate sustainability disclosure practices : Empirical evidence from Indian National Stock Exchange. *Journal of Financial Reporting and Accounting*, 21(2), 300–321. 10.1108/JFRA-01-2021-0023

Kuzey, C., & Uyar, A. (2017). Determinants of sustainability reporting and its impact on firm value : Evidence from the emerging market of Turkey. *Journal of Cleaner Production*, 143, 27–39. 10.1016/j.jclepro.2016.12.153

Line, M., Hawley, H., & Krut, R. (2002). The Development of Global Environmental and Social Reporting. *Corporate Environmental Strategy*, 9(1), 69–78. 10.1016/S1066-7938(01)00159-2

Mette, L. B. (2016). *Mixed Methods Research for Improved Scientific Study*. IGI Global.

Meyer, J. W., & Rowan, B. (1977). Institutionalized Organizations : Formal Structure as Myth and Ceremony. The American Journal of Sociology, 83(2,), 340-363.

Morioka, S. N., Bolis, I., Evans, S., & Carvalho, M. M. (2017). Transforming sustainability challenges into competitive advantage : Multiple case studies kaleidoscope converging into sustainable business models. *Journal of Cleaner Production*, 167, 723–738. 10.1016/j.jclepro.2017.08.118

Muthuri, J. N., & Gilbert, V. (2011). An Institutional Analysis of Corporate Social Responsibility in Kenya. *Journal of Business Ethics*, 98(3), 467–483. 10.1007/s10551-010-0588-9

Omran, M. A., & Ramdhony, D. (2015). Theoretical Perspectives on Corporate Social Responsibility Disclosure : A Critical Review. *International Journal of Accounting and Financial Reporting*, 5(2), 38. 10.5296/ijafr.v5i2.8035

Perrow, C. (1970). *Organizational Analysis : A Sociological View*. Brooks/Cole.

Pin, C. (2023). L'entretien semi-directif. LIEPP Methods Brief / Fiches méthodologiques du LIEPP. https://sciencespo.hal.science/hal-04087897

Pras, B., Evrard, Y., & Roux, E. (2003). Market : Études et recherches en marketing - Fondements, méthodes: Vol. 3ème édition (p. 704). Dunod. https://shs.hal.science/halshs-00150660

Rezaee, Z. (2016). Business sustainability research : A theoretical and integrated perspective. *Journal of Accounting Literature*, 36(1), 48–64. 10.1016/j.acclit.2016.05.003

Rezaee, Z., & Tuo, L. (2017). Voluntary disclosure of non-financial information and its association with sustainability performance. *Advances in Accounting*, 39, 47–59. 10.1016/j.adiac.2017.08.001

Rezaul, I. M. (2018). *Social Research Methodology and New Techniques in Analysis, Interpretation, and Writing*. IGI Global.

Robb, S., & Zarzeski, L. (2001). Nonfinancial Disclosures Across Anglo-American Countries. *Journal of International Accounting, Auditing & Taxation*, 10(1), 71–83. 10.1016/S1061-9518(01)00036-2

Roberts, R. W. (1992). Determinants of corporate social responsibility disclosure : An application of stakeholder theory. *Accounting, Organizations and Society*, 17(6), 595–612. 10.1016/0361-3682(92)90015-K

Roe, M. J. (2003). Political Determinants of Corporate Governance. https://global.oup.com/academic/product/political-determinants-of-corporate-governance-9780199240746?cc=us&lang=en&

Scott, W. (2005). Institutional Theory : Contributing to a Theoretical Research Program.

Seidman, I. (2013). *Interviewing as Qualitative Research : A Guide for Researchers in Education and the Social Sciences*. Teachers College Press.

Suchman, M. C. (1995). Managing legitimacy : Strategic and institutional approaches. *Academy of Management Review*, 20(3), 571–610. 10.2307/258788

Viehs, M., Clark, G., & Feiner, A. (2014). From The Stockholder To The Stakeholder—How Sustainability Can Drive Financial Outperformance. SSRN Electronic Journal. 10.2139/ssrn.2508281

Wacheux, F. (1996). Méthodes qualitatives et recherche en gestion / Frédéric Wacheux. Économica. Paris. https://bibliotheque.tbs-education.fr/Default/doc/SYRACUSE/17765/methodes-qualitatives-et-recherche-en-gestion-frederic-wacheux

Wise, V., & Ali, M. M. (2009). Corporate Governance and Corporate Social Responsibility in Bangladesh with special reference to Commercial Banks. American International University-Bangladesh, Office of Research and Publications (ORP), AIUB Bus Econ Working Paper Series.

Zattoni, A., & Cuomo, F. (2008). Why Adopt Codes of Good Governance? A Comparison of Institutional and Efficiency Perspectives. *Corporate Governance*, 16(1), 1–15. 10.1111/j.1467-8683.2008.00661.x

APPENDIX 1

Table 7. Interview guide

	Themes	Main questions	Secondary questions
	Theme 1: Non-financial disclosure	**Design of concepts related to non-financial disclosure**	
	Familiarisation with non-financial concepts	What do you think of non-financial disclosure	What is your own definition of this concept
	Typology of non-financial information	what are the elements that must be presented in the context of a non financial communication?	What categories of information must be included in a non-financial report ?
	Information channels	What is the best channel for non-financial disclosure?	Should we adopt a single communication channel or opt for diversification ?
	Birth of the problem	How would you rate the level of non-financial communication in Morocco?	How would you rate the quantity and quality of non-financial information published ?
	Theme II: Determinants	**Explanations of the phenomenon (diversification/level of non-financial disclosure)**	
	Influencing factors / Determinants	What are your views on the determinants of our theoretical model (Direct question)	Liste of determinants (**Tableau 2**)
		Are there any other determinants that have not been mentioned that you would like to add?	

APPENDIX 2

Table 8. The participants' attributes

Participants	Age range	Profession	Last qualification obtained	Experience
Participant 1	40 - 50	Researcher - Expert	Phd	10 years
Participant 2	30 - 40	CSR Manager	Bac+5	5 years
Participant 3	40 - 50	Researcher - Expert	Phd	15 years
Participant 4	20 - 30	CSR Project Manager	Bac+5	3 years
Participant 5	30 - 40	Researcher - Expert	Bac+5	6 years
Participant 6	30 - 40	CSR Executive	Bac+5	6 years
Participant 7	30 - 40	Head of CSR and Diversity Committee	Bac+5	1 year
Participant 8	40 - 50	Group Compliance and Sustainable Development Officer	Bac+5	4 years
Participant 9	30 - 40	CSR Manager	Bac+5	5 years

Chapter 10
Qualitative Risk Analysis Based on Avalanche Scenarios in Palandöken Ski Center, Erzurum (NE Turkey)

Tayfun Kurt
https://orcid.org/0000-0002-7279-8854
Independent Researcher, Turkey

ABSTRACT

The chapter proposes a method for regional-scale qualitative risk assessment. To estimate critical snow heights, information was gathered from residents, documents, and back calculations of avalanche simulations based on well-known avalanche events. The elements at risk were extracted from existing digital cadastral databases, and vulnerabilities were identified. The study found that the provincial forest area had the highest annual probability of property loss. The chapter presents a model that offers an effective means of determining standardized responses and controls for common environmental issues based on the level of risk. Additionally, various potential avalanche scenarios are presented to provide an understanding of the potential threats specific to this avalanche-prone area. The resulting map provides reliable and easily understandable information, highlighting areas where avalanches pose a significant risk at the regional level, as well as areas where new avalanche paths may develop under favorable conditions within the defined region.

DOI: 10.4018/979-8-3693-5543-5.ch010

1. INTRODUCTION

Winter tourism is a thriving sector in countries like Turkey, which has emerged as a prominent destination. Offering a diverse range of activities and services tailored to travelers seeking winter experiences, winter tourism has witnessed significant growth in recent years (Evren & Kozak 2018; Pröbstl-Haider 2019; Özçoban, 2019; Evren, 2023). Beyond skiing and snowboarding, Turkey's winter tourism provides various attractions and activities for adventure and enjoyment across all ages (Subaşı & Gür, 2023; Aygün & Baycan, 2023). The country's majestic mountain ranges, picturesque winter resorts, and warm hospitality make it an appealing choice for unforgettable winter vacations (Kurt, 2022; Kurt, 2017). For individuals planning to engage in winter activities in Turkey's mountainous regions, it is crucial to be aware of the potential risks associated with avalanches (Kurt, 2022). Staying informed about weather conditions, following safety guidelines, and respecting the instructions of ski resort authorities significantly contribute to reducing the likelihood of accidents related to avalanches (Tremper, 2013; Nichols et al., 2018;Landrø, 2021; Stetham et al., 2018).

Avalanches at ski resorts can be a serious risk for both visitors and staff. Therefore, managing avalanche risk and conducting risk analyses are essential for resort safety. One method used is qualitative risk analysis, which assesses the likelihood and impact of events (Kurt,2022). For avalanches, this analysis considers various factors and relies on observations and the expertise of experienced individuals. Ski resort staff and local avalanche experts offer valuable insights into past avalanches and current conditions.

While avalanches can occur in Palandöken, as in any mountainous region, ski resorts and local authorities take significant measures to mitigate risks, ensure public safety, and provide efficient rescue operations. Skiers and visitors must be aware of the potential dangers and follow the instructions and guidelines provided by ski resort authorities (Steiger et al., 2019). Staying within designated areas, avoiding off-piste skiing in risky zones, and being equipped with appropriate safety gear such as avalanche transceivers, shovels, and probes are essential precautions to minimize the risk.

Qualitative risk analysis encompasses various methods for identifying and evaluating risks, considering organizational culture, industry dynamics, and external factors (Siraj & Fayej, 2019; Cohen & Palmer, 2014). In Palandöken, we employed this approach to bridge knowledge gaps and support effective avalanche risk management. Because qualitative methodologies can offer valuable insights into the intricate dynamics of avalanche risks in Palandöken (Kurt, 2022; Kurt, 2017). Researchers can delve into the nuances of these hazards, identify vulnerable areas, and develop a holistic understanding of the overall risk landscape (Dvorak et al., 2020;

Costa et al., 2017). This approach can facilitate prioritizing mitigation strategies, enhances decision-making, and ultimately bolsters safety and resilience in avalanche-prone areas (Raihan, 2024). Furthermore, qualitative methods foster collaboration by involving local communities, government agencies, and industry stakeholders (Haimes, 2005). By incorporating diverse perspectives and local knowledge, these assessments contribute to more effective risk management strategies and policy interventions (Logrosa et al., 2022).

This study focused on qualitative risk analysis, a vital approach in environments with limited quantitative data. It leveraged expert opinions, stakeholder input, and descriptive methods to assess risks comprehensively (Ngwenya et al., 2023; Behzadifar et al., 2023). This method delved into the complexities of avalanche hazards, employing techniques like SWOT analysis and expert judgment (Siraj & Fayej, 2019; Cohen & Palmer, 2014). Qualitative risk analysis played a pivotal role in environments where quantitative data were limited or inadequate, emphasizing subjective assessments through expert judgment, stakeholder input, and descriptive methodologies (Ngwenya et al., 2023; Behzadifar et al., 2023). This method allowed for a nuanced exploration of potential risks and their implications, utilizing techniques like SWOT analysis and expert-based assessments to comprehensively evaluate risks from diverse perspectives (Siraj & Fayej, 2019; Cohen & Palmer, 2014).

We especially employed L-type matrix risk analysis, a systematic approach that incorporates an organization's unique management framework in this study. By integrating risk assessments within this framework at Palandöken, interventions for identified avalanche-prone zones were prioritized. The L-type matrix analysis entailed a systematic approach to assessing avalanche risks in mountainous regions such as Palandöken. It commenced with identifying potential hazards, such as avalanche-prone areas, utilizing historical data analysis, expert consultations, and Geographic Information Systems (GIS) for zone mapping. Subsequently, the analysis evaluated the likelihood of avalanche occurrences and their potential impact on the region, integrating qualitative and quantitative data from meteorological observations, terrain analysis, and avalanche forecasting models. Furthermore, organizational factors, including the goals of local authorities and ski resorts, operational strategies for avalanche risk management, and resource allocation for mitigation efforts, were taken into consideration. This integrated approach provided a comprehensive risk profile, highlighting critical areas requiring intervention and guiding effective risk management practices (Karahan & Aydoğmuş, 2023; Cengiz, 2022).

This study highlights the value of qualitative risk analysis in understanding and managing snow avalanche risk in Palandöken. By employing this approach, we can develop a foundation for practitioners and decision-makers to implement appropriate mitigation measures, prioritize resource allocation, and ultimately reduce the impact of avalanches in the region.

2. MATERIALS

The qualitative risk analysis conducted through fieldwork involved several key components in its materials section. It began with the selection of the study area, outlining reasons for its choice and geographical significance. Data collection processes were meticulously detailed, specifying sources and methods utilized for gathering information crucial to risk assessment. Meteorological data played a pivotal role, documenting weather conditions within the study site and their analysis. Threat assessments related to avalanches were conducted using specialized methodologies, integrating local knowledge and expert opinions to enhance accuracy. Geographic Information Systems (GIS) facilitated spatial data handling and mapping tasks, essential for visualizing and analyzing risk factors. Overall, the materials section encapsulated the comprehensive approach adopted in collecting and utilizing diverse data sources to support qualitative risk analysis in the fieldwork.

2.1 Study Area

The study area is located in Palandöken, Erzurum province, which is a prominent hub of winter tourism in Northeast Turkey. It boasts the longest ski resort, highest altitude, and longest winter season in the region (DPT, 2000) (Figure 1). Palandöken Mountain exhibits primarily metamorphic and sedimentary geological features (Fırat & Çalık, 2014). Among these formations are schists, limestones, mica schists, and crystalline rocks. The processes of metamorphism, particularly induced by intense heat and pressure during mountain formation, have led to significant structural changes in the rocks (Parolly, 2020). The mountain's towering peaks often display morphological characteristics influenced by glacial and periglacial dynamics, making it an ideal destination for winter activities in the region.

Figure 1. Overwiev of study area (Source: Google Earth ®, 2024)

Palandöken Ski Resort offers exceptional skiing opportunities but also carries the inherent risk of avalanches. Therefore, evaluating and managing avalanche risk in this region is crucial to ensure visitor safety and prevent accidents. Advanced techniques and methodologies are employed to assess the risk and implement appropriate safety measures. Several factors contribute to avalanche risk, including the geographical features of the mountains, snowfall patterns, weather conditions, and human activities. Steep slopes, snow accumulation, and temperature fluctuations can increase the likelihood of avalanches. Human factors, such as skiing off-piste or disregarding safety guidelines, can also contribute to avalanche incidents. However, the mountainous terrain also poses a widespread hazard risk due to potential avalanches. For instance, Palandöken Ski slopes rank among the world's longest and steepest, with the longest track measuring 12 km and a total length of 28 km. Consequently, guests may face the risk of injury or even death from avalanches (Table 1). The Turkish government officially designated the area as a winter tourism center in 1993, leading to the construction of numerous hotels and ski slopes in close proximity to potential avalanche run-out areas. This underscores the dynamic nature of risk in the area.

Table 1. Hotels and bed capacities in Palandöken

Accommodation place	Bed capacities (Person)
AFAD Observatory	30
Dedeman otel	30
Palan otel	500
Polat hotel	700
Dedeman	100
Sway otel (sanado)	110
Gendarmeria station	30
Total	**2100**

The area covers a total of 3,500 hectares and is located between 2150 and 3175 meters above sea level (asl), featuring steep gradients (Kurt, 2017). The average slope inclination is 18°, and the region receives substantial snowfall during the winter season. As a result, the ski slopes in the area are consistently exposed to the risk of snow avalanches throughout the winter season.

To mitigate the risk, various measures are implemented at Palandöken Ski Resort. These measures may include avalanche forecasting and monitoring systems, controlled release of avalanches in less populated areas, installation of protective structures such as snow fences and barriers, and educating skiers and visitors about avalanche safety. Regular monitoring of snowpack conditions, weather forecasts, and terrain analysis is crucial for determining the current avalanche risk level. These assessments aid in making informed decisions regarding the opening or closure of specific ski slopes or areas within the resort.

2.2 Meteorological Characteristics

Meteorological parameters play a crucial role in assessing the potential for avalanches, but they also introduce uncertainties. Avalanche forecasts are based on weather forecasts, which inherently carry a level of uncertainty. Factors such as the size of the forecast area and the lead time for the forecast contribute to these uncertainties. To minimize such uncertainties, meteorological data is collected from an automated weather observing system (AWOS) located in the ski center. This data is carefully analyzed and utilized by relevant authorities to enhance the accuracy of avalanche forecasts and ensure the safety of winter tourists (Figure 2).

Figure 2. Location of the automated weather observing system (AWOS) in Palandöken ski resort (Photo: Batur Turalıoğlu) (Kurt, 2017)

The region is characterized by a terrestrial climate, which is influenced by its geographical features. Winter seasons in the area are known to be long and harsh, with cold temperatures and abundant snowfall. Historical data from 1921 to 2020 indicates that the region remains covered in snow for approximately 150 days each year (Zeydan and Sevim, 2008).

The climate in the region exhibits diversity, with average air temperatures around 5.6 °C (Irmak et al, 2020). However, extreme temperature variations have been recorded, with the highest temperature reaching 36.5 °C and the lowest temperature dropping as low as -37.2 °C. These extreme temperature ranges highlight the challenging weather conditions that can be encountered in the region based on meteorological records. Additionally, strong winds with speeds of up to 110.2 km/h have been documented, further influencing the overall winter experience.

To ensure the safety of winter tourists and residents, it has been determined that snow heights exceeding 80 cm within a three-day period serve as indicative information for potential avalanche risk in Palandöken. This threshold has been confirmed by officials from AFAD (Disaster and Emergency Management Presidency), which plays a crucial role in monitoring and managing natural disasters, including avalanches, in Turkey.

By considering the accumulated snowfall and the critical threshold of 80 cm, authorities and relevant organizations can make more accurate assessments of the avalanche risk. This information allows them to implement necessary safety measures and precautions to mitigate the potential hazards associated with winter storms and avalanches in Palandöken. Monitoring these snow heights and promptly responding to situations exceeding the threshold is essential for ensuring the well-being and security of individuals in the area.

2.3 Chronology of Avalanche Events

When analyzing and assessing avalanche hazards, it is crucial to examine various factors, including avalanche patterns such as the runout zone and starting zone (Kurt, 2014). These patterns provide valuable insights into the dynamics of avalanche activity and play a significant role in avalanche hazard assessment. To determine these patterns and assess avalanche frequency, multiple sources of information can be utilized. One of the most reliable sources of data for avalanche hazard assessment is documented information about past avalanche events. These records can be collected from formal avalanche observation records as well as informal sources like newspapers, historical records, and photographs. By examining these sources, experts can establish the frequency of avalanche events and gain a comprehensive understanding of their characteristics.

Reports from the local population also contribute to identifying major deposition areas within the region. Combined with on-site assessments of terrain characteristics during fieldwork, this information helps experts identify significant areas where avalanches tend to occur. By studying avalanche patterns and utilizing documented information, experts and authorities can gain a better understanding of the nature and frequency of avalanches in the area. This knowledge is crucial for effective avalanche hazard management and the implementation of appropriate safety measures. Understanding the characteristics of major deposition areas enables the development of strategies to mitigate the potential risks associated with avalanche events in the region (Kurt, 2017).

The authorities are fully aware of the potential damages that can be caused by avalanches and are actively working to prevent future incidents in the region. Extensive investigations and consultations with relevant authorities have helped identify the avalanche-prone areas in Palandöken, which are categorized into three main parts:

- **Northern Ski Slopes**: These slopes have been identified as particularly susceptible to avalanches due to their geographical characteristics and snow accumulation patterns. Special attention and precautionary measures are implemented in this area to mitigate the risk of avalanches.
- **Valley Ski Slopes**: The valley areas within Palandöken are also recognized as potential avalanche zones. The combination of steep slopes and natural terrain features make these areas prone to avalanche activity. Close monitoring and careful assessment of snowpack stability are carried out to ensure the safety of visitors.
- **Avalanche Flow Lines from Tekman Road:** Avalanche flow lines originating from the Tekman Road pose a risk to the surrounding areas. These flow lines are closely studied and mapped to identify potential hazards and establish appropriate safety measures.

Despite the efforts to mitigate avalanche risks, it is important to note that unfortunate incidents have occurred in these regions, resulting in the loss of nine lives (Figure 3). These incidents serve as a reminder of the unpredictable nature of avalanches and the need for continuous monitoring, assessment, and precautionary measures to ensure the safety of residents and visitors.

Figure 3. Avalanche locations in Palandöken skiing centre in Erzurum province (Photo: Batur Turalıoğlu)

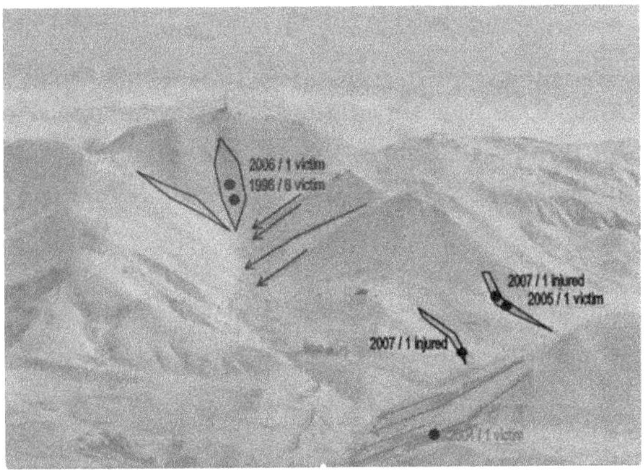

Individuals planning to visit these areas play a crucial role in ensuring their own safety and the safety of others. It is essential for them to stay informed about the current avalanche situation, follow the guidance of local authorities, and make informed decisions regarding their activities in avalanche-prone zones. By promoting education, awareness, and responsible behavior, the aim is to minimize the risk of avalanche-related incidents and ensure the well-being of everyone in the region (Figure 4).

Figure 4. Overview of Valley ski slopes (Photo: Batur Turaloğlu)

2.3.1 Avalanches in North Ski Slopes

The North Ski Slopes in Palandöken are particularly prone to avalanches, with the highest number of incidents occurring during the months of December and January. These avalanches are primarily categorized as powder or dry layer avalanches, posing significant risks to individuals in the area. Among the various slopes in this region, Büyük Ejder hill has been identified as the most dangerous and notorious trigger point for avalanches (Figure 5).

Figure 5. Overview of Büyük Ejder hill and North ski slopes (Photo: Batur Turaloğlu)

Tragically, a total of seven fatalities have been recorded in this region as a result of avalanches. To minimize the potential losses and enhance safety measures, various avalanche control methods are employed. One such method is the use of the Gazex avalanche explosion system, which is implemented at random intervals (Figure 6). This system helps to mitigate the risk by triggering controlled avalanches and reducing the build-up of unstable snow masses.

Figure 6. Overview of Gazex system in Palandöken ski centre (Photo: Batur Turalıoğlı)

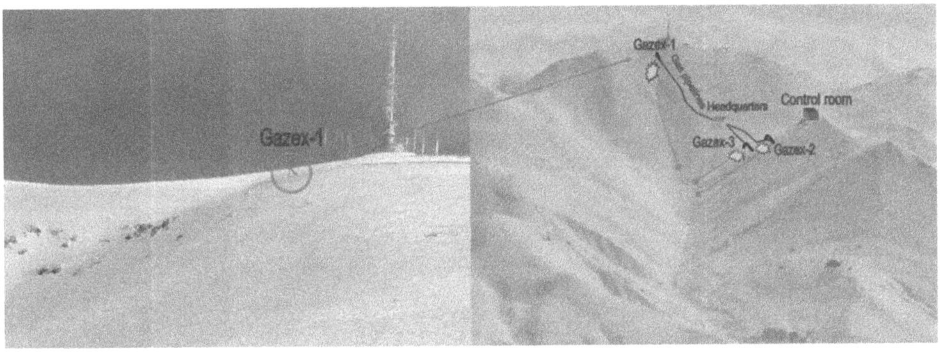

Additionally, to provide further protection, an avalanche catching wall has been constructed in this region. This wall acts as a physical barrier to prevent avalanches from reaching lower-lying areas, thereby reducing the potential damage and danger to individuals. Furthermore, the North Ski Slopes boast the highest concentration of warning signs, strategically placed to alert skiers and snowboarders about the avalanche risks and remind them to exercise caution (Figure 4).

The combination of these avalanche control measures, including the Gazex system, the avalanche catching wall, and the abundance of warning signs, plays a crucial role in ensuring the safety of visitors in the North Ski Slopes of Palandöken. However, it is important to note that avalanches remain unpredictable forces of nature, and it is essential for individuals to remain vigilant, follow safety guidelines, and respect any closures or warnings issued by the authorities.

Continuous monitoring of snow conditions, regular assessments of avalanche risk, and effective communication between relevant stakeholders are key aspects of managing and minimizing avalanche hazards in this region. By prioritizing safety and implementing comprehensive measures, the aim is to provide visitors with a secure environment to enjoy winter sports while mitigating the potential risks associated with avalanches.

2.3.2 Avalanches in Valley Ski Slopes

The Valley ski slopes in Palandöken are considered the most dangerous area in terms of avalanche occurrences. As a result, this area has been officially closed for skiing activities. Despite the closure, there are still some skiers who are determined to ski in this area, disregarding the risks involved. Unfortunately, their actions often lead to triggering avalanches, which can result in loss of life or serious injuries (Figure 7).

Figure 7. Overview of Valley ski slopes (Photo: Batur Turaloğlu)

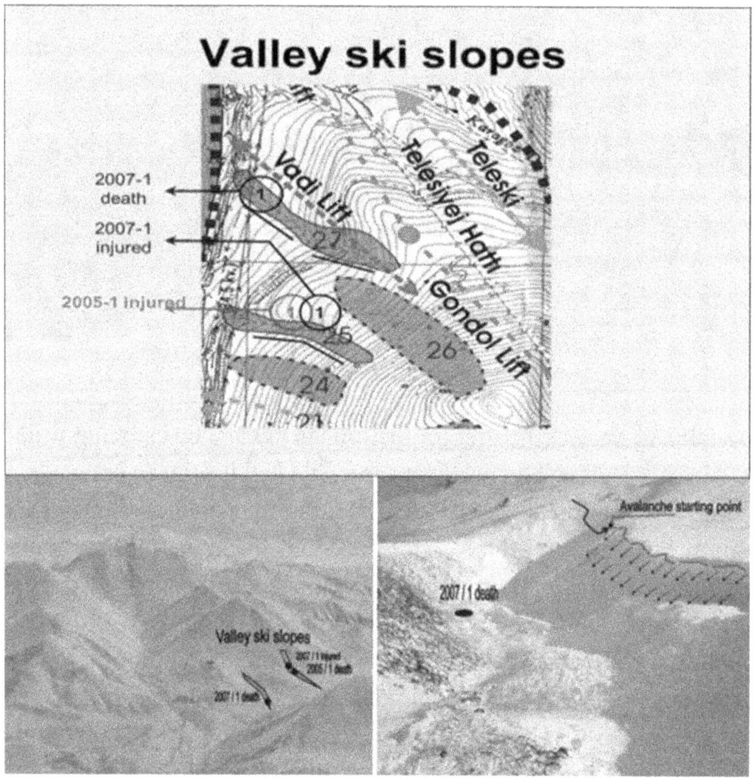

Notably, there have been incidents in the past that highlight the dangers of skiing in the Valley ski slopes. In 2005, a tourist lost their life, while in 2007, two tourists were injured on separate occasions. These incidents serve as a stark reminder of the inherent risks associated with skiing in this area, especially when it is officially closed due to avalanche hazards.

The occurrence of avalanches in the Valley ski slopes is primarily attributed to artificial factors. One such factor is the accumulation of snow in cornices, which are formed by wind transport in the region. These cornices act as precarious overhangs of accumulated snow, and when skiers venture into these areas, their presence can trigger dry layer or block avalanches. This highlights the importance of understanding the dangers associated with cornices and avoiding them to prevent avalanche incidents (Figure 8).

Figure 8. One skier has injured after and avalanche in Palandöken ski centre (Photo: Batur Turalıoğlu)

2.3.3 Avalanches in Tekman Road

The Tekman road region in Palandöken is identified as a hazardous area with a high risk of avalanches, both deadly and beneficial in nature (Figure 9). The region is characterized by rounded inclinations ranging from 300 to 550. Typically, wet snow avalanches are the most common type observed in this area.

Figure 9. Overwiev of Tekman road ski slopes (Photo: Batur Turaloğlu)

It is important to note that the Tekman road region is not intended for skiing activities due to its inherent avalanche risk. Despite this, there have been unfortunate incidents in the past that highlight the dangers associated with venturing into this area. In 2004, a tourist lost their life, emphasizing the severity of the risks involved.

The occurrence of avalanches in the Tekman road region is influenced by various factors. The rounded inclination of the terrain contributes to the accumulation and instability of snow, making it susceptible to avalanches. Additionally, the presence of wet snow further increases the likelihood of avalanches, as it has a higher water content that can affect the stability of the snowpack.

It is crucial for visitors and skiers to be aware of the specific risks associated with the Tekman road region and to strictly adhere to the guidance provided by the authorities. Engaging in skiing or any other activities in this area is strongly discouraged due to the potential for deadly avalanches.

The tragic incident in 2004 serves as a somber reminder of the unpredictable nature of avalanches and the need for caution when exploring such hazardous regions. The authorities continuously work to assess and manage the avalanche risks in the Tekman road region, implementing measures to enhance safety and prevent further accidents (Figure 10).

Figure 10. Overwiev of Tekman road ski slopes (Photo: Batur Turaloğlu)

To mitigate the risks associated with avalanches in this area, comprehensive avalanche forecasting and monitoring systems are employed. These systems aim to provide real-time information about snow conditions, weather patterns, and other relevant factors that can contribute to avalanche formation. By closely monitoring these conditions, authorities can issue timely warnings and take necessary precautions to ensure the safety of individuals in the vicinity.

Raising awareness among tourists and visitors about the potential dangers and risks associated with the Tekman road region is of paramount importance. Clear signage, communication channels, and educational campaigns are utilized to inform the public about the specific hazards and the need to avoid venturing into this high-risk area.

By promoting responsible behavior, providing accurate information, and implementing effective risk management strategies, the authorities strive to create a safer environment and prevent future accidents in the Tekman road region. Visitors are urged to prioritize their safety and adhere to the guidelines and closures set in place to minimize the risks associated with avalanches.

3. METHODOLOGY

Many studies have demonstrated that risk analysis is a multifaceted process that plays a pivotal role in assessing and managing potential risks and threats across various domains (Fjelt et al., 2023; Fuchs et al., 2015; Haimes, 2005). In the specific context of qualitative avalanche risk analysis, the emphasis lies in evaluating the likelihood of avalanche occurrences and comprehensively examining the potential consequences associated with such events, including property damage and endangerment costs (Kurt, 2022; St Clair, 2019; Warrick, 2015; Zweifel and Haegeli, 2014; Eckert et al., 2012).

This study utilized L-type matrix risk analysis. The application of the L Matrix in avalanche risk analysis is of significant importance due to its ability to systematically assess and quantify risk factors. The L Matrix facilitates a detailed examination of potential avalanche hazards by categorizing risks based on their likelihood and potential impact, breaking them down into manageable components. This structured format assists decision-makers in developing targeted risk mitigation strategies, while also providing a common framework for local communities, government agencies, and industry professionals. This promotes collaboration among different stakeholders. By prioritizing risks and optimizing resource allocation, the L Matrix ensures that no significant risk factors are overlooked. Additionally, the iterative nature of the L Matrix allows for continuous monitoring and updating of risk assessments, which is crucial for responding to changing environmental conditions and enhancing community resilience against avalanche threats. In conclusion, the L Matrix is a critical tool in avalanche risk analysis, enabling a systematic, prioritized, and collaborative approach to risk management. This, in turn, enhances the safety and resilience of communities in avalanche-prone areas.

The assessment began by determining the hazard level of snow avalanches through modeling techniques. Factors such as slope inclination, aspect, and vegetation cover, known to influence avalanche occurrences, were analyzed and integrated to generate a hazard map. This map provided an overview of the avalanche-prone areas within the study region. Simultaneously, the elements at risk, including infrastructure such as buildings and roads, as well as the population residing in avalanche-prone areas, were identified and quantified. By considering the vulnerability of these elements and their exposure to potential avalanches, the potential consequences of avalanche events were evaluated.

The hazard and elements at risk were then combined to provide a comprehensive assessment of snow avalanche risk. The assessment results were visualized using GIS tools, allowing for a clear depiction of areas with varying levels of risk.

To validate the findings, available incident information and field surveys were utilized. Incident data from past avalanche events were examined and compared with the risk assessment results to evaluate the accuracy of the qualitative approach. Additionally, field surveys were conducted to gather on-site information and verify the assessment outcomes.

The overall process consists of three main steps: risk identification, risk analysis, and risk evaluation in this paper.

3.1 Risk Identification

In this study, we focus on risk identification, exploring different methods and processes' effectiveness, and evaluating how risk management contributes to overall strategic objectives. Our research aims to aid organizations in enhancing long-term sustainability and performance by refining risk definition and management.

The initial step, risk identification, involved a thorough process. We gathered diverse information to comprehensively understand and document potential avalanche risks and hazards. This process included examining historical avalanche data to assess frequency, severity, and locations, providing insights into potential avalanche-prone areas. Additionally, we analyzed topographic characteristics such as slope angle, aspect, elevation, and vegetation cover to pinpoint susceptible areas. Meteorological conditions like snowfall, temperature fluctuations, wind speed, and humidity levels were also considered, as specific weather conditions can trigger avalanches. Furthermore, we examined human activities such as recreational pursuits, infrastructure development, and land use practices to understand their impact on avalanche risk.

By integrating these sources of information, our risk identification process accurately identified potential avalanche-prone areas and assessed associated risks. This comprehensive analysis is vital for developing effective mitigation strategies and enhancing safety measures in avalanche-prone regions within the study area.

3.2 Risk Analysis

The next step in our analysis was risk assessment, which involved evaluating the likelihood of avalanche occurrences and examining their potential impacts. We utilized scientific methods, expert knowledge, and data analysis to estimate the probability of avalanches, considering factors such as slope angle, snowpack stability, weather conditions, and human activities. Additionally, we assessed the potential magnitude of avalanches and their effects on infrastructure, human lives, and the environment.

After identifying and analyzing the risks, we proceeded to the final step: risk evaluation. This phase entailed prioritizing the identified risks based on their severity and potential impacts. We considered the costs associated with property damage, potential loss of life, activity disruptions, and overall societal and economic consequences. This prioritization enabled decision-makers to focus on implementing risk mitigation measures and allocating resources effectively.

To develop risk maps, we adopted a two-step approach. Firstly, we assessed the likelihood of avalanches for all tracks by determining the probability of specific triggering events, such as intense snowfall, which could initiate avalanches within a specified timeframe. Secondly, we evaluated the frequency of avalanches based on input from the local population and analyzed anomalies in vegetation cover. This information allowed us to categorize avalanches into different frequencies, such as occurring once every year, every 3 years, every 10 years, every 30 years, or every 100 years. These risk maps offer valuable insights into the likelihood and potential impacts of avalanches in the area (Table 2).

Table 2. Evaluation of avalanche occurrence in the study area

Definition of avalanche occurrence	**Frequency (avalanche events) [%]**
Highest	≥ 1.0 (every year)
High	0.1 > to < 1,0 (every 3 years)
Medium	0.02 > to ≤ 0.1 (every 10 years)
Low	0.02 > to ≤ 0.01 (every 30 years)
Lowest	≤ 0.01 (every 100 years)

In the second phase of our risk analysis, we conducted fieldwork and interviews with local residents to assess the potential impacts of snow avalanches. This analysis aimed to evaluate the vulnerability of various structures, including buildings, roads, infrastructure, and the population residing in vulnerable areas. We categorized the assessment into three main areas: health, economy, and environment. Within each category, we further assessed severity levels, ranging from low to high. The objective of this analysis was to comprehensively understand the risks associated with avalanches and their potential consequences. By evaluating the vulnerability of different elements at risk and considering severity levels, decision-makers and stakeholders can effectively prioritize their efforts in mitigating the impacts of avalanches (Table 3).

Table 3. Calculation of risk scores

Risk Score=Occurrence of avalanche x Effect level			**Effect level of avalanche**				
			Highest	**High**	**Medium**	**Low**	**Lowest**
			5	**4**	**3**	**2**	**1**
Occurrence of avalanche	Highest	5	*RS=25*	*RS=20*	*RS=15*	*RS=10*	*RS=5*
	High	4	*RS=20*	*RS=16*	*RS=12*	*RS=8*	*RS=4*
	Medium	3	*RS=15*	*RS=12*	*RS=9*	*RS=6*	*RS=3*
	Low	2	*RS=10*	*RS=8*	*RS=6*	*RS=4*	*RS=2*
	Lowest	1	*RS=5*	*RS=4*	*RS=3*	*RS=2*	*RS=1*
Risk Score (RS)Classification: 0 < Risk Score ≤ 7 = Lowest Risk 7 < Risk Score ≤ 17 = Medium Risk 17 < Risk Score ≤ 25 = Highest Risk							

Risk analysis was conducted to assess the potential impacts of snow avalanches on various elements at risk. By considering the categories of health, economy, and environment, and evaluating severity levels, decision-makers can gain a comprehensive understanding of potential consequences and prioritize efforts in mitigating avalanche risks. For example, the health category focused on assessing the number of people at risk, while the economy category considered exposed buildings and the length of major roads. The environment category accounted for affected vegetation, particularly forest cover, measured in hectares (Table 4). Assigning impact levels to individual avalanche paths and run-out areas within each category facilitated the identification of the most critical avalanche path with the highest risk using Geographic Information Systems (GIS) overlays. This process enabled a comprehensive assessment of potential avalanche consequences on human health, the local economy, and the environment, determined through observational evaluation.

It is important to note that this analysis should be regularly updated to account for changes in infrastructure, population, and other relevant factors. Ongoing collaboration with local communities, experts, and authorities ensures that the assessment remains accurate and up to date, enabling effective risk management strategies to be implemented.

3.3. Risk Evaluation

Risk assessment involved identifying potential risks to an organization and evaluating their possible impacts. This process assessed the likelihood and impact of various risks, considering the organization's objectives, activities, and resources. Typically conducted by experts using a specific methodology, the results aided organizations in strategic decision-making and risk management. This process was crucial for organizations' long-term success and sustainability.

The risk evaluation employed a scoring system, offering a systematic and standardized approach to assess the overall risk level for each element at risk. It's worth noting that the scoring system and risk evaluation methodology outlined in Remondo et al. (2005) served as the basis for developing qualitative risk maps. However, modifications and adaptations specific to the study area may have been made to ensure their applicability and relevance. By considering factors like hazard, vulnerability, and consequences, the scoring system provided a comprehensive understanding of potential avalanche risks (Table 4).

Table 4. Classification and characterisation of risk indicators in the study area

Classification	**EXPOSURE OF AVALANCHE**				
	Lowest	**Low**	**Medium**	**High**	**Highest**
Health (Number of persons who may be affected, number or leisure centres)	**Number of persons ≤ 1** and/ or **Leisure centre** (N) **≤ 1**	**2 ≥ Number of persons ≤ 4** and/ or **1≥ Leisure centre** (N) **≤ 5**	**5 ≥ Number of persons ≤ 9** and/ or **6 ≥ Leisure centre** (N) **≤ 10**	**10 ≥ Number of persons ≤** and/ or **11 ≥ Leisure centre** (N) **≤ 19**	**Number of persons ≥ 21** and/ or **Leisure centre** (N) **≥ 20**
Environment (Protected areas that may be affected)	**Protected areas** (Ha) **≤ 20**	**21≥ Protected areas** (Ha) **≤ 50**	**51 ≥Protected areas** (Ha) **≤ 75**	**76 ≥Protected areas** (Ha) **≤ 150**	**Protected areas** (Ha) **≥ 151**

continued on following page

Table 4. Continued

Classification	EXPOSURE OF AVALANCHE				
	Lowest	Low	Medium	High	Highest
Economy (Length of road that may be affected, number of personal property)	**Property** (N) **≤ 1** and/ or **Length of road** (m) **≤ 100**	**2 ≥ Property** (N) **≤ 5** and/ or **101 ≥ Length of road** (m) **≤ 200**	**6 ≥ Property** (N) **≤ 9** and/ or **201 ≥ Length of road** (m) **≤ 300**	**10 ≥ Property** (N) **≤ 19** and/ or **301 ≥ Length of road** (m) **≤ 400**	**Property** (N) **≥ 20** and/ or **Length of road** (m) **≥ 401**

The occurrence of avalanche values was derived from Table 2, while the effect level values were obtained from Table 4 and utilized for scoring purposes in Table 3. In summary, risk analysis involved assessing the potential impacts of snow avalanches by considering health, economy, and environmental factors. By assigning severity levels to elements at risk and overlaying the results in GIS, the most critical avalanche path was determined.

In conclusion, converting avalanche susceptibility maps into hazard maps, followed by developing qualitative risk maps using a scoring system, provided a comprehensive understanding of avalanche risks in the study area. These risk maps effectively communicated varying degrees of risk across different elements at risk, aiding decision-making and prioritizing mitigation efforts. The methodology outlined in Remondo et al. (2005) served as a framework for the risk evaluation process.

4. RESULTS

The desktop reviews involved an extensive analysis of existing literature, historical records, and available data related to avalanches in the area. This helped establish a foundation of knowledge and provided insights into past avalanche occurrences. Additionally, fieldwork was conducted, involving site visits and direct observations in the study region. This on-site investigation allowed for the identification and verification of avalanche release areas.

The avalanche susceptibility maps were transformed into hazard maps by integrating avalanche trigger probabilities, enabling a more comprehensive assessment of potential avalanche hazards. Through extensive desktop reviews and fieldwork, followed by the development of a susceptibility map, a deeper understanding of avalanche dynamics and associated risks was achieved. This knowledge contributes to overall safety and management of the study area, allowing for effective implementation of mitigation measures to safeguard individuals and infrastructure from potential avalanche risks. In the subsequent step, qualitative risk maps were created using a scoring system to assess hazard, vulnerability, and consequences for each at-risk element.

The qualitative risk maps provided a valuable tool for visualizing and understanding the overall risk landscape in the study area. By assigning scores to different elements at risk based on their susceptibility to avalanches, their vulnerability to avalanche impacts, and the potential consequences of an avalanche event, the risk maps effectively communicated the varying degrees of risk across the region. The information presented in the risk maps facilitated decision-making processes by highlighting areas of higher risk and identifying elements at risk that required prioritized mitigation measures. The risk maps served as a valuable resource for stakeholders and decision-makers involved in disaster management, land-use planning, and infrastructure development in avalanche-prone areas.

To address the knowledge gap in avalanche documentation, a comprehensive approach was taken, combining desktop reviews and fieldwork. The aim was to identify and document avalanche release areas in the study region. Through this process, a total of 35 avalanche release areas were detected, with the majority of them situated at an elevation of approximately 2,500 meters above sea level.

Based on the collected data, a susceptibility map was created, linking the potential avalanche release areas to the possible run-out areas. This map serves as a valuable tool for understanding the dynamics of avalanches in the region and assessing the potential hazards they pose. By delineating the areas susceptible to avalanche activity and their associated run-out zones, it becomes easier to identify areas of high risk and implement appropriate safety measures. Figure 11 showcases the susceptibility map, providing a visual representation of the identified avalanche release areas and their corresponding run-out zones. This map serves as a valuable resource for avalanche forecasting, risk management, and land-use planning in the study region, helping stakeholders and authorities make informed decisions regarding infrastructure development, recreational activities, and public safety measures.

Figure 11. Avalanche depositions zone in Paladöken ski centre (Kurt, 2017)

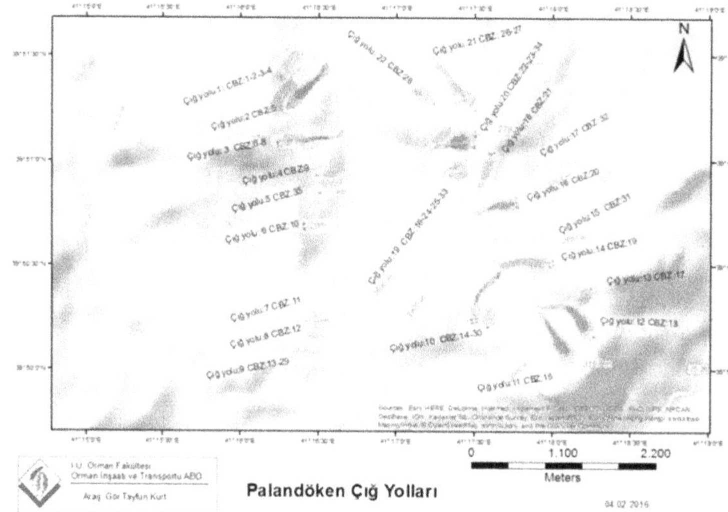

4.1 Interpretation of Avalanche Frequency

The analysis of avalanche frequency and magnitude played a crucial role in hazard assessment. In this study, avalanche frequency was defined as the reciprocal of the avalanche return period, often expressed as avalanches per year. This ratio could range from 1:1 (indicating one avalanche per year) to 1:300 (indicating one avalanche every 300 years). The return period of avalanches was commonly expressed as 1, 3, 10, 30, or 100 years, based on information obtained from authoritative sources.

The occurrence of snow avalanches was influenced by various natural factors, including slope inclination, aspect, and the presence of vegetation. Vegetation analysis was an essential tool for examining past avalanche events and estimating their frequency and intensity. However, in the Palandöken region, herbaceous plants dominated the landscape, and there was no dense vegetation in the starting zones or run-out areas of avalanches. Throughout the research area, species such as alpine tangerines (Festuca sp.), thyme (Thymus sp.), speedwell (Artemisia sp.), agave (Astragalus sp.), and wild clover (Medicago sativa sp.) were commonly observed (Kurt, 2017).

4.2 Determination of Objects at Risk

The following objects and individuals, as presented in Table 5, are at risk within the Palandöken avalanche area. Through the integration of avalanche susceptibility mapping and hazard maps, a qualitative assessment of risk was conducted using the scoring and classification approach described earlier. The findings reveal that multiple avalanche tracks pose a threat to over 1,851 meters of road, and a total of 7 buildings are endangered. Furthermore, based on a thorough review, approximately 2,100 individuals are at risk, and nearly 260 hectares of land are under threat.

Table 5. Assessment of exposed elements at risk in the study area

Number of avalanche path	CBZ (Avalanche path) No	Length of road (Economy)	Number of property (Economy)	Number of persons (Health)		Avalanche occurrence
1	1-2-3-4	600 m	2	690	26 hectares	3
2	5	150 m	2	0	10 hectares	3
3	6-8	150 m	1	30	13 hectares	3
4	9	100 m	0	0	6 hectares	3
5	35	150 m	0	0	5 hectares	3
6	10	100 m	0	0	4 hectares	3
7	11	0	0	0	3 hectares	3
8	12	0	0	0	5 hectares	3
9	13-29	0	0	0	18 hectares	3
10	14-30	300 m	0	0	17 hectares	3
11	15	0	0	0	17 hectares	3
12	18	0	0	0	17 hectares	3
13	17	0	0	0	16 hectares	3
14	19	0	0	0	14 hectares	3
15	31	200 m	0	0	12 hectares	3
16	20	1 m	0	0	8 hectares	3
17	32	0 m	0	0	8 hectares	3

continued on following page

Table 5. Continued

Number of avalanche path	CBZ (Avalanche path) No	Length of road (Economy)	Number of property (Economy)	Number of persons (Health)		Avalanche occurrence
18	21	0 m	0	0	11 hectares	3
19	16-24-25-33	0 m	1	0	10 hectares	3
20	22-23-34	100	1	30	16 hectares	3
21	26-27	0	0	0	15 hectares	3
22	28	0	0	0	10 hectares	3

It is important to note that the information provided in this text is for illustrative purposes only and does not represent the actual extent of risk or specific numbers of individuals or buildings at risk in the Palandöken avalanche area. The figures presented are fictional and used solely for the purpose of demonstrating the concept of risk assessment and its potential implications for land-use planning and risk management.

The qualitative risk assessment conducted on a 1:25,000 scale resulted in the ranking of risks for the study area. A comparison of all the studied avalanche tracks revealed significant variations in the development of risks across the different release zones, as outlined in Table 5. The qualitative risk scores were calculated on a continuous scale, where numerical values indicate the prioritization of risk for each avalanche track.

Based on the calculation of risk scores (Table 6):

- For traffic infrastructure, avalanche track no (CBZ) 1 was identified as having the highest risk score.
- In terms of exposed buildings and the population at risk, avalanche track no (CBZ) 1 also obtained the highest risk score.
- Regarding the environment, specifically the protected area, avalanche track no (CBZ) 1 was found to have the highest risk score.

Table 6. Overview on the avalanche tracks with the highest risk scores in the study region

Avalanche Track no	CBZ No (Avalanche track)	Length of road (Economy)	Avalanche occurrence	Being effected level	Risk Score
1	1, 2, 3, 4	600 m	3	5	**15**

continued on following page

Table 6. Continued

Avalanche Track no	CBZ No (Avalanche track)	Length of road (Economy)	Avalanche occurrence	Being effected level	Risk Score
Avalanche Track no	CBZ No	Number of Building (Economy)	Avalanche occurrence	Being effected level	**Risk Score**
1	1, 2, 3, 4	6	3	5	**15**
Avalanche Track no	CBZ No	Number of human (Health)	Avalanche occurrence	Being effected level	**Risk Score 1**
1	1, 2, 3, 4	690	3	5	**15**
Avalanche Track no	1, 2, 3, 4	Natural park (Environment)	Avalanche occurrence	Being effected level	**Risk Score**
1	1, 2, 3, 4	26 hectares	3	5	**15**

These findings highlight the specific avalanche track (CBZ) 1 as the most critical in terms of risks associated with traffic infrastructure, exposed buildings, citizens at risk, and the environment (protected area) (Figure 12).

Figure 12. Most dangerous avalanche slope in Paladöken ski centre (Kurt, 2017)

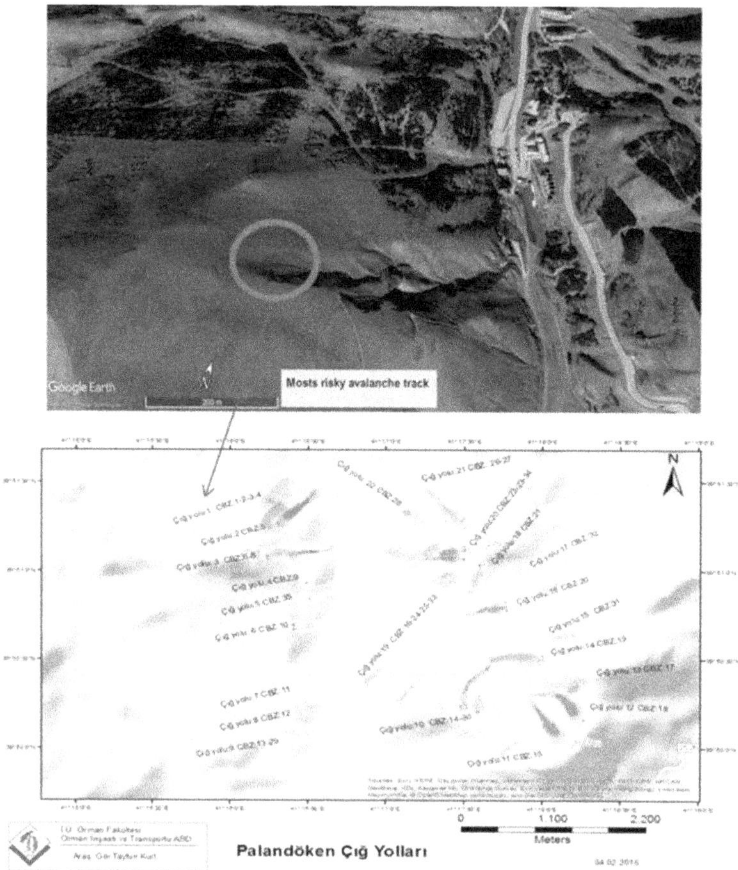

5. CONCLUSION AND DISCUSSION

Due to its nature as a ski resort, Palandöken is prone to avalanche incidents that pose significant risks, resulting in fatalities among both tourists and locals. Previous avalanche events exemplify these dangers. Conducting avalanche risk analyses aims to identify the most hazardous areas in terms of potential loss of life and property. Despite various control measures, the risk of avalanches remains high, necessitating continuous attention and management, as evidenced in other studies (Bentella,

2022). In other words, the prospective nature of our approach requires additional and more in-depth empirical research.

Avalanches can cause severe damage and endanger both infrastructure and human life, making it crucial to prioritize avalanche safety and implement effective mitigation strategies. The avalanche risk at Palandöken ski resort is influenced by natural factors like snow depth, snowpack structure, slope gradient, meteorological conditions, and other environmental variables. Therefore, a systematic and scientific approach is vital to accurately assess avalanche potential. Mitigation measures should align with this assessment to develop tailored safety measures and visitor awareness strategies. By informing skiers, tourists, and locals about high-risk zones, the goal is to minimize casualties and property damage. This study identifies high-risk areas, providing valuable insights for land-use planners and stakeholders involved in hazard and risk management. Utilizing this data can optimize investments in technical protection measures to mitigate potential avalanche impacts in the Palandöken region.

It's important to note that the information presented in the text is based on hypothetical scenarios and doesn't reflect actual risk assessments or specific avalanche tracks in the study area. The mentioned risk scores and track numbers serve illustrative purposes, showcasing the concept of qualitative risk assessment and its implications for the study area.

The significance of conducting qualitative risk analysis in avalanche control cannot be overstated. This approach serves as a pivotal tool for comprehending intricate avalanche risks and devising effective countermeasures. Qualitative analysis involves a thorough evaluation of avalanche probability and potential impacts, drawing upon expert insights and local knowledge. This method facilitates the identification and prioritization of high-risk avalanche zones, thereby optimizing resource allocation. Furthermore, it plays a crucial role in enhancing community awareness of risks, bolstering preparedness efforts, and ensuring overall safety. Ultimately, qualitative risk analysis stands as a foundational element in formulating strategies crucial for mitigating avalanche impacts and safeguarding vulnerable communities.

This study will offer managers information by identifying potential hazards and risks, enabling the prioritization of different risks and effective resource allocation. This facilitates the development of emergency plans and risk reduction strategies. Focusing resources on risks will lead to their mitigation and more efficient utilization. It assists decision-makers in evaluating the effects of specific risk reduction or prevention strategies and enhances awareness among communities and stakeholders, thereby ensuring better preparedness. For instrance; When disseminated on platforms frequented by skiing enthusiasts (utilizing a netnographic approach), this study could establish digital communication and potentially prompt skiers to adopt behaviors that steer clear of risky areas (Alla L & Bentalla, 2022; Bentella & Hmioui, 2022;).

Accuracy, validity, reliability, and data consistency are critical factors in risk mapping, contingent upon data availability and the institutional framework, including legal considerations and economic resources. A multidisciplinary approach, incorporating knowledge in areas such as statistics, geology, and economics, is essential for reliable risk assessment. There's a need to expand the existing avalanche incident system in Turkey, including all types of avalanche events and adhering to standardized protocols. The inconsistencies in historical data highlight the importance of comprehensive and consistent record-keeping.

Qualitative risk mapping approaches have shown the ability to assess snow avalanche risk, particularly in data-scarce mountain regions. These methods, originally quantitative, rely on qualitative information and scoring to evaluate hazard and risk. Subsequent steps should focus on establishing observation networks and collecting reliable data to enhance forecasting and warning services tailored to local conditions.

Ski areas and slopes are highly susceptible to avalanches due to their unique topographical features and the accumulation of snow in steep terrain. Despite ongoing efforts to manage and mitigate avalanche risks, the unpredictable nature of avalanches makes it challenging to completely eliminate the potential for incidents.

Tourists who visit these ski areas are well aware of the inherent risks associated with skiing in avalanche-prone regions. However, their passion for the sport and the allure of challenging slopes often outweigh the potential dangers. Ski resorts and authorities work diligently to implement safety protocols, including avalanche control measures such as snowpack stability assessments, controlled releases of unstable snow, and the use of protective structures. The authorities, in collaboration with local communities, ski resorts, and relevant stakeholders, are actively working to enhance safety measures and raise awareness about avalanche risks. This comprehensive approach includes regular training programs, improved emergency response systems, and the dissemination of up-to-date information regarding avalanche conditions.

However, it is crucial for tourists to exercise caution, adhere to safety guidelines, and stay informed about current avalanche conditions. They should also take advantage of the services provided by ski resorts, such as expert guidance, avalanche education programs, and the availability of safety equipment.

Continuous research and monitoring efforts are vital to further improve avalanche hazard assessment and mitigation strategies in ski areas. This includes advanced snowpack analysis, improved forecasting models, and the development of effective communication channels to promptly inform skiers about changing avalanche conditions.

By acknowledging the potential risks and taking necessary precautions, tourists can enjoy the exhilarating experience of skiing while minimizing the dangers associated with avalanches. Ski resorts and authorities must remain committed to prioritizing the safety of their visitors through ongoing risk management efforts.

The Turkish government's recent investments in winter tourism infrastructure aim to enhance the overall winter tourism experience and attract more visitors worldwide. Winter tourism in Turkey offers natural beauty, thrilling activities, and cultural experiences, establishing Turkey as a prominent player in the global winter tourism industry. By prioritizing safety and implementing effective risk management strategies, Palandöken Ski Resort aims to provide a secure and enjoyable experience for all visitors.

Efforts are continuously made to educate skiers about the risks involved and to enhance safety measures in the Valley ski slopes. It is vital for skiers to be well-informed, undergo proper training, and make responsible decisions while participating in winter sports. Additionally, awareness campaigns, signage, and communication channels are utilized to inform visitors about the dangers of skiing in closed areas and the potential consequences of triggering avalanches.

By fostering a culture of safety, raising awareness, and enforcing regulations, the aim is to prevent accidents and create a secure environment for winter sports enthusiasts in Palandöken. The collaboration between authorities, ski resort management, and visitors plays a pivotal role in mitigating avalanche risks and ensuring the well-being of everyone involved.

DATA AVAILABILITY

The data supporting the findings of this study can be accessed in the doctoral dissertation completed by Tayfun Kurt, available at the following link: Link to the doctoral dissertation with the reference number: 456959.

The author affirms that the data supporting the conclusions of this study are contained within the doctoral dissertation and its supplementary materials.

ACKNOWLEDGMENT

This work was financially supported by the Scientific Research Projects Unit of Istanbul University under Project number 47628. The study incorporates findings from the doctoral dissertation conducted by Tayfun Kurt at the Faculty of Forestry, Istanbul University. The fieldwork, conducted over several weeks, took place in Palandöken, Erzurum Province. I would like to extend my heartfelt gratitude to Batur Turalıoğlu for their valuable assistance and support throughout the research process.

REFERENCES

Alla, , A., Hmioui, A., & Bentalha, B. (2020). La netnographie dans les recherches marketing: La communauté virtuelle comme consom'acteur vecteur d'efficacité marketing. *Alternatives Managériales Economiques*, 2(4), 631–652.

Aygün Oğur, A., & Baycan, T. (2023). Assessing climate change impacts on tourism demand in Turkey. *Environment, Development and Sustainability*, 25(3), 2905–2935. 10.1007/s10668-022-02135-7

Behzadifar, M., Ghanbari, M. K., Azari, S., Bakhtiari, A., Rahimi, S., Ehsanzadeh, S. J., Sharafkhani, N., Moridi, S., & Bragazzi, N. L. (2023). A SWOT analysis of the development of health technology assessment in Iran. *PLoS One*, 18(3), e0283663. 10.1371/journal.pone.028366336996128

Bellaire, S., Jamieson, B., & Stetham, G. (2013) Relating Avalanche Activity to Climate Change and Coupled Ocean-Atmospheric Phenomena. Davos Atmosphere and Cryosphere Assembly DACA-13, Davos, Switzerland.

Bentalha, B. (2022). Smart Technologies for a Sustainable Service Supply Chain: A Prospective Perspective. *Journal of Environmental Issues and Climate Change*, 1(1), 77–89. 10.59110/jeicc.v1i1.81

Bentalha, B., & Hmioui, A. (2022). Smart Service Supply Chain and Just Walk Out Technology: A Netnographic Approach. In Innovations in Smart Cities Applications Volume 5. SCA 2021. Lecture Notes in Networks and Systems, vol 393. Springer. 10.1007/978-3-030-94191-8_18

Bründl, M., Bartelt, P., Schweizer, J., Keiler, M., & Glade, T. (2010). Review and future challenges in snow avalanche risk analysis. In Alcántara-Ayala, I., & Goudie, A. (Eds.), *Geomorphological hazards and disaster prevention* (pp. 49–61). Cambridge University Press. 10.1017/CBO9780511807527.005

Cengiz, Ö. (2022). Analysis of risk factors in fishing activities in Lake Erçek (Van, Turkey) for occupational health and safety. *Momona Ethiopian Journal of Science*, 14(2), 139–148. 10.4314/mejs.v14i2.3

Cohen, M. W., & Palmer, G. R. (2004). Project risk identification and management. AACE International Transactions, IN11.

Coleman, M. E., & Marks, H. M. (1999). Qualitative and quantitative risk assessment. *Food Control*, 10(4-5), 289–297. 10.1016/S0956-7135(99)00052-3

Costa, E. D. F., Cardoso, M. D. I., Kich, J. D., & Corbellini, L. G. (2017). Application Of Qualitative Risk Assessment To Prioritize Hazards In Pork Products In Brazil. In International Symposium On The Epidemiology And Control Of Biological, Chemical And Physical Hazards In Pig And Pork, 12., 2017, Foz Do Iguaçu. Proceedings Book. Concórdia: Embrapa Suínos E Aves.

DPT. (2020) Doğu Anadolu Projesi Ana Planı, Türkiye Cumhuriyeti Başbakanlık Devlet Planlama Teşkilatı Müşteşarlığı, Ankara, 2000. https://www.dap.gov.tr/content/DAP/DAPANAPLAN/PDF/ana_plan.pdf

Dvorak, Z., Rehak, D., David, A., & Cekerevac, Z. (2020). Qualitative approach to environmental risk assessment in transport. *International Journal of Environmental Research and Public Health*, 17(15), 5494. 10.3390/ijerph1715549432751437

Eckert, N., Keylock, C. J., Bertrand, D., Parent, E., Faug, T., Favier, P., & Naaim, M. (2012). Quantitative risk and optimal design approaches in the snow avalanche field: Review and extensions. *Cold Regions Science and Technology*, 79, 1–19. 10.1016/j.coldregions.2012.03.003

Evren, S. (2023). Bir Cumhuriyet Projesi Olarak Türkiye'de Kış Turizmi. Anatolia. *Turizm Araştırmaları Dergisi*, 34(3), 163–175. 10.17123/atad.1307923

Evren, S., & Kozak, N. (2018). Competitive positioning of winter tourism destinations: A comparative analysis of demand and supply sides perspectives–Cases from Turkey. *Journal of Destination Marketing & Management*, 9, 247–257. 10.1016/j.jdmm.2018.01.009

Firat Ersoy, A., & Çalik Sönmez, S. (2014). Hydrogeochemical and isotopic characteristics of the Ilica geothermal system (Erzurum, Turkey). *Environmental Earth Sciences*, 72(11), 4451–4462. 10.1007/s12665-014-3345-z

Fjeld, R. A., DeVol, T. A., & Martinez, N. E. (2023). *Quantitative environmental risk analysis for human health*. John Wiley & Sons.

Fletcher, W. R. J. (2015, March/April). Review and refinement of an existing qualitative risk assessment method for application within an ecosystem-based management framework. *ICES Journal of Marine Science*, 72(3), 1043–1056. 10.1093/icesjms/fsu142

Göymen, K., Demiroglu, O. C., & Tülüş, C. (2017). Snow sports tourism development policies in Turkey. In *Lifestyle sports and public policy* (pp. 63–81). Routledge. 10.4324/9781315624402-5

Haeberli, W., Whiteman, C., & Shroder, F. J. (2015) Snow and ice-related hazards, risks, and disasters, A General Framework, Snow and Ice-related Hazards, Risks and Disasters, Elsevier.

Haimes, Y. Y. (2005). *Risk modeling, assessment, and management*. John Wiley & Sons.

Irmak, M. A., Yilmaz, S., Mutlu, E., & Yilmaz, H. (2020). Analysis of different urban spaces on thermal comfort in cold regions: A case from Erzurum. *Theoretical and Applied Climatology*, 141(3-4), 1593–1609. 10.1007/s00704-020-03289-y

Karahan, V., & Aydoğmuş, E. (2023). Risk Analysis and Risk Assessment in Laboratory Studies. *Avrupa Bilim ve Teknoloji Dergisi*, (49), 55–60. 10.31590/ejosat.1260340

Kazan, H., Hobikoğlu, E. H., Karademir, H., Dalyanci, L., & Turguter, Y. (2015). Economic development of ski industry in experimental innovation: Example of Palandöken Turkey and Alps Switzerland. *Procedia: Social and Behavioral Sciences*, 195, 487–492. 10.1016/j.sbspro.2015.06.245

Kurt, T. (2014). Assessment of the Perchertal avalanche in Tyrol, Austria. *Turkish Journal of Earth Sciences*, 23(3), 339–349. 10.3906/yer-1305-13

Kurt, T. (2017) Improvement of technical capacity in avalanche risk management in Turkey.Dissertation, İstanbul University, Institute of Graduate Studies in Science and Engineering, Departmant of Forest Engineering.

Kurt, T. (2022). Avalanche Hazards with Mitigation in Turkey and Qualitative Risk Assessment for Snow Avalanches in Ayder (Rize, NE Turkey) Using Combination of GIS, Remote Sensing Techniques and Field Studies. In Al Saud, M. M. (Ed.), *Applications of Space Techniques on the Natural Hazards in the MENA Region*. Springer. 10.1007/978-3-030-88874-9_23

Landrø, M. (2021). Why is it safe–enough? Decision-making in avalanche terrain. Dissertation, The Artic University of Norway. https://munin.uit.no/handle/10037/22697

Laternser, M., & Schneebeli, M. (2022). Temporal Trend and Spatial Distribution of Avalanche Activity during the Last 50 Years in Switzerland. *Natural Hazards*, 27(3), 201–230. 10.1023/A:1020327312719

Logrosa, G., Mata, M. A., Lachica, Z. P., Estaña, L. M., & Hassall, M. (2022). Integrating risk assessment and decision-making methods in Analyzing the dynamics of COVID-19 epidemics in Davao City, Mindanao Island, Philippines. *Risk Analysis*, 42(1), 105–125. 10.1111/risa.1377934269475

Ngwenya, S., Mashau, N. S., Mhlongo, E. S., Traoré, A. N., & Mudau, A. G. (2023). Health risk management framework for heavy metals and cyanide in Kwekwe city of Zimbabwe: A mixed-method study protocol. *Journal of Health, Population and Nutrition*, 42(1), 26. 10.1186/s41043-023-00367-537013628

Nichols, T. B., Hawley, A. C., Smith, W. R., Wheeler, A. R.III, & McIntosh, S. E. (2018). Avalanche safety practices among backcountry skiers and snowboarders in Jackson Hole in 2016. *Wilderness & Environmental Medicine*, 29(4), 493–498. 10.1016/j.wem.2018.05.00430213711

Özçoban, E. (2019). Türkiye'de Kış Turizmi ve Erciyes Kayak Merkezi Üzerine Bir Değerlendirme (Winter Tourism in Turkey and a Review on Erciyes Ski Resort). *Journal of Tourism & Gastronomy Studies*, 7(3), 1606–1625. 10.21325/jotags.2019.438

Parolly, G. (2020). The Anatolian high-mountain ranges–Plant diversity between two seas. Plant biogeography and vegetation of high mountains of Central and South-West Asia, 215-286.

Pröbstl-Haider, U., Richens, H., & Türk, S. (Eds.). (2019). *Winter tourism: Trends and challenges*. Cabi International. 10.1079/9781786395207.0000

Raihan, A. (2024). A Systematic Review of Geographic Information Systems (GIS) in Agriculture for Evidence-Based Decision Making and Sustainability. *Global Sustainability Research*, 3(1), 1–24. 10.56556/gssr.v3i1.636

Raihan, A. (2024). A Systematic Review of Geographic Information Systems (GIS) in Agriculture for Evidence- Based Decision Making and Sustainability. *Global Sustainability Research*, 3(1), 1–24. 10.56556/gssr.v3i1.636

Remondo, J., Bonachea, J., & Cendrero, A. (2005). A statistical approach to landslide risk modeling at basin scale; from landslide susceptibility to quantitative risk assessment. *Landslides*, 2(4), 321–328. 10.1007/s10346-005-0016-x

Siraj, N. B., & Fayek, A. R. (2019). Risk identification and common risks in construction: Literature review and content analysis. *Journal of Construction Engineering and Management*, 145(9), 03119004. 10.1061/(ASCE)CO.1943-7862.0001685

St Clair, A. (2019). Exploring the effectiveness of avalanche risk communication: A qualitative study of avalanche bulletin use among backcountry recreationists.

Statham, G., Haegeli, P., Greene, E., Birkeland, K., Israelson, C., Tremper, B., Stethem, C., McMahon, B., White, B., & Kelly, J. (2018). A conceptual model of avalanche hazard. *Natural Hazards*, 90(2), 663–691. 10.1007/s11069-017-3070-5

Steiger, R., Scott, D., Abegg, B., Pons, M., & Aall, C. (2019). A critical review of climate change risk for ski tourism. *Current Issues in Tourism*, 22(11), 1343–1379. 10.1080/13683500.2017.1410110

Subaşı, İ. Ö., & Gür, V. (2023). Recreational skiing-and snowboarding-related extremity injuries: A five-year tertiary trauma center cohort. *Cureus*, 15(7).37649954

Tremper, B. (2013). *Avalanche essentials: A step-by-step system for safety and survival.* Mountaineers Books.

Wilhelm, C. (1998, May). Quantitative risk analysis for evaluation of avalanche protection projects. In *Proceedings of the Anniversary Conference* (Vol. 25, pp. 288-293).

Zeydan, Ö., & Sevim, B. (2008). *İklim Değişikliğinin Kış Turizmine Etkileri*. TMMOB İklim Değişimi Sempozyumu Bildiriler Kitabı.

Zweifel, B., & Haegeli, P. (2014). A qualitative analysis of group formation, leadership and decision making in recreation groups traveling in avalanche terrain. *Journal of Outdoor Recreation and Tourism*, 5, 17–26. 10.1016/j.jort.2014.03.001

Chapter 11
The Adoption of a Qualitative Approach in Management Science:
An Exploratory Survey on the Effects of Big Data on the Performance of Companies

Ahmed El Hammoumi
Polydisciplinary Faculty of Taza, Morocco

Nabil Seghyar
Ibno Tofail Kénitra University, Morocco

Mounir Gouiouez
FLD Fès, Morocco

Meryem Amane
FLDM, Morocco

Abdelaziz Berdi
Faculty of Legal, Economic, and Social Sciences, Morocco

ABSTRACT

This chapter aims to analyze the qualitative method in management science. It examines the theoretical foundations of this approach to achieve its epistemological bases. In this perspective, the authors present an attempt to elucidate the underpinnings on which the assumptions of a knowledge production approach around a defined object in the social sciences are built. They also question why management

DOI: 10.4018/979-8-3693-5543-5.ch011

science, which is traditionally focused on quantitative methods, adopts qualitative methods. Based on a literature review, it can be affirmed that qualitative research aims to deeply understand the context of phenomena and the mechanisms underlying them. Thus, proponents of these methodologies are interested in realities embedded in specific contexts to better grasp the complexity of human interactions within the organization and to address certain challenges, notably the management of cultural diversity and the dynamics of virtual teams. The authors illustrate this research with an exploratory study examining the effects of Big Data on the performance of companies in Morocco.

INTRODUCTION

For a long time, qualitative research was synonymous with a form of methodological impressionism: the researcher conducts interviews, keeps a journal of observations, takes notes on meetings attended, and reads documents. In management sciences, qualitative research aims to deeply understand the context of phenomena and the mechanisms underlying them. Proponents of these methodologies focus on realities embedded in specific contexts. In this vein, Giroux (2003, cited by Notais A, 2009) argues that this methodology allows for the description and understanding of a phenomenon in all its complexity, using a comprehensive approach and considering a large number of factors. Moreover, unlike quantitative methods, qualitative methods offer strong internal validity due to the richness of the data collected.

On one hand, qualitative observation provides the researcher with a deeper understanding and more information about the studied phenomenon, allowing for a more comprehensive view of the different causes of a phenomenon and their potential interactions (Miles and Huberman, 2003). On the other hand, this information and data are more rooted in the reality of practices, allowing for the understanding of the role and impact of the social context in which the phenomenon occurs.

Qualitative methods enable an in-depth and contextual analysis of the study object, facilitated by interviews that allow direct contact with actors who possess relevant information for understanding the phenomenon's reality. In the same sense, Yin (2012) and Wacheux (1996) assert that the qualitative case study approach is more suitable for research questions composed of “how” and “why.” This approach enhances understanding of local perceptions, explaining how people, in a particular context, understand, explain, and manage situations daily (Miles and Huberman, 2003). Similarly, Strauss and Corbin (1990) highlight that these methods are justified for problem categories depending on the actors' experiences. Qualitative researchers are particularly interested in behaviors, life stories, social interactions, organizational functioning, or social movements. This approach grants the researcher a significant degree of freedom in carrying out their project.

The objective of this chapter is to explore the adoption of a qualitative approach in management science, highlighting its theoretical foundations, methodology, and practical applications, and illustrating this research with an exploratory study examining the effects of Big Data on the performance of companies in Morocco.

In the context of modern management, decision-makers face increasingly complex challenges, such as managing cultural diversity, the dynamics of virtual work teams, and adapting to rapid changes in the economic environment. These challenges cannot be fully understood through purely quantitative approaches, which tend to simplify reality by reducing phenomena to measurable variables (Frimousse, S., & Peretti, J. M., 2022). Adopting a qualitative approach allows for an in-depth exploration of the contextual, social, and emotional dimensions underlying organizational behaviors and decisions. This chapter aims to:

- Present the theoretical foundations of the qualitative approach in management sciences, highlighting its differences from quantitative approaches.
- Examine the qualitative research methodology, including data collection and analysis processes.
- Illustrate the practical applications of qualitative research in various management fields, such as understanding consumer needs, evaluating organizational practices, and exploring group dynamics.
- Discuss the challenges and limitations associated with using the qualitative approach, as well as strategies to overcome these obstacles and ensure the validity and reliability of results.

In summary, this chapter seeks to encourage critical reflection on the adoption of the qualitative approach in management science, emphasizing its potential to enrich our understanding of organizational phenomena and inform managerial practices.

1. LITERATURE REVIEW

The qualitative approach is of major interest to researchers seeking to deepen their understanding of a given phenomenon. It is necessary to highlight the analysis techniques and the underlying principles guiding this approach across various fields of study, as well as its advantages compared to the quantitative approach. Before examining these different points, it is essential to define this approach within management science.

1.1. Concepts and Trajectory of the Qualitative Approach in Management Sciences

The qualitative approach in management sciences, initially perceived as an impressionistic and exploratory method, has evolved into a rigorous and systematic practice, thereby offering profound and contextualized perspectives on complex organizational phenomena.

1.1.1. Concepts of the Qualitative Approach in Management Sciences

In management science, unlike the quantitative approach which focuses on the quantification and measurement of variables, the qualitative approach focuses on in-depth and contextual exploration. It aims to capture and interpret organizational phenomena by emphasizing a comprehensive and nuanced understanding of social and organizational realities (Denzin and Lincoln, 2018; Creswell and Poth, 2017).

One of the strengths of this methodology is its flexibility, particularly due to its ability to adapt to the realities of the field and to modify the data access strategy throughout the data production process. This makes the qualitative approach especially valuable for researchers seeking a deeper understanding of a given phenomenon. However, by adopting such an approach, the researcher may face certain criticisms related to objectivity, analysis methods, interaction with the field, and the non-representativeness of situations (Wacheux, 1996). Being aware of these limitations, it seems entirely pertinent to adopt this approach to provide answers to our research questions.

Regarding the conduct of research, once this methodological approach is chosen, several implementation options are available, making the choice sometimes difficult. The following table summarizes the different choices of a qualitative method according to Wacheux (1996, p. 90).

Table 1. Main Choices of a Qualitative Method According to Wacheux (1996)

	Objective and Research Question	Implicit Hypotheses and Control	Investigation et example
Case Study	Exploration, Understanding, and Generating Hypotheses on the "Why" and "How"	Researcher's Ability to Maintain Objectivity During Analysis	The Group, the Organization, Interorganizational Relationships
Action Research	Studying Provoked Changes	Researcher's Ability to Record and Analyze in Real-Time	Organizational Development

continued on following page

Table 1. Continued

	Objective and Research Question	Implicit Hypotheses and Control	Investigation et example
Experimental Research	Simple or Contextual Causalism	Ability to Manipulate Variables and Reproduce the Experiment	Individual and Group Behavior

Source: Frédéric Wacheux (1996, p 90).

The qualitative approach can be employed in the field of management sciences to examine various subjects, such as corporate culture, interpersonal relationships, decision-making processes, or employees' personal experiences. According to Miles, Huberman, and Saldaña (2019), this offers researchers the opportunity to deepen their understanding of organizational phenomena, develop new theories, and provide insightful perspectives to guide management practices.

1.1.2. Trajectory of the Qualitative Approach in Management Science

The qualitative approach in management sciences has significantly evolved over the past decades, transitioning from a method often perceived as impressionistic to a rigorous and systematic approach. This evolution reflects the growing acceptance of qualitative research as an important tool for analyzing complex organizational phenomena. However, the qualitative approach has moved from Impressionism to Rigor:

Methodological Impressionism

Initially, qualitative research was often seen as an impressionistic approach, where methods such as interviews, observations, and document analysis were used relatively informally. These early methods aimed to capture the richness of human experiences and social contexts without necessarily adhering to the rigorous standards often associated with quantitative approaches (Denzin & Lincoln, 2018; Geertz, 1973). This phase marked the beginning of the use of qualitative methods in management sciences, emphasizing a deep understanding of contexts rather than the generalization of results.

The Rise of Rigor: Development of Methodologies

Over time, qualitative research gained methodological rigor. The introduction of Grounded Theory by Glaser and Strauss in the 1960s marked a significant turning point, providing a systematic structure to the collection and analysis of qualitative data (Glaser & Strauss, 1967). This approach allows researchers to develop theories directly based on the collected data, rather than testing predetermined hypotheses, which has helped to enhance the scientific credibility of qualitative research (Corbin & Strauss, 2015).

Academic Acceptance and Recognition

In the 1980s and 1990s, qualitative research began to be widely accepted and recognized in the field of management sciences. Researchers like Miles and Huberman demonstrated how qualitative data could be analyzed rigorously and systematically, contributing to a greater academic recognition of this approach (Miles & Huberman, 1994). This period saw an increase in academic publications and specialized journals including qualitative research, highlighting their ability to offer deep and contextual insights.

The Emergence of Case Studies and Mixed Methods

A major evolution in qualitative research was the rise of case studies and mixed methods. Case studies, popularized by Yin, allowed the exploration of complex phenomena in their real contexts, providing rich and detailed perspectives (Yin, 2018). In parallel, mixed methods, which combine qualitative and quantitative approaches, enabled a more comprehensive and nuanced understanding of management phenomena by leveraging the strengths of each method (Creswell & Plano Clark, 2018).

Contemporary Challenges and New Directions

Today, qualitative research continues to evolve to address the contemporary challenges of management sciences. Researchers are exploring new methods and techniques, such as computer-assisted content analysis, to manage increasingly large volumes of qualitative data (Braun & Clarke, 2021). Additionally, there is a focus on data triangulation, enhancing the validity and reliability of results, and integrating stakeholder perspectives into research processes to strengthen the rigor and relevance of qualitative studies (Flick, 2018; Tracy, 2019).

1.2. Data Analysis Techniques: Coding, Categorization, Grounded Theory

In the process of constructing the research project, the adoption of grounded theory as a theoretical approach quickly emerged as the ideal choice, given its experimental and innovative aim, as well as its flexible approach focusing on developing connections with participants.

The central idea of grounded theory is to emerge theoretical frameworks from the material. Coding is the means by which this process of theorizing from the material is elaborated. In principle, grounded theory in its original form excludes any idea of theoretical coding: coding is there to prevent theory from contaminating the analysis of the material. On the contrary, coding is for theoretical purposes (Point & Voynnet-Fourboul, 2006): it is the central tool through which theory will emerge from the material, like Venus from the waves. At the heart of the idea of grounded theory is the awareness of the risk of circularity: if one approaches material with pre-defined theoretical frameworks, then the temptation is to see in the material only what confirms these theoretical frameworks. There is circularity: the material is pre-structured by the theoretical frameworks mobilized, and it is believed that knowledge has been produced because these theoretical frameworks have been "validated" on empirical material. In doing so, one has eliminated everything that could have been a discovery, all the facts that did not "fit" with the theoretical framework. The risk is to deprive oneself of rich elements hidden in the material and which have every chance of remaining so if this approach is adopted. "The little unexplained facts always contain enough to overturn all the explanations of the big facts." Ayache, M., & Dumez, H. (2011)

Similarly, Saldaña (2016) asserts that it is essential to use data analysis methods in qualitative research, such as coding, categorization, and grounded theory, to obtain relevant information from collected data.

Coding: According to Saldaña (2016), coding involves labeling and classifying data into meaningful units. By grouping data into relevant categories or themes, it helps reduce complexity. The coding process is crucial in grounded theory research. The coding process is not linear: in a circular dynamic, it is marked by constant back-and-forth between the field and the data (Paillé, 1994). In an effort to make the approach accessible to all, Paillé (1994) defined six stages of analysis: initial coding, categorization, linking, integration, modeling, and theorization.

Categorization: After coding the data, researchers have the opportunity to group similar codes into broader categories or themes (Charmaz, 2014).

Categorization often involves comparative analysis of codes to identify similarities and disparities among various data units.

Grounded Theory: According to Strauss and Corbin (1998), grounded theory is an inductive method that involves creating theories or models from the data itself. Scientists study patterns and relationships that form in the data to formulate theoretical ideas.

By combining these methods, scientists have the opportunity to deeply analyze qualitative data to identify hidden trends, patterns, and meanings. This method provides a comprehensive and subtle understanding of the analyzed phenomena, paying particular attention to the participants' perspectives and specific social and organizational realities (Alla, Bentalha, Megzari, & Hmioui, 2022; Bentalha, 2023a, 2023b).

1.3. Underlying Principles of Qualitative Research

The foundations of qualitative research provide an essential conceptual framework to guide the design, implementation, and interpretation of qualitative studies. Here is a more in-depth analysis of these concepts:

Holistic Understanding: The aim of qualitative research is to understand the complexity of the analyzed phenomena in their natural environment. According to Denzin & Lincoln (2018), its goal is to grasp the links between various variables and to explore the social, cultural, and historical aspects influencing these phenomena. This method emphasizes a comprehensive and integrated perspective of social realities, thus providing an in-depth analysis of the dynamics and influences of the context.

Subjectivity and Interpretation: Considering the subjectivity of participants and researchers is an essential principle of qualitative research. According to Creswell & Poth (2017), researchers acknowledge that participants' perceptions and interpretations are influenced by their personal experiences and social context. Thus, qualitative research highlights subjective viewpoints and aims to explore the variety of human experiences.

Social Construction of Reality: From a constructivist perspective, the construction of social reality relies on social interactions and communication exchanges (Charmaz, 2014). Qualitative researchers delve into social structures and participants' interpretations of their experiences. They seek to understand how people make sense of their world and how these processes of meaning-making impact their behaviors and interactions.

Role of the Researcher: Researchers actively participate in qualitative research. They establish exchanges with participants, ask open-ended questions, and encourage discussions to collect valuable and detailed information (Creswell & Poth, 2017). They also acknowledge their own subjectivity and consider their impact on the collected data, ensuring reflexive and rigorous analysis.

Emergence and Flexibility: According to Denzin & Lincoln (2018), qualitative research is often exploratory and promotes the emergence of new topics, ideas, and perspectives during the research process. Researchers are willing to be flexible and adapt their methods and theoretical frameworks based on the collected data. This method provides the opportunity to explore the analyzed phenomena in a comprehensive and subtle manner.

Together, these principles constitute a solid conceptual framework for qualitative research, placing particular emphasis on understanding social realities, considering subjectivity and interpretation, and exploring the processes of constructing the social world.

1.4. Advantages of Qualitative Approach Compared to Quantitative Approach

The qualitative approach offers numerous advantages over the quantitative approach and can play an essential role in many research situations.

In-Depth and Contextual Understanding: The qualitative approach provides a thorough and contextual understanding of the analyzed phenomena. According to Denzin & Lincoln (2018), by emphasizing details, participants' perceptions, and interpretations, it offers rich and nuanced perspectives on social and organizational realities.

Exploration of Motivations and Individual Experiences: The qualitative approach allows for the analysis of participants' motivations, beliefs, and individual experiences using techniques such as in-depth interviews and participant observation. This helps understand subtleties and nuances often overlooked by quantitative methods (Creswell & Poth, 2017).

Flexibility and Adaptation to Complex Contexts: Qualitative research is often more flexible and adaptable to complex and ever-changing contexts. Through it, researchers can adapt their methods and approaches to the specific needs of their study, which is particularly beneficial in areas where variables are difficult to measure or define (Charmaz, 2014).

Exploration of New Areas and Perspectives: The use of the qualitative approach is ideal for exploring new research areas and developing new theoretical perspectives. By fostering the emergence of topics, concepts, and theories from initial data, it encourages innovation and inquiry (Denzin & Lincoln, 2018).

Valuing Participant Voices: The qualitative approach emphasizes participants' experiences and viewpoints, highlighting the opinions of individuals involved in the study. This encourages participant independence and promotes more ethical and balanced research (Creswell & Poth, 2017).

By combining these benefits, the qualitative approach constitutes a complementary and often essential approach for research in social sciences and management sciences, offering profound and relevant insights into human and organizational realities.

2. METHODOLOGY OF QUALITATIVE RESEARCH IN MANAGEMENT SCIENCES

2.1. Data Collection Techniques

Data collection in qualitative research is an essential step that allows for gathering valuable and pertinent information to address the study's objectives. Some commonly employed methods during data collection in qualitative research include:

Semi-Structured Interviews: Semi-structured interviews involve in-depth exchanges with participants, with a set of pre-established open-ended questions in mind. According to Fontana and Frey (2005), this approach allows researchers to delve into participants' experiences, opinions, and perspectives on a specific subject.

Participant Observation: The researcher immerses themselves in the study's context to directly observe participants' behaviors, interactions, and social dynamics during participant observation. Through this approach, valuable contextual data can be obtained, and the studied phenomena can be understood in their natural context (Denzin & Lincoln, 2018).

Document Analysis: Document study involves examining written documents such as reports, journals, archives, or publications to collect relevant information for the study. According to Yin (2013), this approach can provide essential historical, institutional, or contextual insights.

Focus Groups: Focus group discussions bring together a small group of participants to discuss a particular topic under the researcher's supervision.

According to Krueger & Casey (2014), this approach encourages idea exchange and facilitates the collection of different perspectives and social interactions among participants.
Life Stories: The collection of participants' personal stories and life experiences is the goal of life stories. Through this approach, individual journeys, significant events, and participants' interpretations of their own stories can be captured (Atkinson, 1998).

These various data collection methods can be used individually or in combination depending on the research objectives and the nature of the phenomena being evaluated. By carefully selecting data collection methods, researchers can obtain a wealth of rich and diversified data that facilitate in-depth and nuanced analysis of social realities and organizations.

2.2. Consideration of Subjectivity and Interpretation in Qualitative Research

The consideration of subjectivity and interpretation is an essential characteristic of qualitative research, as this approach recognizes that researchers' and participants' perceptions and interpretations play a central role in constructing meaning and understanding the phenomena under study. Here's how subjectivity and interpretation are accounted for in qualitative research:

Acknowledgment of Researchers' Subjectivity: Researchers acknowledge that they have their own subjectivity and influence on the research. According to Charmaz (2014), they are aware that their experiences, beliefs, and viewpoints can impact how they collect, analyze, and interpret data. Researchers can adopt a reflexive approach by being aware of their own position and its consequences on the research through this awareness.
Consideration of Participants' Perspectives: Qualitative research highlights participants' viewpoints and acknowledges that their understanding of events and experiences is influenced by their own social, cultural, and historical context. Thus, scientists consider the variety of voices and experiences of participants when analyzing and interpreting data (Creswell & Poth, 2017).
Interpretative Analysis: Data analysis in qualitative research requires an interpretive approach, where researchers study participants' interpretations of their experiences and the phenomena under study. This study emphasizes the creation of meaning through social exchanges and communication (Denzin & Lincoln, 2018).

Validation and Triangulation: To minimize the consequences of subjectivity and interpretation, scientists frequently employ validation and triangulation methods. These techniques involve comparing the results of different data collection methods, involving independent researchers in data analysis, or seeking convergences between different data sources (Creswell & Poth, 2017).

2.3. Practical Challenges of Qualitative Methods

The adoption of qualitative methods in management science presents several practical challenges. This section explores these challenges, supported by case studies and examples, and highlights the implications and limitations in the conclusion chapter.

2.3.1. Data Collection and Access

One of the primary challenges in qualitative research is obtaining access to relevant and comprehensive data. In the context of studying the effects of Big Data on business performance in Morocco, researchers often face difficulties in accessing proprietary data from companies. For instance, a case study of a Moroccan telecommunications company revealed that gaining access to internal data required extensive negotiations and assurances of confidentiality (Miles & Huberman, 1994). The sensitive nature of the data can limit the scope of the research and the depth of insights.

2.3.2. Data Analysis Complexity

Qualitative data, especially when dealing with Big Data, can be vast and complex. Analyzing this data manually is time-consuming and requires significant expertise. For example, a study involving a Moroccan retail chain used computer-assisted qualitative data analysis software (CAQDAS) to manage and analyze customer feedback data from various digital platforms (Creswell & Plano Clark, 2018). Despite the tools available, researchers faced challenges in coding and categorizing data due to its unstructured nature.

2.3.3. Ensuring Validity and Reliability

Ensuring the validity and reliability of qualitative research can be challenging. In the study of Big Data's impact on Moroccan businesses, researchers must ensure that their interpretations of the data are accurate and consistent. Triangulation, where multiple data sources or methods are used to cross-verify findings, is essential but can be difficult to implement. A case study involving multiple Moroccan SMEs showed that discrepancies between different data sources (e.g., interviews vs. digital analytics) required careful reconciliation (Denzin & Lincoln, 2018).

2.3.4. Cultural and Contextual Factors

Cultural and contextual factors can influence the research process and outcomes. For example, in Morocco, business practices and data governance may differ significantly from those in other regions, affecting how Big Data is utilized and perceived. A study on Moroccan financial institutions found that local regulatory frameworks and cultural attitudes towards data privacy significantly impacted the adoption and implementation of Big Data strategies (Geertz, 1973).

2.3.5. Ethical Considerations

Qualitative research often involves ethical considerations, particularly concerning confidentiality and informed consent. When studying the effects of Big Data on business performance, researchers must navigate the ethical implications of accessing and analyzing potentially sensitive business information. A case study involving a Moroccan tech startup highlighted the importance of obtaining informed consent from all stakeholders and ensuring that data is anonymized to protect identities (Tracy, 2019).

2.4. Implications and Limitations

The implications and limitations of adopting a qualitative approach to study the effects of Big Data on business performance in Morocco are significant.

2.4.1. Implications

Deep Contextual Insights: Qualitative methods provide rich, contextual insights into how Moroccan businesses are leveraging Big Data, uncovering nuances that quantitative methods might miss (Braun & Clarke, 2021).

Policy and Strategy Development: The findings can inform policymakers and business leaders on effective strategies for Big Data adoption and utilization, considering the unique Moroccan context (Flick, 2018).

2.4.2. Limitations

Generalizability: The findings from qualitative studies are often not generalizable to a broader population. The specific context of Moroccan businesses

may limit the applicability of the results to other regions or industries (Miles & Huberman, 1994).
Resource Intensive: Qualitative research, especially involving Big Data, is resource-intensive in terms of time, expertise, and technology. This can limit the scope and scale of the study (Creswell & Plano Clark, 2018).
Subjectivity: The interpretative nature of qualitative research can introduce subjectivity, which may affect the reliability of the findings. Ensuring rigorous methodology and triangulation is essential to mitigate this issue (Denzin & Lincoln, 2018).

2.4.3. Contemporary Challenges and New Directions

Today, qualitative research continues to evolve to meet contemporary challenges in management science. Researchers are exploring new methods and techniques, such as computer-assisted content analysis, to manage increasingly large volumes of qualitative data (Braun & Clarke, 2021). Additionally, there is a strong emphasis on data triangulation, improving the validity and reliability of results, and integrating stakeholder perspectives into the research process to enhance the rigor and relevance of qualitative studies (Flick, 2018; Tracy, 2019).

In conclusion, while the adoption of qualitative methods to study the impact of Big Data on business performance in Morocco offers deep and valuable insights, researchers must navigate significant practical challenges. Addressing these challenges through methodological rigor, ethical considerations, and cultural sensitivity is crucial for producing reliable and impactful research outcomes.

3. EMPIRICAL STUDY: APPLICATIONS OF QUALITATIVE RESEARCH IN MANAGEMENT SCIENCES

This research aims to examine how the use of Big Data enhances corporate performance. This formulation paves the way for an in-depth analysis of the mechanisms and impacts of integrating Big Data into companies' operational and decision-making processes, while maintaining a specific focus on the economic, cultural, and social realities of Morocco. We conducted a qualitative study, collecting data through semi-structured interviews with five companies that use Big Data, to understand its effects on their performance.

3.1. Data and Methods

In this empirical section, we delve into the phase of confrontation between theory and practice. The literature review was dedicated to exploring the scientific literature around the two central themes of our research: the use of Big Data and business performance. We presented the data in a structured manner to facilitate our analysis in the field and establish a comprehensive interview guide covering all aspects of Big Data usage and performance.

3.1.1. Sampling

Our research question explores how the use of Big Data influences performance. We explored the concrete measures that companies using Big Data can take to integrate management practices favoring creativity, innovation, and consequently performance.

To respond as exhaustively as possible to our problematic, our sample is diversified, consisting of IT managers and big data analysts from Moroccan companies. The table below provides an overview of the various interviews.

Table 2. Summary of Interviews Conducted in the Exploratory Phase

Companies	Position	Industry	Duration
Managem	It Manger, Big Data analysts	Mining	35min
Soft Group	Chief Information Officer	Software Development	25min
Cosumar	Financial Manager	Sugar Industry	30min
Ynna Holding	Director of Human Resources	Diverse activities	35min
Attijariwafa bank	Financial analyst	Banking Sector	25min
Attijari Global Research	Financial market analyst	financial sector	25min
National Transport and Logistics Company (NTLC)	Warehouse manager	Transport	30min
Sonasid	Supply Chain Responsible	Steel industry	40min
National Motorway Company of Morocco.	Human Resources Manager	Infrastructure sector	45min
Marjane	Management controller	Commerce and Distribution	40min

3.1.2. Data Collection Method

To conduct our research effectively, we opted for a qualitative analysis. Indeed, the topic revolves around analyzing the effect of using Big Data on performance, which is difficult to measure. Starting in October, I began researching companies using Big Data in Morocco through the internet, social networks, networks of accounting firms and consulting firms, and direct contacts with some officials through our relational network. I sent requests to about ten companies and around twenty individuals. I received a significant number of no responses, and only a few people agreed to answer my questions. This was one of the difficulties encountered during this research.

For the scientific research, we will first establish an interview guide, which will help us conduct our semi-directed individual interviews effectively. However, the semi-directed interview constitutes the main source of data for our practical investigation. In order to demonstrate the importance of this data source, Royer, C. et al. (2009) specify that "the semi-directed or semi-guided interview is certainly the most used in social research. It is semi-directed in the sense that it is neither entirely open nor guided by a large number of specific questions. Generally, the researcher has a series of relatively open guide questions, about which it is imperative that they receive information from the interview. But they do not necessarily ask all the questions in the order they noted them and in the planned formulation. As much as possible, they will let the interviewee speak openly, in the words they wish, and in the order that suits them. The researcher will simply try to refocus the interview on the objectives each time it deviates and to ask the questions that the interviewee does not come up with for themselves, at the most appropriate time and in as natural a manner as possible" MOULAI, S., & BOUKRIF, M. (2018).

A series of meetings with officials of Moroccan companies using Big Data constituted the heart of our empirical base. In this initial exploratory phase, we conducted semi-directed interviews with officials and executives of companies operating in Morocco and using Big Data. However, semi-directed interviews are a qualitative research method used to study social relationships and organizational processes. They involve the use of a set of predefined open questions that are posed to participants, but also allow for a more open exploration of participants' ideas and experiences (Wollesen, B. et al., 2020).

When it comes to understanding the link between Big Data usage and performance, semi-directed interviews can be particularly useful, as they allow researchers to gather detailed information on participants' perceptions and experiences regarding leadership and organizational culture. Semi-directed questions can be adapted to explore different aspects of companies using Big Data, such as data-driven decision-making, performance and productivity management, process and operations optimization, product and service customization, and innovation and development of new products (Bentalha, B., Hmioui, A., & Lhoussaine, A. L. L. A., 2020).

By using this research method, researchers can obtain rich and contextualized qualitative data that can be analyzed to identify emerging themes and patterns, as well as to understand the nuances and complexities of new information and communication technologies.

3.2. Interview Results

The results of the interviews have been compiled into a table. We have listed the categories of questions we addressed with our interviewees. The objective is to compare their responses and highlight similarities.

Since our research question aimed to understand how the use of Big Data by companies in Morocco enhances their performance, it was evident that my questions revolved around three areas: performance and productivity management, process and operations optimization, product and service customization, and innovation and development of new products.

We initiated the interview with an introduction of the interviewee. Then, we explored various topics, starting with understanding Big Data and performance. We then addressed flexibility, the use of advanced technologies, operational efficiency, employee productivity, innovation, quality, profitability, customer satisfaction, market share, and revenue and turnover. We also discussed the company's performance in terms of corporate social responsibility (CSR) and environmental sustainability.

3.3. Exploitation of Interviews

Based on the collected data, we will first seek out similarities between the different responses across various themes:

3.3.1. Understanding of Big Data and Performance

During the interviews, it is evident that all interviewees have a good grasp of the relationship between Big Data and performance in the context of modern businesses. However, Big Data involves collecting, storing, and analyzing vast datasets from various sources. By analyzing large datasets, companies can identify inefficiencies and optimization opportunities in their business processes. This can lead to improvements in productivity, operational efficiency, and overall profitability. Big Data provides companies with factual information and deep insights that can inform strategic decision-making. By using customer data, companies can personalize their offerings and services to meet the specific needs of each customer. This can enhance the customer experience, strengthen brand loyalty, and drive sales. In summary, it can be concluded that Big Data can play a crucial role in enhancing overall business performance by providing actionable insights, optimizing business processes, fostering innovation, and improving the customer experience.

3.3.2. Revenue and Turnover

To assess the health and viability and measure the business performance of a company, it is important to measure companies' performance through their revenues and turnover. However, the use of Big Data enables companies to generate revenue and improve their turnover. By effectively utilizing Big Data, companies can enhance their ability to measure and maximize their financial performance by tracking their revenues, total turnover, and sales growth. By harnessing the information and insights generated by Big Data, companies can make more informed and strategic decisions to achieve their financial goals.

Regarding the companies studied, the interviewees indicate that Big Data allows them to deeply analyze customer purchasing behaviors, preferences, and trends. By analyzing this data, they can better understand their customers and adjust their sales strategies to maximize revenue and turnover. Big Data also enables them to personalize their offerings based on specific customer needs and preferences. This personalization can increase sales by enhancing the appeal of products and services to customers. Additionally, by analyzing large datasets from external sources such as social media, online forums, etc., they can detect emerging market trends. This allows them to anticipate customer demand and react quickly to seize growth opportunities.

3.3.3. Profitability and Operational Efficiency

Profitability refers to a company's ability to generate profits relative to its costs and investments. This can include measures such as gross profit margin, net profit margin, return on investment (ROI), etc. Big Data can have a significant impact on companies' profitability. The interviewees indicate that the use of Big Data enables them to optimize operational processes by analyzing large amounts of data on operational processes. Big Data allows them to identify inefficiencies and optimization opportunities. By improving operational efficiency, they can reduce costs and increase profitability. Indeed, Big Data offers them the opportunity to leverage massive amounts of data to make more informed and strategic decisions. By optimizing operational processes, personalizing offerings, optimizing pricing, managing risks, and forecasting future trends, they can improve profitability and competitiveness in the market.

3.3.4. Quality and Customer Satisfaction

Big Data can play a crucial role in improving product and service quality as well as increasing customer satisfaction. In our research, interviewees indicate that Big Data enables them to analyze data from customer feedback, online reviews, satisfaction surveys, and social media to understand customer needs, preferences, and complaints. This analysis helps identify areas for improvement and take corrective actions to enhance product and service quality. Additionally, Big Data allows them to analyze data from sensors and connected devices integrated into products, enabling real-time monitoring of product performance and reliability. This helps identify potential issues before they become major problems, thereby improving quality and customer satisfaction.

3.3.5. Market Share and Employee Productivity

Big Data can have a significant impact on a company's market share and employee productivity. Indeed, Big Data enables the analysis of competition and market trends, optimization of business processes, training and development of personnel, and assessment of their performance before rewarding them. Concerning the interviewees in our study, they indicate that Big Data enables them to improve their companies' market share by identifying growth opportunities, personalizing offerings to meet customer needs, and optimizing business processes. Additionally, it helps contribute to increasing employee productivity by identifying necessary training and development areas, evaluating performance, and rewarding contributions.

3.3.6. Innovation and Corporate Social Responsibility

Big Data can play an important role in promoting innovation and corporate social responsibility (CSR). It can analyze customer data; personalize products and services to meet the specific needs and preferences of each customer (Alla, L., Jaouhari, E. L., & Bentalha, B., 2022). This personalization can foster innovation by offering unique solutions tailored to individual customer needs. It can also analyze data related to sustainability, environmental footprint, ethical business practices, etc., and assess companies' performance in terms of corporate social responsibility (CSR). This analysis can identify areas for improvement and guide efforts to integrate sustainable and socially responsible business practices. For our research, interviewees indicate that Big Data fosters innovation by providing valuable insights for the development of new products, services, and solutions, as well as improving the customer experience. Additionally, it can support corporate social responsibility by assessing sustainability performance, promoting transparency, and guiding efforts to integrate responsible business practices throughout the organization.

3.4. Discussion

While the results demonstrate the significant impact of big data on various aspects of business performance, the discussion will separate the observations and implications for a clearer understanding.

3.4.1. Understanding of Big Data and Performance

The findings confirm that big data is well understood by business leaders and is seen as a pivotal tool in enhancing business performance. The ability to analyze large datasets to identify inefficiencies and optimize business processes is a recurring theme. This demonstrates the practical application of big data in strategic decision-making and operational improvements.

3.4.2. Revenue and Turnover

Big data's role in generating revenue and increasing turnover is evident from the respondents' insights. By understanding customer behaviors and preferences through data analysis, companies can tailor their strategies to maximize financial outcomes. This personalization and strategic alignment highlight big data's transformative potential in financial performance.

3.4.3. Profitability and Operational Efficiency

Respondents consistently noted the impact of big data on profitability and operational efficiency. The ability to identify inefficiencies and optimize processes through data analysis underscores big data's role in enhancing cost management and profitability. This aligns with existing literature on the strategic benefits of big data (Miles & Huberman, 1994; Braun & Clarke, 2021).

3.4.4. Quality and Customer Satisfaction

The use of big data to improve product and service quality and increase customer satisfaction was a significant observation. By analyzing customer feedback and real-time performance data, companies can proactively address issues and enhance the customer experience. This proactive approach is crucial for maintaining high quality and customer satisfaction levels.

3.4.5. Market Share and Employee Productivity

The analysis of market trends and employee performance through big data is seen as a strategic advantage. Respondents highlighted how big data helps in identifying growth opportunities and optimizing employee productivity. This dual benefit of market and internal efficiency enhancement supports the broader strategic importance of big data.

3.4.6. Innovation and Corporate Social Responsibility

Big data's role in fostering innovation and supporting CSR initiatives was evident. By leveraging data to understand customer needs and sustainability metrics, companies can innovate and align their practices with CSR goals. This demonstrates big data's broader impact beyond immediate business metrics.

CONCLUSION

The adoption of a qualitative approach in management science, illustrated by an exploratory study on the effects of Big Data on the performance of companies in Morocco, highlights the importance of understanding organizational phenomena in all their complexity and contextual specificity.

Qualitative methods allow for a deep immersion in the study context, thus offering a nuanced understanding of internal dynamics and human interactions that often escape quantitative methodologies. Particularly, within the context of Moroccan companies, qualitative exploration of the impacts of Big Data reveals valuable insights into how these technologies influence organizational performance, decision-making, and competitiveness.

The study demonstrates that Big Data, as a strategic tool, not only transforms operational processes but also organizational structures and cultures. The qualitative data collected shows that the effective integration of Big Data relies on a thorough understanding of the organization's specific needs, available skills, and unique cultural and technical challenges for each company.

In conclusion, this research underscores the added value of qualitative approaches in management, especially in studying complex and contextually rooted phenomena such as the impact of Big Data. These approaches offer an enriched and detailed perspective, essential for developing adapted and effective management strategies. For researchers and practitioners in management science, the integration of qualitative methods represents a complementary and indispensable approach for understanding the reality of contemporary businesses, particularly in the face of rapid technological innovations and growing global challenges.

REFERENCES

Alla, L., Bentalha, B., Megzari, G., & Hmioui, A. (2022). Les effets du COVID-19 sur les chaines logistiques: Une analyse qualitative. Reconfigurations du tourisme en contexte de la crise du Covid-19: Quelles démarches pour quelles résiliences au Maroc et ailleurs?, 1(1), 130-153.

Alla, L., Jaouhari, E. L., & Bentalha, B. (2022). Repenser l'entrepreneuriat au sein du secteur informel au Maroc à l'ère du post Covid-19: Pour approche prospective de la formalisation intégrative. *Alternatives Managériales et Economiques*, 4(1), 180–202.

Atkinson, R. (1998). *The life story interview* (Vol. 44). Sage. 10.4135/9781412986205

Ayache, M., & Dumez, H. (2011). Le codage dans la recherche qualitative une nouvelle perspective?. *Le libellio d'Aegis, 7*(2-Eté), 33-46.

Bentalha, B., Hmioui, A., & Lhoussaine, A. L. L. A. (2020). La performance des entreprises de services: Un cadrage théorique d'un concept évolutif. *Alternatives Managériales Economiques*, 2(1), 58–78.

Bentalha, B. (2023). Consumer Perception of Robotic Mobile Fulfillment Systems: A Netnographic Case Study of Amazon. In Innovation, Strategy, and Transformation Frameworks for the Modern Enterprise (pp. 243-264). IGI Global. 10.4018/979-8-3693-0458-7.ch010

Bentalha, B. (2023). A Framework for Exploring the Global Performance of Service Supply Chain Management: A Qualitative Study. Government Impact on Sustainable and Responsible Supply Chain Management, 106-129.

Braun, V., & Clarke, V. (2021). *Thematic Analysis: A Practical Guide*. SAGE Publications.

Charmaz, K. (2014). Constructing grounded theory (introducing qualitative methods series). *Constr. grounded theory*.

Corbin, J., & Strauss, A. (2015). *Basics of Qualitative Research: Techniques and Procedures for Developing Grounded Theory*. Sage Publications.

Creswell, J. W., & Plano Clark, V. L. (2018). *Designing and Conducting Mixed Methods Research*. SAGE Publications.

Creswell, J. W., & Poth, C. N. (2017). *Qualitative inquiry and research design: Choosing among five approaches*. Sage Publications.

Denzin, N. K., & Lincoln, Y. S. (2018). *The SAGE Handbook of Qualitative Research*. SAGE Publications.

Flick, U. (2018). *An Introduction to Qualitative Research*. SAGE Publications.

Fontana, A., & Frey, J. H. (2005). The interview: From structured questions to negotiated text. In Denzin, N. K., & Lincoln, Y. S. (Eds.), *The SAGE handbook of qualitative research* (pp. 695–727).

Frimousse, S., & Peretti, J. M. (2022). Pour des organisations post-Covid inclusives. *Question(s) de management, 38*(1), 187-262.

Geertz, C. (1973). *The Interpretation of Cultures: Selected Essays*. Basic Books.

Glaser, B. G., & Strauss, A. L. (1967). *The Discovery of Grounded Theory: Strategies for Qualitative Research*. Aldine Publishing.

Krueger, R. A., & Casey, M. A. (2014). *Focus groups: A practical guide for applied research*. Sage Publications.

Miles, M. B., & Huberman, A. M. (1994). *Qualitative Data Analysis: An Expanded Sourcebook*. SAGE Publications.

Miles, M. B., & Huberman, A. M. (2003). *Analyse des données qualitatives*. De Boeck Supérieur.

Miles, M. B., Huberman, A. M., & Saldaña, J. (2019). *Qualitative data analysis: A methods sourcebook*. Sage Publications.

Moulai, S., & Boukrif, M. (2018). Le choix du successeur dans les PME familiales. *Revue africaine de management, 1*(3).

Paillé, P. (1994). L'analyse par théorisation ancrée. *Cahiers de Recherche Sociologique*, (23), 147–181. 10.7202/1002253ar

Point, S., & Fourboul, C. V. (2006). Le codage à visée théorique. *Recherche et Applications en Marketing*, 21(4), 61–78. 10.1177/076737010602100404

Royer, C., Baribeau, C., & Duchesne, A. (2009). Les entretiens individuels dans la recherche en sciences sociales au Québec: Où en sommes-nous? Un panorama des usages. *Recherches Qualitatives*, 7, 64–79.

Saldaña, J. (2021). The coding manual for qualitative researchers.

Strauss, A., & Corbin, J. (1990). *Basics of qualitative research*. Sage publications.

Strauss, A., & Corbin, J. (1998). *Basics of qualitative research: Techniques and procedures for developing grounded theory*. Sage Publications.

Tracy, S. J. (2019). *Qualitative Research Methods: Collecting Evidence, Crafting Analysis, Communicating Impact*. Wiley-Blackwell.

Wacheux, F. (1996). Méthodes qualitatives de recherches en gestion.

Wollesen, B., Wildbredt, A., van Schooten, K. S., Lim, M. L., & Delbaere, K. (2020). The effects of cognitive-motor training interventions on executive functions in older people: A systematic review and meta-analysis. *European Review of Aging and Physical Activity*, 17(1), 1–22. 10.1186/s11556-020-00240-y32636957

Yin, R. K. (2013). *Case study research: Design and methods*. Sage Publications.

Yin, R. K. (2018). *Case Study Research and Applications: Design and Methods*. SAGE Publications.

APPENDIX

Table 3. Presentation of Various Companies in the Sample

Companies	Presentation
Managem	Managem, a Moroccan mining and metals company, has been a pioneer in digital transformation since 1930. It uses advanced technologies like artificial intelligence and automation to optimize processes, enhance productivity, reduce costs, and ensure agile management of activities.
Soft Group	Soft Group, a leading provider of customized software solutions, is a pioneer in digitalization and New Methods of Operations Management (NMMO), enhancing operational efficiency and competitiveness by optimizing processes, automating repetitive tasks, and utilizing data for informed decision-making.
Cosumar	Cosumar, a pioneering Moroccan sugar company, has grown significantly since 1929 and is utilizing digital transformation to enhance product quality, production processes, and meet international standards, playing a significant role in Morocco's sugar supply.
Ynna Holding	Ynna Holding, a Moroccan conglomerate, is a leader in various sectors, integrating digitalization into all aspects of its operations, including supply chain, customer relations, and human resources management, to optimize processes and increase operational efficiency.
Attijariwafa Bank	Attijariwafa Bank, Morocco's top bank, merges Wafabank and Banque Commerciale du Maroc, serving 10.6 million clients worldwide. Emphasizing modernization and digital solutions.
Attijari Global Research	Attijari Global Research is a research company based in the MENA (Middle East and North Africa) region specializing in financial analysis, investment, and wealth management. Founded on strong principles of research and analysis, Attijari Global Research strives to provide valuable insights and informed advice to its clients, which often include financial institutions, asset management companies, institutional investors, and businesses. This company utilizes big data to analyze financial markets, forecast economic trends, and provide strategic advice to investors.
National Transport and Logistics Company (NTLC)	The National Company of Transport and Logistics (SNTL) is a company specialized in transportation and logistics, based in Morocco. With a solid experience in the field, SNTL offers a comprehensive range of transportation and logistics services tailored to the needs of its clients, both nationally and internationally. SNTL utilizes big data to optimize transportation routes, manage warehouses more efficiently, and track shipments in real-time.
National Motorway Company of Morocco.	The Moroccan National Highway Company (ADM) is a public enterprise responsible for the design, construction, operation, and maintenance of the country's highway network. Its main mission is to develop and operate the Moroccan highway network, thus contributing to improving connectivity, mobility, and economic development in the country. It may be involved in the planning and implementation of new highway network extension projects to accommodate the country's economic and demographic growth. ADM utilizes big data to analyze traffic patterns, forecast congestions, enhance road safety, and plan investments in road infrastructure.
Sonasid	Sonasid, or the National Company of Iron and Steel, is a Moroccan company specialized in the production and marketing of steel products. Founded in 1962, Sonasid has become one of the leading companies in the steel industry in North Africa, playing a crucial role in Morocco's industrial development. Sonasid could utilize big data to optimize its production processes, enhance supply chain management, anticipate market demand, and maintain product quality.

continued on following page

Table 3. Continued

Companies	Presentation
Marjane	Marjane is a Moroccan retail company, part of the Moroccan group ONA (Omnium Nord-Africain). Founded in 1990, Marjane has quickly become one of the leading supermarket and hypermarket chains in Morocco, offering a wide range of food and non-food products at competitive prices. Marjane utilizes big data to optimize inventory management, personalize offers for customers, analyze purchasing trends, and improve operational efficiency.

Compilation of References

Aamoum, H., & Soufiane, N. (2021). L'étude de la divulgation volontaire d'informations à propos du capital immatériel à la lumière de la théorie de l'agence et la théorie du signal : Quels déterminants pour quels objectifs? The study of the voluntary disclosure of information about intangible capital in the light of agency theory and signal theory: Which determinants for which objectives? 2, 2658-8455. 10.5281/zenodo.5531367

Abarar, I., Elhamma, A., & Razgani, H. (2014). De l'observation participante comme méthode de recherché. *Revue Marocaine des Sciences de Management*, 3, 72–84.

Abdolmohammadi, M., Simnett, R., Thibodeau, J. C., & Wright, A. M. (2006). Sell-Side Analysts' Reports and the Current External Reporting Model. *Accounting Horizons*, 20(4), 375–389. 10.2308/acch.2006.20.4.375

Abdul Rahman, R., & Alsayegh, M. F. (2021). Determinants of corporate environment, social and governance (ESG) reporting among Asian firms. *Journal of Risk and Financial Management*, 14(4), 167. 10.3390/jrfm14040167

Ackerly, B., & True, J. (2010). *Doing feminist research in political and social science*. Palgrave Macmillan. 10.1007/978-1-137-05442-5

Adams, C. A. (2004). The ethical, social and environmental reporting-performance portrayal gap. *Accounting, Auditing & Accountability Journal*, 17(5), 731–757. 10.1108/09513570410567791

Adams, C., & van Manen, M. (2008). Phenomenology. In Given, L. M. (Ed.), *The SAGE Encyclopedia of Qualitative Research Methods* (pp. 615–620). Sage Publications.

Aftiss, A. (2019). L'entrepreneuriat au Maroc : Les contraintes et les conditions de développement. *Revue de l'Entrepreneuriat et de l'Innovation*, II(8), 1–9.

Aggarwal, P., & Singh, A. K. (2019). CSR and sustainability reporting practices in India : An in-depth content analysis of top-listed companies. *Social Responsibility Journal*, 15(8), 1033–1053. 10.1108/SRJ-03-2018-0078

Ajemba, M. N., Arene, E. C., Ajemba, M. N., & Arene, E. C. (2022). Research gaps for future research and their identification. *World Journal of Advanced Research and Reviews*, 16(1), 1. Advance online publication. 10.30574/wjarr.2022.16.1.1062

Akkermans, J., Brenninkmeijer, V., Huibers, M., & Blonk, R. W. B. (2013). Competencies for the Contemporary Career : Development and Preliminary Validation of the Career Competencies Questionnaire. *Journal of Career Development*, 40(3), 245–267. 10.1177/0894845312467501

Al Shammre, A. S., Alshebami, A. S., Ali Seraj, A. H., Elshaer, I. A., & Al Marri, S. H. (2023). Unleashing environmental performance: The impact of green entrepreneurial motivation on small enterprises. *Frontiers in Environmental Science*, 11(April), 1–12. 10.3389/fenvs.2023.1176804

Alla, L., Bentalha, B., Megzari, G., & Hmioui, A. (2022). Les effets du COVID-19 sur les chaines logistiques: Une analyse qualitative. Reconfigurations du tourisme en contexte de la crise du Covid-19: Quelles démarches pour quelles résiliences au Maroc et ailleurs?, 1(1), 130-153.

Alla, L., Bentalha, B., Megzari, G., & Hmioui, A. (2022). The effects of COVID-19 on supply chains: A qualitative analysis. Reconfigurations du tourisme en contexte de la crise du Covid-19: Quelles démarches pour quelles résiliences au Maroc et ailleurs, 1(1), 130-153.

Alla, L., El Jaouhari, S., & Bentalha, B. (2022). Repenser l'entrepreneuriat au sein du secteur informel au Maroc à l'ère du post Covid-19 : Pour approche prospective de la formalisation intégrative. *Alternatives Managériales et Economiques, 4*(1), 180–202. 10.48374/IMIST.PRSM/ame-v4i1.30472

Alla, L., Hmioui, A., & Bentalha, B. (2020). La netnographie dans les recherches marketing: La communauté virtuelle comme consom'acteur vecteur d'efficacité marketing. *Alternatives Managériales Economiques*, 2(4), 631–652.

Alla, L., Jaouhari, E. L., & Bentalha, B. (2022). Repenser l'entrepreneuriat au sein du secteur informel au Maroc à l'ère du post Covid-19 : Pour approche prospective de la formalisation intégrative. *Alternatives Managériales et Economiques*, 4(1), 180–202.

Alla, L., Jaouhari, E. L., & Bentalha, B. (2022). Repenser l'entrepreneuriat au sein du secteur informel au Maroc à l'ère du post Covid-19: Pour approche prospective de la formalisation intégrative. *Alternatives Managériales et Economiques*, 4(1), 180–202.

Allard-Poesi, F., & Maréchal, G. (2014). Construction de l'objet de la recherche. In T. R.-A. et al (Éd.), Méthodes de recherche en management (p. 48-76). Dunod. https://hal.science/hal-01123768

Al-Tuwaijri, S. A., Christensen, T. E., & Hughes, K. E.II. (2004). The relations among environmental disclosure, environmental performance, and economic performance : A simultaneous equations approach. *Accounting, Organizations and Society*, 29(5), 447–471. 10.1016/S0361-3682(03)00032-1

Amaeshi, K., Adi, A. B. C., Ogbechie, C., & Amao, O. O. (2006). Corporate Social Responsibility in Nigeria : Western Mimicry or Indigenous Influences? (SSRN Scholarly Paper 896500). 10.2139/ssrn.896500

Amir, E., Lev, B., & Sougiannis, T. (2003). Do financial analysts get intangibles? *European Accounting Review*, 12(4), 635–659. Advance online publication. 10.1080/0963818032000141879

Anghel, G. A., & Anghel, M. A. (2022). Green Entrepreneurship among Students—Social and Behavioral Motivation. *Sustainability (Basel)*, 14(14), 1–19. 10.3390/su14148730

Argyris, C. (1985). *Action Science*. Jossey-Bass.

Aria, M., & Cuccurullo, C. (2017). bibliometrix : An R-tool for comprehensive science mapping analysis. *Journal of Informetrics*, 11(4), 959–975. 10.1016/j.joi.2017.08.007

Asdiou, A., & Mokhtari, B. (2019). The Indicators of the Corporate Social Responsibility : A Search for Efficiency or a Concern for Legitimacy Towards Stakeholders? A Proposed Explanatory Model. [IJSSMET]. *International Journal of Service Science, Management, Engineering, and Technology*, 10(1), 62–76. 10.4018/IJSSMET.2019010105

Ashforth, B. E., & Gibbs, B. W. (1990). The Double-Edge of Organizational Legitimation. *Organization Science*, 1(2), 177–194. 10.1287/orsc.1.2.177

Aspers, P., & Corte, U. (2019). *What is Qualitative in Qualitative Research*. https://link.springer.com/article/10.1007/s11133-019-9413-7

Aspers, P., & Corte, U. (2019). What is qualitative in qualitative research. *Qualitative Sociology*, 42(2), 139–160. 10.1007/s11133-019-9413-731105362

Asselineau, A., Combaudon, C., & Fouré-Joopen, H. (2022). In Validation scientifiqued'une nouvelle approche du management par la recherche-intervention:Cas de 12 entreprises « libérées ». Recherche en Sciences de Gestion-ManagementSciences-Ciencias de Gestión, 150, 31-55.

Atkinson, R. (1998). *The life story interview* (Vol. 44). Sage. 10.4135/9781412986205

Attanasio, O., Kugler, A., & Meghir, C. (2015). Subsidizing vocational training for disadvantaged youth in Colombia: Evidence from a randomized experiment (NBER Working Paper No. 21390). National Bureau of Economic Research.

Avenier, M.-J. (2010). Shaping a Constructivist View of Organizational Design Science. *Organization Studies*, 31(9-10), 1229–1255. 10.1177/0170840610374395

Ayache, M., & Dumez, H. (2011). Le codage dans la recherche qualitative une nouvelle perspective?. *Le libellio d'Aegis, 7*(2-Eté), 33-46.

Aygün Oğur, A., & Baycan, T. (2023). Assessing climate change impacts on tourism demand in Turkey. *Environment, Development and Sustainability*, 25(3), 2905–2935. 10.1007/s10668-022-02135-7

Azam, O., Zulfiqar, H., & Lillah, Z. I. (2023). An Ethnographic Study on Female Entrepreneurship in Patriarchal Structure. *Motivation and Challenges An Ethnographic Study on Female Entrepreneurship in Patriarchal Structure : Motivation and Challenges*, XIII(April), 108–118.

Aziz, O. E., & Asdiou, A. (2024). Les déterminants de la communication extra-financière : Cas des entreprises Marocaines cotées à la bourse de valeurs mobilières de Casablanca. Alternatives Managériales Economiques, 6(1), Article 1.

Baethge C, Goldbeck-Wood S, Mertens S. (2019). SANRA-a scale for the quality assessment of narrative review articles. Res Integr Peer Rev. 26;4:5. .10.1186/s41073-019-0064-8

Bajpai, N. (2018). Business Research Methods. *Pearson Education*, 362. https://www.amazon.com/Business-Research-Methods-2nd-Edition/dp/1741032539

Baker, L., Phelan, S., Snelgrove, R., Varpio, L., Maggi, J., & Ng, S. (2016). Recognizing and Responding to Ethically Important Moments in Qualitative Research. *Journal of Graduate Medical Education*, 8(4), 607–608. 10.4300/JGME-D-16-00384.127777677

Balmer, D. F., Varpio, L., Bennett, D., & Teunissen, P. W. (2021). Longitudinal qualitative research in medical education : Time to conceptualise time. *Medical Education*, 55(11), 1253–1260. 10.1111/medu.1454233847408

Bansal, P. (2005). Evolving sustainably : A longitudinal study of corporate sustainable development. https://onlinelibrary.wiley.com/doi/abs/10.1002/smj.441

Bardin. (1977). L'Analyse du contenu, https://www.persee.fr/doc/colan_0336-1500_1977_num_35_1_4449

Barratt, M., Choi, T. Y., & Li, M. (2011). Qualitative case studies in operations management : Trends, research outcomes, and future research implications. *Journal of Operations Management*, 29(4), 329–342. 10.1016/j.jom.2010.06.002

Bassen, A., & Kovács, A. M. (2008). Environmental, Social and Governance Key Performance Indicators from a Capital Market Perspective.

Beck, S., Bercovitz, J., Bergenholtz, C., Brasseur, T.-M., D'Este, P., Dorn, A., Doser, M., Dosi, C., Effert, A., Furtuna, R., Goodyear, M., Grimpe, C., Haeussler, C., Hans, F., Heinisch, B., Katona, N., Kleinberger-Pierer, H., Kokshagina, O., LaFlamme, M., & Zyontz, S. (2021). Experimenting with Open Innovation in Science (OIS) practices : A novel approach to co-developing research proposals. *CERN ideaSquare Journal of Experimental Innovation*, 5(2), 28–49. 10.23726/cij.2021.1328

Behzadifar, M., Ghanbari, M. K., Azari, S., Bakhtiari, A., Rahimi, S., Ehsanzadeh, S. J., Sharafkhani, N., Moridi, S., & Bragazzi, N. L. (2023). A SWOT analysis of the development of health technology assessment in Iran. *PLoS One*, 18(3), e0283663. 10.1371/journal.pone.028366336996128

Bellaire, S., Jamieson, B., & Stetham, G. (2013) Relating Avalanche Activity to Climate Change and Coupled Ocean-Atmospheric Phenomena. Davos Atmosphere and Cryosphere Assembly DACA-13, Davos, Switzerland.

Bellanca, S., & Colot, O. (2014). Motivations et typologie d'entrepreneurs. *La Revue Des Sciences de Gestion*, 266(2), 63. 10.3917/rsg.266.0063

Benoteau, I. (2015). Quels effets du recrutement en contrat aidé sur la trajectoire professionnelle ? Document d'études de la DARES, 192.

Benoteau, I. (2015). The effects of subsidized contracts on the career trajectories of employees facing difficulties in accessing employment: Evidence from propensity score matching. *Economie & Statistique*, 477(1), 167–198. 10.3406/estat.2015.10589

Bentalha, B. (2023). A Framework for Exploring the Global Performance of Service Supply Chain Management: A Qualitative Study. Government Impact on Sustainable and Responsible Supply Chain Management, 106-129.

Bentalha, B. (2023a). Influence des actions sociales de la RSE sur les valeurs perçues des employés: Étude qualitative des œuvres sociales de l'entreprise tajmouati. Repères et Perspectives Economiques, 7(2).

Bentalha, B. (2023b). A Framework for Exploring the Global Performance of Service Supply Chain Management: A Qualitative Study. Government Impact on Sustainable and Responsible Supply Chain Management, 106-129.

Bentalha, B. (2023b). Consumer Perception of Robotic Mobile Fulfillment Systems: A Netnographic Case Study of Amazon. In Innovation, Strategy, and Transformation Frameworks for the Modern Enterprise (pp. 243-264). IGI Global. 10.4018/979-8-3693-0458-7.ch010

Bentalha, B. (2023c). A Framework for Exploring the Global Performance of Service Supply Chain Management: A Qualitative Study. Government Impact on Sustainable and Responsible Supply Chain Management, 106-129.

Bentalha, B. (2023d). Influence des actions sociales de la RSE sur les valeurs perçues des employés: Étude qualitative des œuvres sociales de l'entreprise Taajmouati. Repères et Perspectives Economiques, 7(2).

Bentalha, B. (2024a). Contribution des randonnées au développement d'un tourisme durable dans la région Fès Meknès. Numérisation et durabilité, Vers une économie responsable et innovante, 226-245.

Bentalha, B. (2024b). Employees and Robots in Amazon's Robotic Mobile Fulfillment Systems: A Netnographic Analysis of a Supply Chain Transformation. In Blockchain Applications for Smart Contract Technologies (pp. 188-207). IGI Global. 10.4018/979-8-3693-1511-8.ch009

Bentalha, B., & Hmioui, A. (2022). *Smart Service Supply Chain and Just Walk Out Technology : A Netnographic Approach* (p. 223-236). 10.1007/978-3-030-94191-8_18

Bentalha, B. (2022). Smart Technologies for a Sustainable Service Supply Chain: A Prospective Perspective. *Journal of Environmental Issues and Climate Change*, 1(1), 77–89. 10.59110/jeicc.v1i1.81

Bentalha, B. (2023). Influence des actions sociales de la RSE sur les valeurs perçues des employés : Étude qualitative des œuvres sociales de l'entreprise Taajmouati. *Repères et Perspectives Economiques, 7*(2), 183–204. 10.34874/IMIST.PRSM/RPE/43586

Bentalha, B. (2023a). A Framework for Exploring the Global Performance of Service Supply Chain Management : A Qualitative Study. In *Government Impact on Sustainable and Responsible Supply Chain Management* (pp. 106–129). IGI Global., 10.4018/978-1-6684-9062-4.ch007

Bentalha, B. (2023a). Les fondements théoriques du supply chain management: Une analyse critique. *Alternatives Managériales Economiques*, 5(2), 1–22.

Bentalha, B., Hmioui, A., & Lhoussaine, A. L. L. A. (2020). La performance des entreprises de services: Un cadrage théorique d'un concept évolutif. *Alternatives Managériales Economiques*, 2(1), 58–78.

Bentalha, B., & Hmioui, A. (2021). Smart service supply chain and Just Walk Out technology: a netnographic approach. In *The Proceedings of the International Conference on Smart City Applications* (pp. 223-236). Cham: Springer International Publishing.

Bentalha, B., Hmioui, A., & Alla, L. (2020). La performance des entreprises de services : Un cadrage théorique d'un concept évolutif. *Alternatives Managériales et Economiques*, 2, 58–78.

Bentalha, B., Hmioui, A., & Alla, L. (Eds.). (2023). *Integrating Intelligence and Sustainability in Supply Chains*. IGI Global. 10.4018/979-8-3693-0225-5

Berger, P. L., & Luckmann, T. (1966). *The social construction of reality: A treatise in the sociology of knowledge*. Anchor.

Bergson, H. (2014). *Time and free will : An essay on the immediate data of consciousness* (Routledge, T.Francis Group, Trans.). F. L. Pogson. 10.4324/9781315830254

Bernard, C. (1865). *An Introduction to the Study of Experimental Medicine* (Green, H. C., Trans.). Dover Publications.

Bernardi, L. (2021). Qualitative longitudinal research in family sociology. In Schneider, N. F., & Kreyenfeld, M. (Eds.), *Research Handbook on the Sociology of the Family*. Edward Elgar Publishing., 10.4337/9781788975544.00015

Berrone, P., & Gomez-Mejia, L. R. (2009). Environmental performance and executive compensation : An integrated agency-institutional perspective. *Academy of Management Journal*, 52(1), 103–126. 10.5465/amj.2009.36461950

Blanchet, A., & Gotman, A. (1992). *L'enquête et ses méthodes : L'entretien*. Nathan.

Blossfeld, H.-P., & Hofmeister, H. A. (Eds.). (2008). *Globalization, uncertainty and women's careers : An international comparison* (Paperback ed). Edward Elgar.

Boatright, J. R. (2014). Ethics and the conduct of business (7. ed., Pearson new internat. ed). Pearson.

Boddy, C. R. (2016). Sample size for qualitative research. *Qualitative Market Research*, 19(4), 426–432. 10.1108/QMR-06-2016-0053

Bong, W. K., & Chen, W. (2024). Increasing faculty's competence in digital accessibility for inclusive education : A systematic literature review. *International Journal of Inclusive Education*, 28(2), 197–213. 10.1080/13603116.2021.1937344

Borg, W. R. (1963). *Educational research: An introduction.* David McKay Company, Inc.

Borman, W. C., Ilgen, D. R., & Klimoski, R. J. (2003). *Handbook of psychology, volume twelve : Industrial and organizational psychology* (New York:).

Bouhtati, N., El Yaacoubi, Y., Bentalha, B., & Lhoussaine, A. L. L. A. (2024). Leviers logistiques de promotion du commerce équitable des produits de terroir: Une esquisse de modèle d'analyse. Alternatives Managériales Economiques, 6(Spécial 1), 17-35.

Bourner, T., & Brook, C. (2019). Comparing and contrasting action research and action learning. In Mertler, C. A. (Ed.), *The Wiley Handbook of Action Research in Education* (pp. 185–206). Wiley-Blackwell. 10.1002/9781119399490.ch9

Bradbury, H. (2015). *The SAGE Handbook of Action Research.* SAGE Publications. 10.4135/9781473921290

Brammer, S., & Pavelin, S. (2004). Voluntary social and environmental disclosures by large UK companies. *Business Ethics (Oxford, England)*, 23(2/3), 86–99. 10.1111/j.1467-8608.2004.00356.x

Brannen, J. (2006). Cultures of Intergenerational Transmission in Four-Generation Families. *The Sociological Review*, 54(1), 133–154. 10.1111/j.1467-954X.2006.00605.x

Braun, V., & Clarke, V. (2006). Using thematic analysis in psychology. *Qualitative Research in Psychology*, 3(2), 77–101. 10.1191/1478088706qp063oa

Braun, V., & Clarke, V. (2021). *Thematic Analysis: A Practical Guide.* SAGE Publications.

Brewer, J. D. (2000). *Ethnography.* Open University Press.

Broadus, R. N. (1987). Toward a definition of "bibliometrics". *Scientometrics*, 12(5), 373–379. 10.1007/BF02016680

Broecke, S. (2012). Decomposing Wage Inequality in Tunisia. African Development Bank (AfDB).

Bründl, M., Bartelt, P., Schweizer, J., Keiler, M., & Glade, T. (2010). Review and future challenges in snow avalanche risk analysis. In Alcántara-Ayala, I., & Goudie, A. (Eds.), *Geomorphological hazards and disaster prevention* (pp. 49–61). Cambridge University Press. 10.1017/CBO9780511807527.005

Bruyat, C., & Julien, P. A. (2001). Defining the field of research in entrepreneurship. *Journal of Business Venturing*, 16(2), 165–180. 10.1016/S0883-9026(99)00043-9

Bryman, A. (2016). *Social research methods*. Oxford University Press.

Buchanan, D. A., Boddy, D., & McCalman, J. (1988). Getting. In *Getting Out, Getting Back: the art of the possible. In A. Bryman (dir.), Doing Research inOrganisations*. Routledge.

Buffington, C., Foster, L., Jarmin, R. S., & Ohlmacher, S. (2016). *The Management and Organizational Practices Survey (MOPS) : An Overview* (SSRN Scholarly Paper 2793858). 10.2139/ssrn.2793858

Buono, A., Savall, H., & Cappelletti, L. (dir.) (2018). Intervention-Research: from-Conceptualization to Publication. Charlotte: IAP.

Burnard, K. J. (2023). Developing a robust case study protocol. *Management Research Review*, 47(2), 204–225. 10.1108/MRR-11-2021-0821

Burney, V., Arnold-Saritepe, A., & McCann, C. M. (2023). Rethinking the place of qualitative methods in behavior analysis. *Perspectives on Behavior Science*, 46(1), 185–200. 10.1007/s40614-022-00362-x37006599

Busse, C., & August, E. (2021). How to Write and Publish a Research Paper for a Peer-Reviewed Journal. *Journal of Cancer Education*, 36(5), 909–913. 10.1007/s13187-020-01751-z32356250

Cahuc, P., Carcillo, S., & Zylberberg, A. (2014). *Labor Economics*. MIT Press.

Campbell, J. L. (2007). Why would corporations behave in socially responsible ways? An institutional theory of corporate social responsibility. *Academy of Management Review*, 32(3), 946–967. 10.5465/amr.2007.25275684

Campbell, J., Dunnette, M., Lawler, E., & Weick, K. (1970). *Managerial Behavior Performance And Effectiveness*.

Campos, M., Peixoto, F., Bártolo-Ribeiro, R., & Almeida, L. S. (2022). Adapting as I Go : An Analysis of the Relationship between Academic Expectations, Self-Efficacy, and Adaptation to Higher Education. *Education Sciences*, 12(10), 10. Advance online publication. 10.3390/educsci12100658

Cappelletti, L. (2009). Performing an internal control function to sustain SOX 404and improve risk management: Evidence from Europe. Management Account-ingQuaterly, été, 10(4), 17-27.

Cappelletti, L., Voyant, O., Savall, H., & Noguera, F. (2018). 40 Years of Socio-Economic Approach of Management (SEAM): What We Know and Where WeGo? Academy of Management Annual Meeting, Chicago.

Cappelletti, L. (2010). *La recherche-intervention en contrôle de gestion. Actes duCongrès de l'AFC*. CD-Rom.

Cappelletti, L., & Baker, R. (2010). Measuring and developing human capital througha pragmatic action research: A French case study. *Action Research*, 8(2), 211–232. 10.1177/1476750309349976

Cappelletti, L., Heurtel, P., & Lefebvre, P. (Dir.). (2022). Crise de la connaissance etconnaissance de la crise. Les points de vue du CNAM. E. *Ms*.

Carsrud, A., and M. B. (2009). *Understanding the Entrepreneurial Mind: Opening the Black Box* (H. Springer (ed.)).

Cassell, C., & Bishop, V. (2019). Qualitative Data Analysis : Exploring Themes, Metaphors and Stories. *European Management Review*, 16(1), 195–207. 10.1111/emre.12176

Cavalcanti, M. F. R. (2017). Guidelines for Qualitative Research in Organization Studies : Controversy and Possibilities. *Administração: Ensino e Pesquisa*, 18(3), 457–488. 10.13058/raep.2017.v18n3.522

Cengiz, Ö. (2022). Analysis of risk factors in fishing activities in Lake Erçek (Van, Turkey) for occupational health and safety. *Momona Ethiopian Journal of Science*, 14(2), 139–148. 10.4314/mejs.v14i2.3

CESE. (2016). L'auto-entreprise, levier de développement et d'intégration du secteur informel. Auto - Saisine n°27.

CESE. (2021). *Une approche intégrée pour résorber l'économie informelle au MarocConseil Economique Social e Environnemental*.

Chakroun, R., Matoussi, H., & Mbirki, S. (2017). Determinants of CSR disclosure of Tunisian listed banks : A multi-support analysis. *Social Responsibility Journal,* 13(3), 552–584. 10.1108/SRJ-04-2016-0055

Charmaz, K. (2006). Constructing Grounded Theory : A Practical Guide Through Qualitative Analysis. In *Introducing Qualitative Methods* (Vol. 1).

Charmaz, K. (2014). Constructing grounded theory (introducing qualitative methods series). *Constr. grounded theory.*

Charmaz, K. (2014). *Constructing Grounded Theory.* https://uk.sagepub.com/en-gb/eur/constructing-grounded-theory/book235960

Charmaz, K. (2014). *Constructing grounded theory* (2nd ed.). Sage Publications.

Charreire, S., & Huault, I. (2001). Le constructivisme dans la pratique de re- cherche : Une évaluation à partir de seize thèses de doctorat.

Château-Terrisse, P., Codello, P., Béji-Bécheur, A., Jougleux, M., Chevrier,S. et-Vandangeon-Derumez, I. (2016). Réflexivité et éthique du chercheur dans laconduite d'une recherche-intervention. La revue des sciences de gestion, 277, 45-56.

Chelly, A. (2004). *L'identification et développement d'opportunités entrepreneuriales.*

Chen, C., J., Patten, D. M., & Roberts, R. W. (. (2008). Corporate Charitable Contributions : A Corporate Social Performance or Legitimacy Strategy? *Journal of Business Ethics*, 82(1), 131–144. 10.1007/s10551-007-9567-1

Cherkaoui, A. & Belgaid, A. (2018). Effectivité des pratiques RSE communiquées par les entreprises marocaines au prisme de la gestion de la diversité: enseignements d'une étude de cas. Revue internationale des sciences de l'organisation, 6, 141-181. 10.3917/riso.006.0141

Cherkaoui, A. & El Aabar, A. (2022). Contribution à l'étude d'impact des pratiques RSE des entreprises au Maroc: vers une création de valeur durable et partagée – Cas de l'OCP. Question(s) de management, 39, 13-40. 10.3917/qdm.219.0013

Cherkaoui, A. (2016). Pratiques RSE des PME au Maroc : Une analyse perceptuelle auprès des dirigeants casablancais. Question(s) de management, 14(3), 13-26. 10.3917/qdm.163.0013

Cherkaoui, A., & Belgaid, A. (2021). Les nouveaux défis des managers de proximité face à la décentralisation de la fonction RH au Maroc – Cas d'un groupe financier panafricain. *Question(s) de management, 32*(2), 43-68. 10.3917/qdm.212.0043

Cherkaoui, A., & Jallal, R. (2017). Spécificités des pratiques RH socialement responsables au contexte des PME au Maroc : Les enseignements d'une étude de cas. *Question(s) de management, 18*(3), 63-77. 10.3917/qdm.173.0063

Cherkaoui, A. (2020). Du reporting de la performance globale à la stratégie de reporting ESG : Pratiques des entreprises faisant appel public à l'épargne au Maroc. *Journal Of Social Science and Organization Management*, 1(1), 1. Advance online publication. 10.48434/IMIST.PRSM/jossom-v1i1.22348

Chiapello, E., & Gilbert, P. (2019). *Management Tools. A Social Science Perspective*. Cambridge University Press. 10.1017/9781108553858

Chigbu, U. E. (2019). Visually Hypothesising in Scientific Paper Writing:Confirming and Refuting Qualitative Research Hypotheses Using Diagrams. *Publications / MDPI*, 7(22), 22. Advance online publication. 10.3390/publications7010022

Chigbu, U. E., Atiku, S. O., & Du Plessis, C. C. (2023). The Science of Literature Reviews : Searching, Identifying, Selecting, and Synthesising. *Publications/MDPI*, 11(1), 1. Advance online publication. 10.3390/publications11010002

Clandinin, D. J., & Connelly, F. M. (2000). *Narrative Inquiry: Experience and Story in Qualitative Research*. Jossey-Bass.

Clarkson, P. M., Overell, M. B., & Chapple, L. (2011). Environmental Reporting and its Relation to Corporate Environmental Performance. *Abacus*, 47(1), 27–60. 10.1111/j.1467-6281.2011.00330.x

Cobb, P., Confrey, J., Disessa, A., Lehrer, R., & Schauble, L. (2003). Design Experiments in Educational Research. *Educational Researcher*, 32(1), 9–13. 10.3102/0013189X032001009

Cobo, M., López-Herrera, A., Herrera-Viedma, E., & Herrera, F. (2011). Science mapping software tools : Review, analysis, and cooperative study among tools. *Journal of the American Society for Information Science and Technology*, 62(7), 1382–1402. 10.1002/asi.21525

Cohen, M. W., & Palmer, G. R. (2004). Project risk identification and management. AACE International Transactions, IN11.

Cohen, J. R., Krishnamoorthy, G., & Wright, A. M. (2000). Evidence on the Effect of Financial and Nonfinancial Trends on Analytical Review. *Auditing*, 19(1), 27–48. 10.2308/aud.2000.19.1.27

Coleman, M. E., & Marks, H. M. (1999). Qualitative and quantitative risk assessment. *Food Control*, 10(4-5), 289–297. 10.1016/S0956-7135(99)00052-3

Condon, L.A. & Coulson, N.S. (2017). Designing and Delivering Interventions for Health Behavior Change in Adolescents Using Multitechnology Systems: From Identification of Target Behaviors to Implementation. *Behavior Change Research and Theory Psychological and Technological Perspectives*, 27-45. 10.1016/B978-0-12-802690-8.00001-3

Conrad, P. (2001). Health Research, Qualitative, in International Encyclopedia of the Social &. *Behavioral Science*, 6608–6612. Advance online publication. 10.1016/B0-08-043076-7/03904-8

Cooke, A., Smith, D., & Booth, A. (2012). Beyond PICO : The SPIDER tool for qualitative evidence synthesis. *Qualitative Health Research*, 22(10), 1435–1443. 10.1177/1049732312452938822829486

Corbin, J., & Strauss, A. (2015). Basics of qualitative research (4e éd.). Sage.

Corbin, J., & Strauss, A. (2015). *Basics of Qualitative Research: Techniques and Procedures for Developing Grounded Theory*. Sage Publications.

Costa, E. D. F., Cardoso, M. D. I., Kich, J. D., & Corbellini, L. G. (2017). Application Of Qualitative Risk Assessment To Prioritize Hazards In Pork Products In Brazil. In International Symposium On The Epidemiology And Control Of Biological, Chemical And Physical Hazards In Pig And Pork, 12., 2017, Foz Do Iguaçu. Proceedings Book. Concórdia: Embrapa Suínos E Aves.

Cowen, S. S., Ferreri, L. B., & Parker, L. D. (1987). The impact of corporate characteristics on social responsibility disclosure : A typology and frequency-based analysis. *Accounting, Organizations and Society*, 12(2), 111–122. 10.1016/0361-3682(87)90001-8

Crépon, B., Duflo, E., Gurgand, M., Rathelot, R., & Zamora, P. (2014). Do labor market policies have displacement effects: Evidence from a clustered randomized experiment. *The Quarterly Journal of Economics*, 128(2), 531–580. 10.1093/qje/qjt001

Creswell, J. W. (2007). *Qualitative inquiry and research design : Choosing among five approaches, 2nd ed* (p. xvii, 395). Sage Publications, Inc.

Creswell, J. W. (2008). *Research Design : Qualitative, Quantitative, and Mixed Methods Approaches* (3e édition). SAGE Publications Inc.

Creswell, J. W., & Poth, C. N. (2018). Qualitative inquiry and research design (4e éd.). Sage.

Creswell, J. W., & Clark, V. L. P. (2017). *Designing and conducting mixed methods research*. Sage publications.

Creswell, J. W., & Creswell, J. D. (2018). *Research design: Qualitative, quantitative, and mixed methods approaches* (5th ed.). Sage Publications.

Creswell, J. W., & Plano Clark, V. L. (2018). *Designing and Conducting Mixed Methods Research*. SAGE Publications.

Creswell, J. W., & Poth, C. N. (2016). Qualitative inquiry and research design: Choosing among five approaches. *Sage (Atlanta, Ga.)*.

Creswell, J. W., & Poth, C. N. (2017). *Qualitative inquiry and research design: Choosing among five approaches*. Sage Publications.

Cristina García Vivar, Anne McQueen, & Navidad Canga. (2011). *Qualitative research proposal : A model to help novice researchers*.

Cuervo-Cazurra, A., & Aguilera, R. (2004). *The Worldwide Diffusion of Codes of Good Governance*. 10.1093/acprof:oso/9780199269761.003.0015

Cunliffe, A. L. (2003). Intersubjective voices: The role of the "theorist". *Administration & Society*, 46(4), 504–533.

da Silva, O. D. L., & de Sousa, Á. S. T. (2016). Gestion de la qualité dans la recherche en sciences sociales et humaines : articulation entre les méthodes qualitatives et quantitatives. *Recherches Qualitatives, 18*, 171–181. http://www.recherche-qualitative.qc.ca/Revue.html

da Silva, J. J., & Silva Cirani, C. B. (2020). The capability of organizational innovation : Systematic review of literature and research proposals. *Gestão & Produção*, 27(4), e4819. Advance online publication. 10.1590/0104-530x4819-20

Danjou, I. (2002). l'entrepreneuriat : Un champ fertile à la recherche de son Unité. *Revue Française de Gestion*, 28(138), 109–125.

David, A. (1999). Intervention methodologies in management research. Track:Collaborative Management / Research Approaches.

De Massis, A., & Kotlar, J. (2014). The case study method in family business research : Guidelines for qualitative scholarship. *Journal of Family Business Strategy*, 5(1), 15–29. 10.1016/j.jfbs.2014.01.007

Deb, D., Dey, R., & Balas, V. E. (2019). Literature Review and Technical Reading. In Deb, D., Dey, R., & Balas, V. E. (Eds.), *Engineering Research Methodology : A Practical Insight for Researchers* (pp. 9–21). Springer. 10.1007/978-981-13-2947-0_2

Deci & Ryan. (1985). intrinsic motivation and self- determinaton in human behaviour. In *Plenum Press* (Springer S).

Deci, E. L., & Ryan, R. M. (1985). The general causality orientations scale: Self-determination in personality. *Journal of Research in Personality*, 19(2), 109–134. 10.1016/0092-6566(85)90023-6

Deci, E. L., & Ryan, R. M. (2000). The "What" and "Why" of Goal Pursuits: Human Needs and the Self-Determination of Behavior. *Psychological Inquiry*, 11(4), 227–268. 10.1207/S15327965PLI1104_01

Deegan, C. (2002). The Legitimising Effect of Social and Environmental Disclosures – A Theoretical Foundation. *Accounting, Auditing & Accountability Journal*, 15, 282–311. 10.1108/09513570210435852

Del Río-Rama, M. de la C., Álvarez-García, J., Mun, N. K., & Durán-Sánchez, A. (2021). Influence of the Quality Perceived of Service of a Higher Education Center on the Loyalty of Students. *Frontiers in Psychology*, 12, 671407. Advance online publication. 10.3389/fpsyg.2021.67140734163410

DeMunck, V. C., & Sobo, E. J. (1998). *Using methods in the field: a practical introduction and casebook*. AltaMira Press.

Dentchev, N. (2004). Corporate Social Performance as a Business Strategy. *Journal of Business Ethics*, 55(4), 395–410. 10.1007/s10551-004-1348-5

Denzin, N. K., & Lincoln, Y. S. (2018). The Sage handbook of qualitative research (5e éd.). Sage.

Denzin, N., & Lincoln, Y. (2000). Grounded Theory : Objectivist and Constructivist Methods. In *Handbook of Qualitative Research* (p. 509-535).

Denzin, N. K. (2008). Strategies of Qualitative Inquiry. *Sage (Atlanta, Ga.)*.

Denzin, N. K., & Lincoln, Y. S. (2011). *The SAGE handbook of qualitative research*. Sage Publications.

Denzin, N. K., & Lincoln, Y. S. (2018). *The SAGE Handbook of Qualitative Research*. SAGE Publications.

Derntl, M. (2014). Basics of research paper writing and publishing. *International Journal of Technology Enhanced Learning*, 6(2), 105–123. 10.1504/IJTEL.2014.066856

Derrington, M. L. (2019). *Qualitative Longitudinal Methods : Researching Implementation and Change*. SAGE Publications, Inc., 10.4135/9781071814277

Derviş, H. (2020). Bibliometric Analysis using Bibliometrix an R Package. *Journal of Scientometric Research*, 8(3), 156–160. 10.5530/jscires.8.3.32

DeWalt, K. M., & DeWalt, B. R. (2002). *Participant observation: a guide for field-workers*. AltaMira Press.

Dibi, A. (2022). Le pilotage de la performance globale des établissements éducatifs: Des pistes pour faire mieux avec peu. *ACCRA*, 15(3), 61–91. 10.3917/accra.015.0061

DiMaggio, P. J., & Powell, W. W. (1983). The Iron Cage Revisited : Institutional Isomorphism and Collective Rationality in Organizational Fields. *American Sociological Review*, 48(2), 147. 10.2307/2095101

Ding, F., & Curtis, F. (2021). 'I feel lost and somehow messy' : A narrative inquiry into the identity struggle of a first-year university student. *Higher Education Research & Development*, 40(6), 1146–1160. 10.1080/07294360.2020.1804333

Diop, P. M. (2012). *La motivation entrepreneuriale dans le contexte sub-saharien francophone*.

Dissanayake, D., Tilt, C., & Xydias-Lobo, M. (2016). Sustainability reporting by publicly listed companies in Sri Lanka. *Journal of Cleaner Production*, 129, 169–182. 10.1016/j.jclepro.2016.04.086

Dodgson, J. E. (2019). Reflexivity in Qualitative Research. *Journal of Human Lactation*, 35(2), 220–222. 10.1177/089033441983099030849272

Dollhausen, K., Lattke, S., Scheliga, F., Andrä, W., Spexard-Heinemann, A., Geffers, J., & Banscherus, U. (2013). *Opening Higher Education to Adults. Final Report.*

Donaldson, T., & Preston, L. E. (1995). The Stakeholder Theory of the Corporation : Concepts, Evidence, and Implications | Academy of Management Review. https://journals.aom.org/doi/abs/10.5465/amr.1995.9503271992

Donath Olomi, B. R. (2001). Evolution Of Entrepreneurial Motivation: The Transition From Economic Necessity to Entrepreneurship. *Business Management Review*, 7(2), 107–136.

Donthu, N., Kumar, S., Mukherjee, D., Pandey, N., & Lim, W. M. (2021). How to conduct a bibliometric analysis : An overview and guidelines. *Journal of Business Research*, 133, 285–296. 10.1016/j.jbusres.2021.04.070

DPT. (2020) Doğu Anadolu Projesi Ana Planı, Türkiye Cumhuriyeti Başbakanlık Devlet Planlama Teşkilatı Müşteşarlığı, Ankara, 2000. https://www.dap.gov.tr/content/DAP/DAPANAPLAN/PDF/ana_plan.pdf

Drews, C. (2015). *Understanding Motivations for Entrepreneurship A Review of Recent Research Evidence Ute Stephan, Mark Hart and Cord-. February*, 1–54. 10.13140/RG.2.1.3343.2165

Dul, J., & Hak, T. (2008). *Case Study Methodology in Business Research.*

Dunn, W. N. (2011). *Public Policy Analysis. Pearson* (5th ed.). Graduate School of Public and International Affairs.

Dvorak, Z., Rehak, D., David, A., & Cekerevac, Z. (2020). Qualitative approach to environmental risk assessment in transport. *International Journal of Environmental Research and Public Health*, 17(15), 5494. 10.3390/ijerph1715549432751437

Dye, R. A. (2001). An evaluation of « essays on disclosure » and the disclosure literature in accounting. *Journal of Accounting and Economics*, 32(1-3), 181–235. 10.1016/S0165-4101(01)00024-6

Dyer, W. G., & Wilkins, A. L. (1991). Better Stories, Not Better Constructs, to Generate Better Theory : A Rejoinder to Eisenhardt. *Academy of Management Review*, 16(3), 613–619. 10.2307/258920

Eccles, R. G., & Krzus, M. P. (2010). *One Report : Integrated Reporting for a Sustainable Strategy*. John Wiley & Sons.

Eckert, N., Keylock, C. J., Bertrand, D., Parent, E., Faug, T., Favier, P., & Naaim, M. (2012). Quantitative risk and optimal design approaches in the snow avalanche field: Review and extensions. *Cold Regions Science and Technology*, 79, 1–19. 10.1016/j.coldregions.2012.03.003

Edmondson, A. C., & Mcmanus, S. E. (2007). Methodological fit in management field research. *Academy of Management Review*, 32(4), 1155–1179. 10.5465/amr.2007.26586086

Eisenhardt, K. M. (1989). Building Theories from Case Study Research. *Academy of Management Review*, 14(4), 532–550. 10.2307/258557

Eisenhardt, K. M., & Graebner, M. E. (2007). Theory building from cases : Opportunities and challenges. *Academy of Management Journal*, 50(1), 25–32. 10.5465/amj.2007.24160888

Eisenhardt, K. M., Graebner, M. E., & Sonenshein, S. (2016). Grand Challenges and Inductive Methods : Rigor Without Rigor Mortis. *Academy of Management Journal*, 59(4), 1113–1123. 10.5465/amj.2016.4004

El Rhaz, L., & Bouzineb, Y. (2021). Le secteur informel au Maroc: Principales caractéristiques et tendances générales d'évolution. *Les Brefs Du Plan*, 212(0), 1–5.

Elder, G. H., Johnson, M. K., & Crosnoe, R. (2003). The Emergence and Development of Life Course Theory. In Mortimer, J. T., & Shanahan, M. J. (Eds.), *Handbook of the Life Course* (pp. 3–19). Springer US., 10.1007/978-0-306-48247-2_1

Elhamma, A. (2012). The Relationship between Firm Size, Activity Based Costing and Performance: An Application on Moroccan Enterprises. *Journal of Accounting, Business & Management,* 19(1).

Elhamma, A. (2023). Impact of mandatory IFRS adoption on foreign direct investment: the moderating role of conflict of interest regulation. *Journal of Financial Reporting and Accounting*. 10.1108/JFRA-04-2022-0145

Elhamma, A. (2015). The relationship between budgetary evaluation, firm size and performance. *Journal of Management Development*, 34(8), 973–986. 10.1108/JMD-06-2014-0053

Elhamma, A. (2024). Determinants of national IFRS adoption: Evidence from the Middle East and North Africa region. *International Journal of Accounting. Auditing and Performance Evaluation*, 20(1-2), 69–90. 10.1504/IJAAPE.2024.135535

Elhamma, A., & Zhang, Y. F. (2013). The relationship between activity based costing, business strategy and performance in Moroccan enterprises. *Accounting and Management Information Systems*, 12(1), 22–38.

Elkhder, E. E., Al-Zubair, B. T., Mohammed, E.-R. H., & Onia, S. I. (2022). Application of Knowledge Management in The Sudanese Universities : An Analytical Study. *Indonesian Journal of Education and Social Studies*, 1(1), 1. Advance online publication. 10.33650/ijess.v1i1.3411

Estay, C., Durrieu, F., & Diop, P. M. (2012). Motivation entrepreneuriale et logique d'action du créateur. *Revue Internationale P.M.E (Norwalk, Conn.)*, 24(1), 135–165. 10.7202/1012554ar

Evren, S. (2023). Bir Cumhuriyet Projesi Olarak Türkiye'de Kış Turizmi. Anatolia. *Turizm Araştırmaları Dergisi*, 34(3), 163–175. 10.17123/atad.1307923

Evren, S., & Kozak, N. (2018). Competitive positioning of winter tourism destinations: A comparative analysis of demand and supply sides perspectives–Cases from Turkey. *Journal of Destination Marketing & Management*, 9, 247–257. 10.1016/j.jdmm.2018.01.009

Fang Cai, M. W. (2008). A Counterfactual Analysis on Unlimited Surplus Labor in Rural China. *China & World Economy*.

Farquhar, J., Michels, N., & Robson, J. (2020). Triangulation in industrial qualitative case study research : Widening the scope. *Industrial Marketing Management*, 87, 160–170. 10.1016/j.indmarman.2020.02.001

Farrat, O., & Hajji, Z. (2022). Contribution à l'analyse des déterminants de la divulgation des informations RSE sur les sites web au Maroc : Cas des organismes financiers cotés dans la bourse de Casablanca. 3(1).

Farrell, E. (2020). Researching Lived Experience in Education : Misunderstood or Missed Opportunity? *International Journal of Qualitative Methods*, 19, 160940692094206. 10.1177/1609406920942066

Fellenz, M. R. (2006). Toward Fairness in Assessing Student Groupwork : A Protocol for Peer Evaluation of Individual Contributions. *Journal of Management Education*, 30(4), 570–591. 10.1177/1052562906286713

Fenouillet, F. (2009). Chapitre 16. Vers une approche intégrative des théories de la motivation. In *Traité de psychologie de la motivation* (pp. 305–338). 10.3917/dunod.carre.2019.01.0305

Fernando, A., & Pandey, I. (2012). Corporate social responsibility reporting : A survey of listed Sri Lankan companies. *Journal for International Business and Entrepreneurship Development*, 6(2), 172–187. 10.1504/JIBED.2012.048569

Filion, L. J. (1997). Le champ de l ' entrepreneuriat : historique, évolution, tendances Le champ de l ' entrepreneuriat : historique, évolution, tendances. *Revue Internationale P.M.E. Économie et Gestion de La Petite et Moyenne Entreprise, 10*(2), 129–172. https://id.erudit.org/iderudit/1009026ar

Fine, G. A. (2001). Participant Observation. in International Encyclopedia of the Social &. *Behavioral Science*, 2001, 11073–11078. 10.1016/B0-08-043076-7/00771-3

Firat Ersoy, A., & Çalik Sönmez, S. (2014). Hydrogeochemical and isotopic characteristics of the Ilica geothermal system (Erzurum, Turkey). *Environmental Earth Sciences*, 72(11), 4451–4462. 10.1007/s12665-014-3345-z

Fjeld, R. A., DeVol, T. A., & Martinez, N. E. (2023). *Quantitative environmental risk analysis for human health*. John Wiley & Sons.

Flagstad, I., & Johnsen, S. Å. K. (2022). The psychology of green entrepreneurship: Founder-driven development of green climate in small-scale companies. *Cogent Business and Management*, 9(1), 2079245. Advance online publication. 10.1080/23311975.2022.2079245

Fletcher, W. R. J. (2015, March/April). Review and refinement of an existing qualitative risk assessment method for application within an ecosystem-based management framework. *ICES Journal of Marine Science*, 72(3), 1043–1056. 10.1093/icesjms/fsu142

Flick, U. (2011). Designing Qualitative Research. *Sage (Atlanta, Ga.).*

Flick, U. (2018). *An introduction to qualitative research* (6th ed.). Sage Publications.

Flick, U. (2018). *An Introduction to Qualitative Research.* SAGE Publications.

Flyvbjerg, B. (2006). Five Misunderstandings About Case-Study Research. *Qualitative Inquiry*, 12(2), 219–245. Advance online publication. 10.1177/1077800405284363

Fortanier, F., Kolk, A., & Pinkse, J. (2011). Harmonization in CSR reporting : MNEs and global CSR standards. *MIR. Management International Review*, 51(5), 665–696. 10.1007/s11575-011-0089-9

Franzoni, C., & Stephan, P. (2023). Last Word on the Discussion regarding the paper : "Uncertainty and risk-taking in science : Meaning, measurement and management in peer review of research proposals". *Research Policy*, 52(3), 104710. Advance online publication. 10.1016/j.respol.2022.104710

Freeman, R. E. (2010). *Strategic Management : A Stakeholder Approach.* Cambridge University Press. 10.1017/CBO9781139192675

Frimousse, S., & Peretti, J. M. (2022). Pour des organisations post-Covid inclusives. *Question(s) de management, 38*(1), 187-262.

Gabarret, I., & Vedel, B. (2015). Pour une nouvelle approche de la motivation entrepreneuriale. *La Revue Des Sciences de Gestion*, 271(1), 13. 10.3917/rsg.271.0013

Galdas, P. (2017). *Revisiting Bias in Qualitative Research : Reflections on Its Relationship With Funding and Impact.* https://journals.sagepub.com/doi/10.1177/1609406917748992

Gallais, T. L. (2008). Wherever I go there I am: Reflections on reflexivity and the research stance. *Reflective Practice*, 9(2), 145–155. 10.1080/14623940802005475

Gallo, P., & Jones-Christensen, L. (2011). Firm Size Matters : An Empirical Investigation of Organizational Size and Ownership on Sustainability-Related Behaviors. Business & Society -. *Business & Society*, 50(2), 315–349. 10.1177/0007650311398784

Gamerschlag, R., Möller, K., & Verbeeten, F. (2011). Determinants of voluntary CSR disclosure : Empirical evidence from Germany. *Review of Managerial Science*, 5(2-3), 233–262. 10.1007/s11846-010-0052-3

Gangl, M. (2002). Changing Labour Markets and Early Career Outcomes : Labour Market Entry in Europe Over the Past Decade. *Work, Employment and Society*, 16(1), 67–90. 10.1177/09500170222119254

Gao, F., Dong, Y., Ni, C., & Fu, R. (2016). Determinants and economic consequences of non-financial disclosure quality. *European Accounting Review*, 25(2), 287–317. 10.1080/09638180.2015.1013049

García-Arca, J., González-Portela Garrido, A. T., & Prado-Prado, J. C. (2024). Deploying Sustainability Through Employee Participation. An Action Research Proposal. *Systemic Practice and Action Research*, 37(1), 19–40. 10.1007/s11213-023-09639-x

Gartner, W. B. (1988). "Who is an entrepreneur?" is (still) the wrong question. *American Journal of Small Business*, 12(4), 11–32. 10.1177/104225878801200401

Gaus, N. (2017). Selecting research approaches and research designs : A reflective essay. *Qualitative Research Journal*, 17(2), 99–112. 10.1108/QRJ-07-2016-0041

Gavard-Perret, M.-L., Gotteland, D., Haon, C., & Jolibert, A. (2008). Méthodologie de la recherche.

Geertz, C. (1973). *The Interpretation of Cultures: Selected Essays*. Basic Books.

Gerring, J. (2004). What Is a Case Study and What Is It Good for? *The American Political Science Review*, 98(2), 341–354. 10.1017/S0003055404001182

Ghirelli, C., Havari, E., Santangelo, G., & Scettri, M. (2019). Evidence from an Italian region. *International Journal of Manpower*, 40(3), 500–524. 10.1108/IJM-02-2018-0062

Gibbert, M., Ruigrok, W., & Wicki, B. (2008). What passes as a rigorous case study? *Strategic Management Journal*, 29(13), 1465–1474. 10.1002/smj.722

Giordano, Y. (2003). Conduire un projet de recherche. Une perspective qualitative (p. 318). Editions Management et Société. https://shs.hal.science/halshs-00440011

Glaser, B. G., & Strauss, A. L. (1967). *The Discovery of Grounded Theory: Strategies for Qualitative Research*. Aldine Tra., 10.1037/h0083159

Glaser, B., & Strauss, A. (2017). *Discovery of Grounded Theory : Strategies for Qualitative Research*. Routledge., 10.4324/9780203793206

Goldkuhl, G. (2012). Pragmatism vs interpretivism in qualitative information systems research. *European Journal of Information Systems*, 21(2), 135–146. 10.1057/ejis.2011.54

Gold, R. L. (1958). Roles in sociological field observations. *Social Forces*, 36(3), 217–223. 10.2307/2573808

González-Sánchez, R., Alonso-Muñoz, S., & Medina-Salgado, M. S. (2023). Circularity in waste management : A research proposal to achieve the 2030 Agenda. *Operations Management Research : Advancing Practice Through Research*, 16(3), 1520–1540. 10.1007/s12063-023-00373-0

Göymen, K., Demiroglu, O. C., & Tülüş, C. (2017). Snow sports tourism development policies in Turkey. In *Lifestyle sports and public policy* (pp. 63–81). Routledge. 10.4324/9781315624402-5

Gravemeijer, K., & Cobb, P. (2006). Design research from a learning design perspective. *Educational design research.*

Gray, R. H., Owen, D. L., & Adams, C. A. (1996). Accounting and Accountability : Changes and Challenges in Corporate Social and Environmental Reporting. https://durham-repository.worktribe.com/output/1123501/accounting-and-accountability-changes-and-challenges-in-corporate-social-and-environmental-reporting

Groh, M., Krishnan, N., McKenzie, D., & Vishwanath, T. (2016). *Do wage subsidies provide a stepping-stone to employment for recent college graduates? Evidence from a randomized experiment in Jordan.* World Bank. 10.1162/REST_a_00584

Grosemans, I., & De Cuyper, N. (2021). Career competencies in the transition from higher education to the labor market : Examining developmental trajectories. *Journal of Vocational Behavior*, 128, 103602. 10.1016/j.jvb.2021.103602

Grossoehme, D. H. (2014). Overview of Qualitative Research. *Journal of Health Care Chaplaincy*, 20(3), 109–122. 10.1080/08854726.2014.92566024926897

Guba, E. G., & Lincoln, Y. S. (2005). Paradigmatic controversies, contradictions, and emerging confluences. In The Sage handbook of qualitative research (3e éd., pp.191-215). Sage.

Guba, E. G., & Lincoln, Y. S. (1994). Competing paradigms in qualitative research. In Denzin, N. K., & Lincoln, Y. S. (Eds.), *Handbook of qualitative research* (pp. 105–117). Sage Publications.

Guillemette, F. (2006). L'approche de la Grounded Theory; pour innover? *Recherches Qualitatives*, 26(1), 32–50. 10.7202/1085397ar

Gummesson, E. (2007). Case study research and network theory : Birds of a feather. *Qualitative Research in Organizations and Management*, 2(3), 226–248. 10.1108/17465640710835373

Guthrie, J., & Parker, L. D. (1989). Corporate Social Reporting : A Rebuttal of Legitimacy Theory. *Accounting and Business Research*, 19(76), 343–352. 10.1080/00014788.1989.9728863

Hackston, D., & Milne, M. J. (1996). Some determinants of social and environmental disclosures in New Zealand companies. *Accounting, Auditing & Accountability Journal*, 9(1), 77–108. 10.1108/09513579610109987

Haeberli, W., Whiteman, C., & Shroder, F. J. (2015) Snow and ice-related hazards, risks, and disasters, A General Framework, Snow and Ice-related Hazards, Risks and Disasters, Elsevier.

Haimes, Y. Y. (2005). *Risk modeling, assessment, and management.* John Wiley & Sons.

Haller, A., Link, M., & Groß, T. (2017). The Term 'Non-financial Information' – A Semantic Analysis of a Key Feature of Current and Future Corporate Reporting. *Accounting in Europe*, 14(3), 407–429. 10.1080/17449480.2017.1374548

Hammond, Z., & Jackson, Y. (2015). *Culturally responsive teaching and the brain : Promoting authentic engagement and rigor among culturally and linguistically diverse students.* Corwin, a SAGE company.

Hannah, S. (2019). Literature review as a research methodology: An overview and guidelines. *Journal of Business Research*, 104, 333–339. 10.1016/j.jbusres.2019.07.039

Hassanein, A., & Hussainey, K. (2015). Is forward-looking financial disclosure really informative? Evidence from UK narrative statements. *International Review of Financial Analysis*, 41(C), 52–61. 10.1016/j.irfa.2015.05.025

HCP. (2006-2007). Enquête nationale sur le secteur informel - Rapport de synthèse. Rabat: HCP.

HCP. (2014). Enquête nationale sur le secteur informel 2013 - 2014: Rapport de synthèse.

HCP. (2022). La situation du marché du travail en 2021. Rabat: Haut-Commissariat au Plan. Récupéré sur https://www.hcp.ma/La-situation-du-marche-du-travail-en-2021_a2810.html

Healy, M., Hammer, S., & McIlveen, P. (2022). Mapping graduate employability and career development in higher education research : A citation network analysis. *Studies in Higher Education*, 47(4), 799–811. 10.1080/03075079.2020.1804851

Heidegger, M., Macquarrie, J., & Robinson, E. S. (2008). *Being and time*. Harper-Perennial/Modern Thought.

Hejaji, I., & Cherkaoui, A. (2022). L'influence du profil du dirigeant-propriétaire sur les pratiques RH au sein des PME marocaines : Enseignements de deux études de cas. *Revue Management & Innovation*, 6(2), 46–69. 10.3917/rmi.206.0046

Hessels, J., Van Gelderen, M., & Thurik, R. (2008). Entrepreneurial aspirations, motivations, and their drivers. *Small Business Economics*, 31(3), 323–339. 10.1007/s11187-008-9134-x

Hill, M. J. (2005). *The Public Policy Process*. Pearson Longman.

Hillman, A. J., & Keim, G. D. (2001). Shareholder Value, Stakeholder Management, and Social Issues : What's the Bottom Line? *Strategic Management Journal*, 22(2), 125–139. 10.1002/1097-0266(200101)22:2<125::AID-SMJ150>3.0.CO;2-H

Hlady Rispal, M., & Jouison-Laffitte, E. (2015). La contribution des méthodes qualitatives au développement du champ de l'entrepreneuriat. *Revue de l'Entrepreneuriat*, 14(1), 15–40. 10.3917/entre.141.0015

Hlady-Rispal, M., Fayolle, A., & Gartner, W. B. (2021). In search of creative qualitative methods to capture current entrepreneurship research challenges. *Journal of Small Business Management*, 59(5), 887–912. 10.1080/00472778.2020.1865541

Hmioui, A., Alla, L., & Bentalha, B. (2019). Évaluation des formations en ligne : Etude qualitative des perceptions des utilisateurs des MOOC au Maroc. *ICT Day's19*.

Hmioui, A., Alla, L., & Bentalha, B. (2019). Évaluation des formations en ligne: Etude qualitative des perceptions des utilisateurs des MOOC au Maroc. In ICT-Days' 19.

Hmioui, A., Alla, L., & Bentalha, B. (2019). Évaluation des formations en ligne: Étude qualitative des perceptions des utilisateurs des MOOC au Maroc. In ICT-Days' 19.

Hmioui, A., Alla, L., & Bentalha, B. (2019, avril). Évaluation des formations en ligne : Etude qualitative des perceptions des utilisateurs des MOOC au Maroc. ICT-Days'19. https://hal.science/hal-02943698

Hmioui, A., Alla, L., & Bentalha, B. (2022). Perception du tourisme durable par les clients étrangers: Cas du séjour dans un établissement hôtelier. Reconfigurations du tourisme en contexte de la crise du Covid-19: Quelles démarches pour quelles résiliences au Maroc et ailleurs?, 1(1), 5-38.

Hmioui, A., & Bentalha, B. (2021). Service supply chain management: A literature review. *International Journal of Logistics Systems and Management*, 40(3), 332–353. 10.1504/IJLSM.2021.120253

Hmioui, A., Bentalha, B., & Alla, L. (2020). *La netnographie dans les recherches marketing : La communauté virtuelle comme consom'acteur vecteur d'efficacité marketing. 4*, 631-652. 10.48374/IMIST.PRSM/ame-v2i4.23559

Hollweck, T. (2016). Robert K. Yin. (2014). Case Study Research Design and Methods (5th ed.). Thousand Oaks, CA : Sage. 282 pages. *The Canadian Journal of Program Evaluation, 30*. 10.3138/cjpe.30.1.108

Hoon, C. (2013). Meta-Synthesis of Qualitative Case Studies : An Approach to Theory Building. *Organizational Research Methods*, 16(4), 522–556. 10.1177/1094428113484969

Huberman, A. (2014). Qualitative data analysis a methods sourcebook.

Hughes, K. D. (2003). Pushed or pulled? Women's entry into self-employment and small business ownership. *Gender, Work and Organization*, 10(4), 433–454. 10.1111/1468-0432.00205

Hummel, K., & Schlick, C. (2016). The relationship between sustainability performance and sustainability disclosure – Reconciling voluntary disclosure theory and legitimacy theory. *Journal of Accounting and Public Policy*, 35(5), 455–476. 10.1016/j.jaccpubpol.2016.06.001

Irmak, M. A., Yilmaz, S., Mutlu, E., & Yilmaz, H. (2020). Analysis of different urban spaces on thermal comfort in cold regions: A case from Erzurum. *Theoretical and Applied Climatology*, 141(3-4), 1593–1609. 10.1007/s00704-020-03289-y

Jabar, M. A., Sidi, F., Selamat, M. H., Ghani, A. A. A., & Ibrahim, H. (2009). An Investigation into Methods and Concepts of Qualitative Research in Information System Research. *Computer and Information Science*, 2(4), 4. Advance online publication. 10.5539/cis.v2n4p47

Jackson, G., & Apostolakou, A. (2010). Corporate Social Responsibility in Western Europe : An Institutional Mirror or Substitute? *Journal of Business Ethics*, 94(3), 371–394. 10.1007/s10551-009-0269-8

Jacob, E. (1987). Qualitative research traditions: A review. *Review of Educational Research*, 57(1), 1–50. 10.3102/00346543057001001

Jacob, E. (1988). Clarifying qualitative research: A focus on traditions. *Educational Researcher*, 17(1), 16–24. 10.2307/1175461

Jacobs, T., De Maeyer, S., & Beck, M. (1999). Family Formation and the Divisions of Labour in Belgian Couple Households. *Quality & Quantity*, 33(3), 291–304. 10.1023/A:1004600921053

Jamali, D. (2010). The CSR of MNC Subsidiaries in Developing Countries : Global, Local, Substantive or Diluted? *Journal of Business Ethics*, 93(S2), 181–200. 10.1007/s10551-010-0560-8

Javadian, G., Dobratz, C., Gupta, A., Gupta, V. K., & Martin, J. A. (2020, juillet). *Qualitative Research in Entrepreneurship Studies : A State-of-Science*. 10.1177/0971355720930564

Jennings, G. R. (2005). Business, Social Science Methods Used. Encyclopedia of Social Measurement, 219-230. 10.1016/B0-12-369398-5/00270-X

Jensen, M. C., & Meckling, W. H. (1976). Theory of the firm : Managerial behavior, agency costs and ownership structure. *Journal of Financial Economics*, 3(4), 305–360. 10.1016/0304-405X(76)90026-X

Johansson, E. (2023). Reconceptualizing Time to Understand How to Lead Digitalization in Education. *Forskning Og Forandring*, 6(1), 3–21. 10.23865/fof.v6.5344

Jorgensen, D. (1989). The methodology of participant observation. In Pogrebin, M. (Ed.), *Qualitative Approaches to Criminal Justice: Perspectives from the Field* (pp. 12–26). Sage.

Joseph, F. I., Earland, J., & Ahmed, M. A. (2021). Experience of Conducting Sensitive Qualitative Research as a Cultural Outsider : Formulation of a Guide for Reflexivity. *International Journal of Qualitative Methods*, 20, 16094069211058616. 10.1177/16094069211058616

Judge, T. A., & Klinger, R. (2008). Job satisfaction : Subjective well-being at work. In *The science of subjective well-being* (pp. 393–413). The Guilford Press.

Judge, T. A., Piccolo, R. F., Podsakoff, N. P., Shaw, J. C., & Rich, B. L. (2010). The relationship between pay and job satisfaction : A meta-analysis of the literature. *Journal of Vocational Behavior*, 77(2), 157–167. 10.1016/j.jvb.2010.04.002

Kahneman, D. T. (1979). Prospect Theory: An analysis of decision under risk. Econometrica, 47, pp. p. 263-291.

Kapiszewski, D., & Karcher, S. (2021). Transparency in Practice in Qualitative Research. *PS, Political Science & Politics*, 54(2), 285–291. 10.1017/S1049096520000955

Karahan, V., & Aydoğmuş, E. (2023). Risk Analysis and Risk Assessment in Laboratory Studies. *Avrupa Bilim ve Teknoloji Dergisi*, (49), 55–60. 10.31590/ejosat.1260340

Kawulich, B. (2012). *Collecting data through observation, In book: Doing Social Research: A global context. McGraw Hill.*

Kawulich, B. B. (2005). Participant Observation as a Data Collection Method. *Forum Qualitative Social Research*, 6(2), 43. http://nbn-resolving.de/urn:nbn:de:0114-fqs0502430

Kazan, H., Hobikoğlu, E. H., Karademir, H., Dalyanci, L., & Turguter, Y. (2015). Economic development of ski industry in experimental innovation: Example of Palandöken Turkey and Alps Switzerland. *Procedia: Social and Behavioral Sciences*, 195, 487–492. 10.1016/j.sbspro.2015.06.245

Kedia, B. L., Nordtvedt, R., & Pérez, L. M. (2002). International business strategies, decision-making theories, and leadership styles : An integrated framework. *Competitiveness Review*, 12(1), 38–52. 10.1108/eb046433

Kellogg, B. T. (2024). An Asset-Based Approach to Understanding College Students as Sources of Support. *Journal of College Student Retention*, 25(4), 820–845. 10.1177/15210251211025212

Kemmis, S., Wilkinson, J., Edwards-Groves, C., Hardy, I., Grootenboer, P., & Bristol, L. (2014). *Changing Practices, Changing Education*. Springer Singapore., 10.1007/978-981-4560-47-4

Ketele, J.-M. D., & Roegiers, X. (1996). *Méthodologie du recueil d'informations : Fondements des méthodes d'observations, de questionnaires, d'interviews et d'études de documents*. De Boeck Université.

Khan, A., Muttakin, M., & Siddiqui, J. (2013). Corporate Governance and Corporate Social Responsibility Disclosures : Evidence from an Emerging Economy. *Journal of Business Ethics*, 114(2), 207–223. 10.1007/s10551-012-1336-0

Kincheloe, J. L., McLaren, P., & Steinberg, S. R. (2018). Critical pedagogy and qualitative research: Advancing the bricolage. In Denzin, N. K., & Lincoln, Y. S. (Eds.), *The landscape of qualitative research* (5th ed., pp. 235–260). Sage Publications.

Kirkwood, J., & Campbell-Hunt, C. (2007). Using Multiple Paradigm Research Methodologies To Gain New Insights Into Entrepreneurial Motivations. *Journal of Enterprising Culture*, 15(03), 219–241. 10.1142/S0218495807000137

Kirkwood, J., & Walton, S. (2010). What motivates ecopreneurs to start businesses? *International Journal of Entrepreneurial Behaviour & Research*, 16(3), 204–228. 10.1108/13552551011042799

Kirova, V. (2011). *La direction marketing stratégique comme vecteur de resserrement du lien marketing stratégie dans l'organisation: Etude du cas Laser-Confinoga*. Thèse Sciences de Gestion. Paris: Bordeaux – Montesquieu.

Kirzner, I. (1973). *Competition and Entrepreneurship.*

Klopper, H. (2008). The qualitative research proposal. *Curationis*, 31(4), 4. Advance online publication. 10.4102/curationis.v31i4.106219653539

Knust, M., & Hanft, A. (2009). *Continuing Higher Education and Lifelong Learning.* Springer Netherlands., 10.1007/978-1-4020-9676-1

Kolk, A. (2002). The Effectiveness of Self-regulation. European Management Journal. https://www.academia.edu/33220513/The_Effectiveness_of_Self_regulation

Kontos, M. (2008). The biographical embeddedness of women's self-employment. Motivations, strategies and policies. In Apitzsch, U. (Ed.), *K. M.* Self-Employment Activities of Women and Minorities. 10.1007/978-3-531-90816-8_5

Kontos, M. (Ed.). (2008). *Self-employment activities of women and minorities: Their success or failure in relation to social citizenship policies.* VS Verlag für Sozialwissenschaften., 10.1007/978-3-531-90532-7

Krause, N. (1995). Negative interaction and satisfaction with social support among older adults. *The Journals of Gerontology. Series B, Psychological Sciences and Social Sciences*, 50(2), 59–P73. 10.1093/geronb/50B.2.P597757834

Krippendorff, K. (2018). *Content analysis: An introduction to its methodology.* Sage publications.

Krueger, R. A., & Casey, M. A. (2014). *Focus groups: A practical guide for applied research.* Sage Publications.

Kumar, K., Kumari, R., Poonia, A., & Kumar, R. (2021). Factors influencing corporate sustainability disclosure practices : Empirical evidence from Indian National Stock Exchange. *Journal of Financial Reporting and Accounting*, 21(2), 300–321. 10.1108/JFRA-01-2021-0023

Kurt, T. (2017) Improvement of technical capacity in avalanche risk management in Turkey.Dissertation, İstanbul University, Institute of Graduate Studies in Science and Engineering, Departmant of Forest Engineering.

Kurt, T. (2014). Assessment of the Perchertal avalanche in Tyrol, Austria. *Turkish Journal of Earth Sciences*, 23(3), 339–349. 10.3906/yer-1305-13

Kurt, T. (2022). Avalanche Hazards with Mitigation in Turkey and Qualitative Risk Assessment for Snow Avalanches in Ayder (Rize, NE Turkey) Using Combination of GIS, Remote Sensing Techniques and Field Studies. In Al Saud, M. M. (Ed.), *Applications of Space Techniques on the Natural Hazards in the MENA Region.* Springer. 10.1007/978-3-030-88874-9_23

Kuzey, C., & Uyar, A. (2017). Determinants of sustainability reporting and its impact on firm value : Evidence from the emerging market of Turkey. *Journal of Cleaner Production*, 143, 27–39. 10.1016/j.jclepro.2016.12.153

Lamsiah, A., & Bentalha, B. (2023a). 2022 Qatar World Cup: A Netnographic Analysis of the Relationship Between Sport, Media and Politic . Réflexions sportives, 1(3), 69-95.

Lamsiah, A., & Bentalha, B. (2023b). Moroccan Jews And Economic Intelligence: The Implicit Power Of Belonging. Le concept de l'intéligence en sciences juridiques, économiques et sociales, 71-90.

Lamsiah, A., Benabderrazik, Y., & Bentalha, B. (2021). The Economic and political components of diplomatic discourse: State actors in Morocco-UK Free Trade Agreement as a case study. *Alternatives Managériales Economiques*, 3(4), 496–517.

Lamsiah, A., & Bentalha, B. (2022). Economic Diplomacy and Economic Intelligence Amalgamation : The Moroccan-Nigerian Economic Exchange as a Case Study. *Alternatives Managériales Economiques, 4*(4), Article 4. 10.48374/IMIST.PRSM/ame-v4i4.35571

Lamsiah, A., & Bentalha, B. (2022). Economic Diplomacy and Economic Intelligence Amalgamation: The Moroccan-Nigerian Economic Exchange as a Case Study. *Alternatives Managériales Economiques*, 4(4), 582–603.

Landrø, M. (2021). Why is it safe–enough? Decision-making in avalanche terrain. Dissertation, The Artic University of Norway. https://munin.uit.no/handle/10037/22697

Laternser, M., & Schneebeli, M. (2022). Temporal Trend and Spatial Distribution of Avalanche Activity during the Last 50 Years in Switzerland. *Natural Hazards*, 27(3), 201–230. 10.1023/A:1020327312719

Leo, S. (2019). Pitfalls of tourism graduate students in presenting the ingredients of research proposals. *Journal of Hospitality, Leisure, Sport and Tourism Education*, 24, 178–189. 10.1016/j.jhlste.2019.03.003

Lincoln, Y. S., Lynham, S. A., & Guba, E. G. (2018). Paradigmatic controversies, contradictions, and emerging confluences, revisited. Dans N.K. Denzin & Y.S. Lincoln (Eds.), The Sage handbook of qualitative research (5e éd., pp.108-150). Sage.

Lincoln, Y. S., & Guba, E. G. (1986). But is it rigorous? Trustworthiness and authenticity in naturalistic evaluation. *New Directions for Program Evaluation*, 1986(30), 73–84. 10.1002/ev.1427

Lincoln, Y. S., & Guba, E. G. (2013). *The constructivist credo*. Left Coast Press.

Lindgreen, A., Di Benedetto, C. A., Verdich, C., Vanhamme, J., Venkatraman, V., Pattinson, S., Clarke, A. H., & Khan, Z. (2019). How to write really good research funding applications. *Industrial Marketing Management*, 77, 232–239. 10.1016/j.indmarman.2019.02.015

Line, M., Hawley, H., & Krut, R. (2002). The Development of Global Environmental and Social Reporting. *Corporate Environmental Strategy*, 9(1), 69–78. 10.1016/S1066-7938(01)00159-2

Liu, Z.-Y., Lomovtseva, N., & Korobeynikova, E. (2020). Online Learning Platforms : Reconstructing Modern Higher Education. *International Journal of Emerging Technologies in Learning*, 15(13), 4–21. 10.3991/ijet.v15i13.14645

Li, Y., Mohd Nordin, N. R., Akhter, S., Kumar, T., & Shaheen, M. (2023). Does green entrepreneurial behavior enhance through entrepreneurship education, perceived-ability to use technology, and commitment to environment? Understanding the contribution of entrepreneurial motivation and university support. *Ekonomska Istrazivanja*, 36(3), 2153375. Advance online publication. 10.1080/1331677X.2022.2153375

Logrosa, G., Mata, M. A., Lachica, Z. P., Estaña, L. M., & Hassall, M. (2022). Integrating risk assessment and decision-making methods in Analyzing the dynamics of COVID-19 epidemics in Davao City, Mindanao Island, Philippines. *Risk Analysis*, 42(1), 105–125. 10.1111/risa.1377934269475

Louche, C. (2019). Chapitre 6. La motivation au travail. In *Psychologie sociale des organisations*. Paris: Dun.

Louitri, A., & Sahraoui, D. (2014). Contexte et contextualisation dans les recherches en Management au Maroc. *Revue Marocaine Des Sciences De Management*, 85–100.

Low, M. B., & MacMillan, I. C. (1988). Ent past research and future challenges. *Journal of Management*, 14(2), 139–161. 10.1177/014920638801400202

Maibom, J. (2014). *The Effects of Debt Relief on Labor Market Outcom*.

Malinowski, B. (1922). *Argonauts Of The Western Pacific*. Routledge & Kegan Paul.

Marican, N., Ali, A., Aziz, R., & Yusof, M. (2023). *Designing Value Proposition for Indigenous Inspired Spa Strategic Marketing : A Research Proposal | SpringerLink*. https://link.springer.com/chapter/10.1007/978-3-031-08093-7_22

Martin, P. Y., & Turner, B. A. (1986). Grounded Theory and Organizational Research. *The Journal of Applied Behavioral Science*, 22(2), 141–157. 10.1177/002188638602200207

Ma, S. Z. (2023). Considerations in the Area of Management : Directions and Currents. *East African Scholars Journal of Economics. Business and Management*, 6(08), 239–247. 10.36349/easjebm.2023.v06i08.001

Maurand-Valet, A. (2011). Choix méthodologiques en Sciences de Gestion: pourquoi tant de chiffres?. Revue management et avenir, (3), 289-302.

Maxwell, J. (2012). *Qualitative research design : An interactive approach.*

McCarney, R., Warner, J., Iliffe, S., van Haselen, R., Griffin, M., & Fisher, P. (2007). The Hawthorne Effect: A randomised, controlled trial. *BMC Medical Research Methodology*, 7(1), 30. 10.1186/1471-2288-7-3017608932

McKee-Ryan, F. M., & Harvey, J. (2011). "I have a job, but...": A review of underemployment. *Journal of Management*, 37(4), 962–996. 10.1177/0149206311398134

Mcmullen, J. S., Bagby, D. R., & Palich, L. E. (2008). Economic freedom and the motivation to engage in entrepreneurial action. *Entrepreneurship Theory and Practice*, 32(5), 875–895. 10.1111/j.1540-6520.2008.00260.x

McNabb, D. E. (2017, septembre). How to Write a Research Proposal. In *Routledge eBooks* (p. 100-108). 10.4324/9781315181158-9

McTaggart, R. (1991). Principles for Participatory Action Research. *Adult Education Quarterly*, 41(3), 168–187. 10.1177/0001848191041003003

Merriam, S. B., & Grenier, R. S. (2019). *Qualitative research in practice: Examples for discussion and analysis*. Jossey-Bass.

Merriam, S. B., & Tisdell, E. J. (2016). *Qualitative research: A guide to design and implementation* (4th ed.). Jossey-Bass.

Mette, L. B. (2016). *Mixed Methods Research for Improved Scientific Study*. IGI Global.

Meyer, J. W., & Rowan, B. (1977). Institutionalized Organizations : Formal Structure as Myth and Ceremony. The American Journal of Sociology, 83(2,), 340-363.

Mia, M. M., Rizwan, S., Zayed, N. M., Nitsenko, V., Miroshnyk, O., Kryshtal, H., & Ostapenko, R. (2022). The Impact of Green Entrepreneurship on Social Change and Factors Influencing AMO Theory. *Systems*, 10(5), 132. Advance online publication. 10.3390/systems10050132

Miles, A. (2017). *Research Methods and Strategies Workshop : A Taxonomy of Research Gaps : Identifying and Defining the Seven Research Gaps*. https://www.researchgate.net/publication/319244623_ARTICLE_Research_Methods_and_Strategies_Workshop_A_Taxonomy_of_Research_Gaps_Identifying_and_Defining_the_Seven_Research_Gaps

Miles, M. B., & Huberman, A. M. (1994). Qualitative data analysis: An expanded sourcebook. Sage.

Miles, M. B., & Huberman, A. M. (1994). *Qualitative Data Analysis: An Expanded Sourcebook*. SAGE Publications.

Miles, M. B., & Huberman, A. M. (2003). *Analyse des données qualitatives*. De Boeck Supérieur.

Miles, M. B., Huberman, A. M., & Saldaña, J. (2019). *Qualitative data analysis: A methods sourcebook*. Sage Publications.

Miles, M. B., & Huberman, M. A. (1994). *Qualitative Data Analysis*. Sage Publi.

Millar, J., & Ridge, T. (2013). Lone mothers and paid work : The 'family-work project'. *International Review of Sociology*, 23(3), 564–577. 10.1080/03906701.2013.856161

Miller, D., & Friesen, P. H. (1984). A Longitudinal Study of the Corporate Life Cycle. *Management Science*, 30(10), 1161–1183. 10.1287/mnsc.30.10.1161

Mir, R., & Watson, A. (2000). Strategic Management and the Philosophy of Science : The Case for a Constructivist Methodology. *Strategic Management Journal*, 21(9), 941–953. 10.1002/1097-0266(200009)21:9<941::AID-SMJ141>3.0.CO;2-D

Morioka, S. N., Bolis, I., Evans, S., & Carvalho, M. M. (2017). Transforming sustainability challenges into competitive advantage : Multiple case studies kaleidoscope converging into sustainable business models. *Journal of Cleaner Production*, 167, 723–738. 10.1016/j.jclepro.2017.08.118

Moulai, S., & Boukrif, M. (2018). Le choix du successeur dans les PME familiales. *Revue africaine de management, 1*(3).

Moustakas, C. (1994). *Phenomenological research methods*. Sage Publications. 10.4135/9781412995658

Muthuri, J. N., & Gilbert, V. (2011). An Institutional Analysis of Corporate Social Responsibility in Kenya. *Journal of Business Ethics*, 98(3), 467–483. 10.1007/s10551-010-0588-9

Myers, M. D. (2009). Qualitative Research in Bsiness & Management. In *Revista Brasileira de Linguística Aplicada* (SAGE Publi, Vol. 5, Issue 1). https://revistas.ufrj.br/index.php/rce/article/download/1659/1508%0Ahttp://hipatiapress.com/hpjournals/index.php/qre/article/view/1348%5Cnhttp://www.tandfonline.com/doi/abs/10.1080/09500799708666915%5Cnhttps://mckinseyonsociety.com/downloads/reports/Educa

Nanuru, R. F., & Utama, M. D. (2020). Dynamics of proposition and funding of lecturers' research proposal at private universities in higher education service institutes region XII – Indonesia. *International Journal of Scientific and Technology Research*, 9(1), 3085–3090. 10.4108/eai.18-7-2019.2289201

Nasution, N., Hasnati, H., Hasan, M. A., & Yandra, A. (2019). E-Service Application For Online Research Proposal Guidance. *International Journal of Scientific and Technology Research*, 8(7), 71–76.

Naylor, J. C., Pritchard, R. D., & Ilgen, D. R. (1980). Theory of Behavior in Organizations. In *Organizational Behavior 6*. Academic P. 10.4324/9781315701967-22

Ngwenya, S., Mashau, N. S., Mhlongo, E. S., Traoré, A. N., & Mudau, A. G. (2023). Health risk management framework for heavy metals and cyanide in Kwekwe city of Zimbabwe: A mixed-method study protocol. *Journal of Health, Population and Nutrition*, 42(1), 26. 10.1186/s41043-023-00367-537013628

Nichols, T. B., Hawley, A. C., Smith, W. R., Wheeler, A. R.III, & McIntosh, S. E. (2018). Avalanche safety practices among backcountry skiers and snowboarders in Jackson Hole in 2016. *Wilderness & Environmental Medicine*, 29(4), 493–498. 10.1016/j.wem.2018.05.00430213711

Norum, K. E. (2008). Artifact analysis. In Given, L. M. (Ed.), *The SAGE encyclopedia of qualitative research methods* (pp. 25–26). Sage Publications.

Nowell, L. S., Norris, J. M., White, D. E., & Moules, N. J. (2017). Thematic Analysis : Striving to Meet the Trustworthiness Criteria. *International Journal of Qualitative Methods*, 16(1), 1609406917733847. 10.1177/1609406917733847

Oddershede, A. M., Watkins, F. J., & Quezada, L. E. (2018). Methodology proposal for appraising health service ICT system implication. *2018 7th International Conference on Computers Communications and Control (ICCCC)*, 58-62. 10.1109/ICCCC.2018.8390437

Omran, M. A., & Ramdhony, D. (2015). Theoretical Perspectives on Corporate Social Responsibility Disclosure : A Critical Review. *International Journal of Accounting and Financial Reporting*, 5(2), 38. 10.5296/ijafr.v5i2.8035

Orlikowski, W. J., & Yates, J. (2002). It's About Time : Temporal Structuring in Organizations. *Organization Science*, 13(6), 684–700. 10.1287/orsc.13.6.684.501

Oswald, D., Sherratt, F., & Smmith, S. (2014). Handling the Hawthorne effect: The challenges surrounding a participant observer. *Review of Social Studies*, 1(1), 53–73. 10.21586/ross0000004

Ottrey, E., Rees, C. E., Kemp, C., Brock, T. P., Leech, M., Lyons, K., Monrouxe, L. V., Morphet, J., & Palermo, C. (2021). Exploring health care graduates' conceptualisations of preparedness for practice : A longitudinal qualitative research study. *Medical Education*, 55(9), 1078–1090. 10.1111/medu.1447533617656

Özçoban, E. (2019). Türkiye'de Kış Turizmi ve Erciyes Kayak Merkezi Üzerine Bir Değerlendirme (Winter Tourism in Turkey and a Review on Erciyes Ski Resort). *Journal of Tourism & Gastronomy Studies*, 7(3), 1606–1625. 10.21325/jotags.2019.438

Paillé, P. (1994). L'analyse par théorisation ancrée. *Cahiers de Recherche Sociologique*, (23), 147–181. 10.7202/1002253ar

Park, R. E., Burgess, E. W., & McKenzie, R. D. (1967). *The city*. University of Chicago Press.

Parolly, G. (2020). The Anatolian high-mountain ranges–Plant diversity between two seas. Plant biogeography and vegetation of high mountains of Central and South-West Asia, 215-286.

Patton, M. Q. (2002). Qualitative research and evaluation methods (3e éd.). Sage.

Patton, M. Q. (2015). *Qualitative research & evaluation methods: Integrating theory and practice* (4th ed.). Sage Publications.

Patton, M. Q., & Patton, M. Q. (2002). *Qualitative research and evaluation methods* (3rd ed.). Sage Publications.

Pearce, P. F., & Ferguson, L. A. (2017). How to write abstracts for manuscripts, presentations, and grants : Maximizing information in a 30-s sound bite world. *Journal of the American Association of Nurse Practitioners*, 29(8), 452–460. 10.1002/2327-6924.1248628660736

Perrow, C. (1970). *Organizational Analysis : A Sociological View*. Brooks/Cole.

Pin, C. (2023). L'entretien semi-directif. LIEPP Methods Brief / Fiches méthodologiques du LIEPP. https://sciencespo.hal.science/hal-04087897

Piperopoulos, P. (2010). Qualitative research in SMEs and entrepreneurship : A literature review of case study research. *International Journal of Economics and Business Research*, 2(6), 494. 10.1504/IJEBR.2010.035701

Point, S., & Fourboul, C. V. (2006). Le codage à visée théorique. *Recherche et Applications en Marketing*, 21(4), 61–78. 10.1177/076737010602100404

Pope, C., & Mays, N. (1995). Qualitative Research: Reaching The parts other methods cannot reach: An introduction to qualitative methods in health and health services research. *BMJ (Clinical Research Ed.)*, 311(6996), 42–45. 10.1136/bmj.311.6996.427613329

Prabowo, H., Ikhsan, R. B., & Yuniarty, Y. (2022). Drivers of Green Entrepreneurial Intention: Why Does Sustainability Awareness Matter Among University Students? *Frontiers in Psychology*, 13(March), 1–17. 10.3389/fpsyg.2022.87314035418917

Pras, B., Evrard, Y., & Roux, E. (2003). Market : Études et recherches en marketing - Fondements, méthodes: Vol. 3ème édition (p. 704). Dunod. https://shs.hal.science/halshs-00150660

Prediger, S., & Schnell, S. (2014). *Investigating the Dynamics of Stochastic Learning Processes : A Didactical Research Perspective*. Its Methodological and Theoretical Framework, Illustrated for the Case of the Short Term–Long Term Distinction., 10.1007/978-94-007-7155-0_29

Pröbstl-Haider, U., Richens, H., & Türk, S. (Eds.). (2019). *Winter tourism: Trends and challenges*. Cabi International. 10.1079/9781786395207.0000

Prudence, J. (2010). *La maîtrise du slack budgétaire: Une analyse autour des théories de la déviance et de la criminologie: le cas d'une entreprise de distribution*. Thèse Sciences de Gestion. Paris: Université Paris Dauphine.

Puranik, S. (2020). *Innovative teaching methods in higher education*. BSSS Journal of Education., 10.51767/je0907

Rahman, H., Hasibuan, A. F., Syah, D. H., Sagala, G. H., & Prayogo, R. R. (2022). Intrapreneurship : As the outcome of entrepreneurship education among business students. *Cogent Education*, 9(1), 2149004. 10.1080/2331186X.2022.2149004

Raihan, A. (2024). A Systematic Review of Geographic Information Systems (GIS) in Agriculture for Evidence-Based Decision Making and Sustainability. *Global Sustainability Research*, 3(1), 1–24. 10.56556/gssr.v3i1.636

Ratan, S. K., Anand, T., & Ratan, J. (2019). Formulation of Research Question – Stepwise Approach. *Journal of Indian Association of Pediatric Surgeons*, 24(1), 15. 10.4103/jiaps.JIAPS_76_1830686882

Ravitch, S. M., & Carl, N. M. (2021). Qualitative research: Bridging the conceptual, theoretical, and methodological (2e éd.). Sage.

Razgani, H. (2011). *Les réseaux sociaux et la compétence de l'auditeur – Cas du cabinet Audit &Co*. Thèse Sciences de Gestion. Paris: Université Paris Dauphine.

Reid, A.-M., Brown, J. M., Smith, J. M., Cope, A. C., & Jamieson, S. (2018). Ethical dilemmas and reflexivity in qualitative research. *Perspectives on Medical Education*, 7(2), 69–75. 10.1007/S40037-018-0412-229536374

Remondo, J., Bonachea, J., & Cendrero, A. (2005). A statistical approach to landslide risk modeling at basin scale; from landslide susceptibility to quantitative risk assessment. *Landslides*, 2(4), 321–328. 10.1007/s10346-005-0016-x

Rethlefsen, M. L., Kirtley, S., Waffenschmidt, S., Ayala, A. P., Moher, D., Page, M. J., Koffel, J. B., Blunt, H., Brigham, T., Chang, S., Clark, J., Conway, A., Couban, R., de Kock, S., Farrah, K., Fehrmann, P., Foster, M., Fowler, S. A., Glanville, J., & Young, S. (2021). PRISMA-S : An extension to the PRISMA Statement for Reporting Literature Searches in Systematic Reviews. *Systematic Reviews*, 10(1), 39. 10.1186/s13643-020-01542-z33499930

Rezaee, Z. (2016). Business sustainability research : A theoretical and integrated perspective. *Journal of Accounting Literature*, 36(1), 48–64. 10.1016/j.acclit.2016.05.003

Rezaee, Z., & Tuo, L. (2017). Voluntary disclosure of non-financial information and its association with sustainability performance. *Advances in Accounting*, 39, 47–59. 10.1016/j.adiac.2017.08.001

Rezaul, I. M. (2018). *Social Research Methodology and New Techniques in Analysis, Interpretation, and Writing*. IGI Global.

Robb, S., & Zarzeski, L. (2001). Nonfinancial Disclosures Across Anglo-American Countries. *Journal of International Accounting, Auditing & Taxation*, 10(1), 71–83. 10.1016/S1061-9518(01)00036-2

Robertson, S. P. (2008). The Qualitative Research Process as a Journey - Mapping Your Course with Qualitative Research Software. *Qualitative Research Journal*, 8(2), 81–90. 10.3316/QRJ0802081

Roberts, R. W. (1992). Determinants of corporate social responsibility disclosure : An application of stakeholder theory. *Accounting, Organizations and Society*, 17(6), 595–612. 10.1016/0361-3682(92)90015-K

Roe, M. J. (2003). Political Determinants of Corporate Governance. https://global.oup.com/academic/product/political-determinants-of-corporate-governance-9780199240746?cc=us&lang=en&

Rousseau, D. M., & ten Have, S. (2022). Evidence-based change management. *Organizational Dynamics*, 51(3), 100899. 10.1016/j.orgdyn.2022.100899

Rowley, J. (2012). Conducting research interviews. *Management Research Review*, 35(3/4), 260–271. 10.1108/01409171211210154

Royer, C., Baribeau, C., & Duchesne, A. (2009). Les entretiens individuels dans la recherche en sciences sociales au Québec: Où en sommes-nous? Un panorama des usages. *Recherches Qualitatives*, 7, 64–79.

Rubin, H. J., & Rubin, I. S. (2012). *Qualitative interviewing: The art of hearing data* (3rd ed.). Sage Publications.

Ruth, K. (1990). Motivation theory and industrial and organizational psychology. *Handbook of Industrial and Organizational Psychology, 1*(2), 75–130.

Saldaña, J. (2021). The coding manual for qualitative researchers.

Saldaña, J. (2021). The coding manual for qualitative researchers. *Sage (Atlanta, Ga.)*.

Sarkis-Onofre, R., Catalá-López, F., Aromataris, E., & Lockwood, C. (2021). How to properly use the PRISMA Statement. *Systematic Reviews*, 10(1), 117. 10.1186/s13643-021-01671-z33875004

Savall, H. et Zardet, V. (2014). Recherche qualimétrique. Observer l'objet complexe. Economica.

Schatzki, T. R. (2013). *The timespace of human activity : On performance, society, and history as indeterminate teleological events* (First paperback edition). Lexington Books.

Schensul, J., Schensul, S., & Lecompte, M. (1999). *Essential ethnographic methods: observations, interviews, and questionnaires (Book 2 in Ethnographer's Toolkit).* AltaMira Press.

Schonfeld D.J. & Dreyer B.P. (2008). Research Foundations, Methods, and Issues in Developmental-Behavioral Pediatrics, Developmental-Behavioral Pediatrics Evidence and Practice, 47-56. https://doi.org/10.1016/B978-0-323-04025-9.50006-4

Schwandt, T. A. (2000). Three epistemological stances for qualitative inquiry. In Handbook of qualitative research (2e éd., pp.189-213). Sage.

Schwandt, T. A. (2007). The SAGE Dictionary of Qualitative Inquiry. In *The SAGE Dictionary of Qualitative Inquiry*. 10.4135/9781412986281

Scott, W. (2005). Institutional Theory : Contributing to a Theoretical Research Program.

Seidman, I. (2013). *Interviewing as Qualitative Research : A Guide for Researchers in Education and the Social Sciences*. Teachers College Press.

Shane, S., Kolvereid, L., & Westhead, P. (1991). An exploratory examination of the reasons leading to new firm formation across country and gender. *Journal of Business Venturing*, 6(6), 431–446. 10.1016/0883-9026(91)90029-D

Shepherd, D. A. (2015). Party On! A call for entrepreneurship research that is more interactive, activity based, cognitively hot, compassionate, and prosocial. *Journal of Business Venturing*, 30(4), 489–507. 10.1016/j.jbusvent.2015.02.001

Shoaib, M., Ali, N., Anwar, B., & Abdullah, F. (2022). Library services and facilities in higher education institutions during coronavirus disease (COVID-19) in Pakistan. *Journal of Information Science*. Advance online publication. 10.1177/01655515221141035

Shrout, P. E., & Lane, S. P. (2012). Reliability. In Cooper, H., Camic, P. M., Long, D. L., Panter, A. T., Rindskopf, D., & Sher, K. J. (Eds.), *APA handbook of research methods in psychology, 1. Foundations, planning, measures, and psychometrics* (pp. 643–660). American Psychological Association. 10.1037/13619-034

Siggelkow, N. (2007). Persuasion with case studies. *Academy of Management Journal*, 50(1), 20–24. 10.5465/amj.2007.24160882

Silyn-Roberts, H. (2013, novembre). *A Research Proposal*. https://www.sciencedirect.com/science/article/pii/B9780080982854000054

Siraj, N. B., & Fayek, A. R. (2019). Risk identification and common risks in construction: Literature review and content analysis. *Journal of Construction Engineering and Management*, 145(9), 03119004. 10.1061/(ASCE)CO.1943-7862.0001685

Siswanto, Armanu, Setiawan, M., & Nimran, U. (2013). Entrepreneurial Motivation in Pondok Pesantren. *International Journal of Business and Behavioral Sciences*, 3(2), 42–54.

Sivula, A., Shamsuzzoha, A., & Helo, P. (2020). Requirements for Blockchain Technology in Supply Chain Management : An Exploratory Case Study. *Operations and Supply Chain Management: An International Journal*, 14(1), 39–50. 10.31387/oscm0440284

Solymossy, E. (1998). *Entrepreneurial dimensions : The relationship of individual, venture, and environmental factors to success: Vol.* Academia.e.

Sørheim, R., & Landström, H. (2001). Informal investors - A categorization, with policy implications. *Entrepreneurship and Regional Development*, 13(4), 351–370. 10.1080/08985620110067511

Sousa, D. (2014). Validation in Qualitative Research : General Aspects and Specificities of the Descriptive Phenomenological Method. *Qualitative Research in Psychology*, 11(2), 211–227. 10.1080/14780887.2013.853855

Spradley, J. P. (1980). Participant Observation. New York: Holt, Rinehart and Winston.

St Clair, A. (2019). Exploring the effectiveness of avalanche risk communication: A qualitative study of avalanche bulletin use among backcountry recreationists.

Stake, R. E. (1995). *The art of case study research.* Sage Publications, Inc.

Statham, G., Haegeli, P., Greene, E., Birkeland, K., Israelson, C., Tremper, B., Stethem, C., McMahon, B., White, B., & Kelly, J. (2018). A conceptual model of avalanche hazard. *Natural Hazards*, 90(2), 663–691. 10.1007/s11069-017-3070-5

Stefanovic, I., Prokic, S., & Rankovic, L. (2010). Motivational and success factors of entrepreneurs: The evidence from a developing country. *Zbornik Radova Ekonomskog Fakultet Au Rijeci*, 28(2), 251–269.

Steiger, R., Scott, D., Abegg, B., Pons, M., & Aall, C. (2019). A critical review of climate change risk for ski tourism. *Current Issues in Tourism*, 22(11), 1343–1379. 10.1080/13683500.2017.1410110

Stevenson, H. H. (1984). A Perspective on Entrepreneurship. *Harvard Business School, 9*(384–131), 85–94. https://books.google.de/books?hl=de&lr=&id=jqWuD7dF3GcC&oi=fnd&pg=PA7&dq=A+Perspective+On+Entrepreneurship&ots=IfMZFRw4e4&sig=NaodtCg2v8fPel2eVhspBHlGmmM

Stevenson, H. H., & Jarillo, J. C. (1990). A paradigm of entrepreneurship: Entrepreneurial management. *Strategic Management Journal, 11*, 17–27. 10.1007/978-3-540-48543-8_7

Stewart, D. W. (2022). *Qualitative research methods*. University Press.

Strauss, A., & Corbin, J. (1990). *Basics of qualitative research*. Sage publications.

Strauss, A., & Corbin, J. (1990). *Basics of qualitative research: Grounded theory proceduresand techniques*. Sage.

Strauss, A., & Corbin, J. (1998). *Basics of qualitative research: Techniques and procedures for developing grounded theory* (2nd ed.). Sage Publications.

Subaşı, İ. Ö., & Gür, V. (2023). Recreational skiing-and snowboarding-related extremity injuries: A five-year tertiary trauma center cohort. *Cureus*, 15(7).37649954

Suchman, M. C. (1995). Managing legitimacy : Strategic and institutional approaches. *Academy of Management Review*, 20(3), 571–610. 10.2307/258788

Sulastiningsih & Wikaningtyas, S. U. (2023). Green Entrepreneurship Motivation Model. *Academic Journal of Interdisciplinary Studies, 12*(2), 262. 10.36941/ajis-2023-0047

Sułkowski, Ł. (2009). Interpretative approach in management sciences. *Argumenta Oeconomica*, 23, 127–149.

Suryani, A. W., & Utami, H. (2020). Rigour in Qualitative Studies : Are we on track? *Jurnal Akuntansi Dan Keuangan*, 22(2), 2. Advance online publication. 10.9744/jak.22.2.47-58

Suyadi, Husnaini, & Elvina. (2020). Undergraduate students' difficulties in writing a research proposal : A case study. *International Journal of Scientific and Technology Research*, 9(3), 838–843.

Takahashi, A. R. W., & Araujo, L. (2019). Case study research : Opening up research opportunities. *RAUSP Management Journal*, 55(1), 100–111. 10.1108/RAUSP-05-2019-0109

Taquette, S. R., & Borges da Matta Souza, L. M. (2022). Ethical Dilemmas in Qualitative Research : A Critical Literature Review. *International Journal of Qualitative Methods*, 21. 10.1177/16094069221078731

Tashakkori, A., & Teddlie, C. (2010). Putting the human back in "human research methodology": The researcher in mixed methods research. *Journal of Mixed Methods Research*, 4(4), 271–277. 10.1177/1558689810382532

Taylor, C. J. (2020). *A qualitative longitudinal evaluative study of individual acceptance of leader concepts*.

Tedlock, B. (2003). Ethnography and ethnographic representation. In Denzin, N. K., & Lincoln, Y. S. (Eds.), *Strategies of qualitative inquiry* (pp. 165–213). Sage Publications.

Tedlock, B. (2020). Avances en etnografía de la observación participante. *Revista Colombiana de Antropología*, 56(2), 91–115.

Teehankee, B. (2018). Critical Realist Action Research and Humanistic Management Education. *Humanistic Management Journal*, 3(1), 71–90. 10.1007/s41463-018-0033-2

Thiétart, R.-A. (2014). Méthodes de recherche en management. Dunod. 10.3917/dunod.thiet.2014.01

Thissen, W. A., & Walker, W. E. (2013). *Public Policy Analysis: New Developments. Springer.*

Thomson, R., & Holland, J. (2003). Hindsight, foresight and insight : The challenges of longitudinal qualitative research. *International Journal of Social Research Methodology*, 6(3), 233–244. 10.1080/1364557032000091833

Tijan, E., Jović, M., Aksentijević, S., & Pucihar, A. (2021). Digital transformation in the maritime transport sector. *Technological Forecasting and Social Change*, 170, 120879. 10.1016/j.techfore.2021.120879

Tjitalaksana, S. M. (2014). Identification of Five Motivational Factors of Ecopreneurship for Start-Up Business. *Universitas Telkom*, 1–8. https://openlibrary.telkomuniversity.ac.id/pustaka/files/100142/jurnal_eproc/identification-of-five-motivational-factors-of-ecopreneurship-for-start-up-business-a-case-study-of-enigami-papers-.pdf

Tortora, A. M. R., Di Pasquale, V., Riemma, S., & Iannone, R. (2021). Towards a tool for SMEs to Maintenance Management process self-assessment and improvement : Literature overview and research proposal. *Proc. Summer Sch. Francesco Turco*. Proceedings of the Summer School Francesco Turco. https://www.scopus.com/inward/record.uri?eid=2-s2.0-85124673905&partnerID=40&md5=19b7115dddc283a3f4f6db1591876661

Touzani, A. (2024). L ' Entrepreneuriat et développement économique au Maroc : Défis et Perspectives. *Revue Franaise d'Economie et de Gestion*, 5(5), 218–235.

Tracy, S. J. (2019). *Qualitative Research Methods: Collecting Evidence, Crafting Analysis, Communicating Impact*. Wiley-Blackwell.

Tracy, S. J. (2020). *Qualitative research methods: Collecting evidence, crafting analysis, communicating impact.* Wiley-Blackwell.

Tran, G., & Yeomans, J. S. (2009). Constructions of Strategies Toward Success in English Literacy in Children by Dual-Language Families : A Qualitative Research Proposal. *International Journal of Learning*, 16(1), 295–308. 10.18848/1447-9494/CGP/v16i01/45943

Tremper, B. (2013). *Avalanche essentials: A step-by-step system for safety and survival.* Mountaineers Books.

Treviño-Eberhard, D., & Kaufmann-Kuchta, K. (2022). Does It Make a Difference? Relations of Institutional Frameworks and the Regional Provision of Continuing Higher Education in England and Spain. *Education Sciences*, 12(2), 132. 10.3390/educsci12020132

Tunku Ahmad, T. B. (2020). Learning from Others : Their Experiences, Challenges and Difficulties. *IIUM Journal of Educational Studies*, 7(2), 1–2. 10.31436/ijes.v7i2.294

Tuononen, T., Hyytinen, H., Kleemola, K., Hailikari, T., Männikkö, I., & Toom, A. (2022). Systematic Review of Learning Generic Skills in Higher Education—Enhancing and Impeding Factors. *Frontiers in Education*, 7, 885917. Advance online publication. 10.3389/feduc.2022.885917

Turegun, O. (2020). Motivational Factors of Ecopreneurs. *Uluslararası Akademik Yönetim Bilimleri Dergisi*, 6(9), 10–19.

Udas, K. (1998). Participatory action research as critical pedagogy. *Systemic Practice and Action Research*, 11(6), 599–628. 10.1023/A:1022184120067

Ulrich Mayer, K. (2004). Whose Lives? How History, Societies, and Institutions Define and Shape Life Courses. *Research in Human Development*, 1(3), 161–187. 10.1207/s15427617rhd0103_3

Vagle, M. D. (2018). Crafting phenomenological research (2e éd.). Routledge.

Van de Ven, A. H., & Johnson, P. E. (2006). Knowledge for Theory and Practice. *TheAcademy of Management Review*, 31(4), 802–821.

Vasanthakumari, S. (2021, avril). *Writing research proposal.* https://wjarr.com/sites/default/files/WJARR-2021-0138.pdf

Venegas-Muggli, J. I. (2020). Higher education dropout of non-traditional mature freshmen : The role of sociodemographic characteristics. *Studies in Continuing Education*, 42(3), 316–332. 10.1080/0158037X.2019.1652157

Verbruggen, M., Van Emmerik, H., Van Gils, A., Meng, C., & De Grip, A. (2015). Does early-career underemployment impact future career success? A path dependency perspective. *Journal of Vocational Behavior*, 90, 101–110. 10.1016/j.jvb.2015.08.002

Verina, N., Titko, J., & Shina, I. (2021). Digital transformation outcomes in higher education : Pilot study in Latvia. *International Journal of Learning and Change*, 13(4/5), 459. 10.1504/IJLC.2021.116661

Verleye, K. (2019). Designing, writing-up and reviewing case study research : An equifinality perspective. *Journal of Service Management*, 30(5), 549–576. 10.1108/JOSM-08-2019-0257

Vermunt, J. (2005). Relations between student learning patterns and personal and contextual factors and academic performance. *Higher Education*, 49(3), 205–234. 10.1007/s10734-004-6664-2

Viehs, M., Clark, G., & Feiner, A. (2014). From The Stockholder To The Stakeholder—How Sustainability Can Drive Financial Outperformance. SSRN Electronic Journal. 10.2139/ssrn.2508281

Vinten, G. (1994). Participant Observation: A Model for Organizational Investigation? *Journal of Managerial Psychology*, 9(2), 30–38. 10.1108/02683949410059299

Vollero, A., Siano, A., & Della Volpe, M. (2019). A Systems Perspective for Conceptualizing Sustainability in Long-lived Family Businesses. Research Proposals on Risk Taking and Innovativeness. *Systems Research and Behavioral Science*, 36(1), 111–127. 10.1002/sres.2548

Von Krogh, G. (2020). Building capacity for empirical discovering in management and organization studies. *Academy of Management Discoveries*, 6(2), 159–164. 10.5465/amd.2020.0053

Voss, C., Tsikriktsis, N., & Frohlich, M. (2002). Case research in operations management. *International Journal of Operations & Production Management*, 22(2), 195–219. 10.1108/01443570210414329

Wacheux, F. (1996). Méthodes qualitatives de recherches en gestion.

Wacheux, F. (1996). Méthodes qualitatives et recherche en gestion / Frédéric Wacheux. Économica. Paris. https://bibliotheque.tbs-education.fr/Default/doc/SYRACUSE/17765/methodes-qualitatives-et-recherche-en-gestion-frederic-wacheux

Wacheux, F. (1996). Méthodes qualitatives et recherche en gestion. *Economica.*

Waltman, L., van Eck, N. J., & Noyons, E. C. M. (2010). A unified approach to mapping and clustering of bibliometric networks. *Journal of Informetrics*, 4(4), 629–635. 10.1016/j.joi.2010.07.002

Wang, W., Cao, Q., Zhuo, C., Mou, Y., Pu, Z., & Zhou, Y. (2021). COVID-19 to Green Entrepreneurial Intention: Role of Green Entrepreneurial Self-Efficacy, Optimism, Ecological Values, Social Responsibility, and Green Entrepreneurial Motivation. *Frontiers in Psychology*, 12(October), 732904. Advance online publication. 10.3389/fpsyg.2021.73290434721199

Weick, K. (1995). *Sensemaking in Organizations*. Sage.

Weigold, I. K., Weigold, A., Ling, S., & Jang, M. (2021). College as a Growth Opportunity : Assessing Personal Growth Initiative and Self-determination Theory. *Journal of Happiness Studies*, 22(5), 2143–2163. 10.1007/s10902-020-00312-x

Wiersma, W. (2000). Research Methods in Education: An Introduction. Allyn and Bacon.

Wilhelm, C. (1998, May). Quantitative risk analysis for evaluation of avalanche protection projects. In *Proceedings of the Anniversary Conference* (Vol. 25, pp. 288-293).

Williams, C. C. (2005). The undeclared sector, self-employment and public policy. *International Journal of Entrepreneurial Behaviour & Research*, 11(4), 244–257. 10.1108/13552550510603289

Wilson, T. D. (2005). Participant Observation. Encyclopedia of Social Measurement, 19-24. 10.1016/B0-12-369398-5/00398-4

Wise, V., & Ali, M. M. (2009). Corporate Governance and Corporate Social Responsibility in Bangladesh with special reference to Commercial Banks. American International University-Bangladesh, Office of Research and Publications (ORP), AIUB Bus Econ Working Paper Series.

Wollesen, B., Wildbredt, A., van Schooten, K. S., Lim, M. L., & Delbaere, K. (2020). The effects of cognitive-motor training interventions on executive functions in older people: A systematic review and meta-analysis. *European Review of Aging and Physical Activity*, 17(1), 1–22. 10.1186/s11556-020-00240-y32636957

Worley, C. G., Zardet, V., Bonnet, M., & Savall, A. (2015). *Becoming Agile - How the SEAM Approach to Management Builds Adaptability*. Wiley. 10.1002/9781119182351

Xing, Y., Hu, Y., & Zhou, J.-Z. (2021). Higher education and family background : Which really matters to individual's socioeconomic status development in China. *International Journal of Educational Development*, 81, 102334. 10.1016/j.ijedudev.2020.102334

Yadav, D. (2022). Criteria for Good Qualitative Research : A Comprehensive Review. *The Asia-Pacific Education Researcher*, 31(6), 679–689. 10.1007/s40299-021-00619-0

Yin, R. K. (2018). *Case Study Research and Applications* (6e éd.). https://uk.sagepub.com/en-gb/eur/case-study-research-and-applications/book250150

Yin, R. K. (1990). Case Study Research - Design and methods. *Sage (Atlanta, Ga.)*.

Yin, R. K. (1994). Case study research and applications: Design and methods. In *Journal of Hospitality & Tourism Research*. Sage Publi. 10.1177/109634809702100108

Yin, R. K. (2009). *Case stud research: design and methods* (4th ed.). Sage Publications.

Yin, R. K. (2013). *Case study research: Design and methods*. Sage Publications.

Yin, R. K. (2014). *Case Study Research*. SAGE Publications.

Yin, R. K. (2018). *Case Study Research and Applications: Design and Methods*. SAGE Publications.

Yosso, T. J. (2005). Whose culture has capital? A critical race theory discussion of community cultural wealth. *Race Ethnicity and Education, 8*(1), 69-91. 10.1080/1361332052000341006

Zacharias, M. E., Saldaña, J., & Saldana, J. (2002). Longitudinal Qualitative Research : Analyzing Change through Time. *Canadian Journal of Education / Revue canadienne de l'éducation, 27*(2/3), 341. 10.2307/1602237

Zattoni, A., & Cuomo, F. (2008). Why Adopt Codes of Good Governance? A Comparison of Institutional and Efficiency Perspectives. *Corporate Governance*, 16(1), 1–15. 10.1111/j.1467-8683.2008.00661.x

Zerbino, P., Stefanini, A., & Aloini, D. (2021). Process Science in Action : A Literature Review on Process Mining in Business Management. *Technological Forecasting and Social Change*, 172, 121021. 10.1016/j.techfore.2021.121021

Zeydan, Ö., & Sevim, B. (2008). *İklim Değişikliğinin Kış Turizmine Etkileri*. TMMOB İklim Değişimi Sempozyumu Bildiriler Kitabı.

Zhampeissova, K., Gura, A., Vanina, E., & Egorova, Z. (2020). Academic Performance and Cognitive Load in Mobile Learning. *International Journal of Interactive Mobile Technologies*, 14(21), 78. 10.3991/ijim.v14i21.18439

Zhang, J. J. (2020). Ethical Issues in Research. In Kobayashi, A. (Ed.), *International Encyclopedia of Human Geography* (2nd ed., pp. 299–302). Elsevier. 10.1016/B978-0-08-102295-5.10176-3

Zikmund, W., Babin, B., Carr, J., & Griffin, M. (2010). Business Research Methods Eight Edition. *South Western Educ Pub*, 668.

Zupic, I., & Čater, T. (2015). Bibliometric Methods in Management and Organization. *Organizational Research Methods*, 18(3), 429–472. 10.1177/1094428114562629

Zweifel, B., & Haegeli, P. (2014). A qualitative analysis of group formation, leadership and decision making in recreation groups traveling in avalanche terrain. *Journal of Outdoor Recreation and Tourism*, 5, 17–26. 10.1016/j.jort.2014.03.001

About the Contributors

Badr Bentalha investigates the structural dynamics and control of complex networks, applying his findings to supply chain management, Industry 4.0, risk analysis, and digital supply chains. His pioneering research has established seminal academic and practical approaches for sustainable and intelligent supply chain optimization. Using optimization, simulation, control theory, and AI, he tackles real-world supply chain and operations challenges. His work emphasizes the intersection of supply chain management, operations research, industrial engineering, and digital technology. As a professor, he teaches undergraduate, graduate, and, doctoral courses in operations management, supply chain management, logistics, management information systems, and strategic management. Through guest lectures, webinars, and scholarly presentations, he engages students and fosters an active learning environment. He aims to equip future industry leaders with the management knowledge and technological skills to build more resilient, adaptable, and sustainable supply chain operations. Dr. Bentalha's academic background includes management and economics, operations research, and supply chain management. He studied management operations and supply chain management, graduating with honors. He earned PhD degrees and completed an aggregation in economics and management. Before entering academia, he primarily worked in industry and consulting, focusing on process optimization in manufacturing, logistics, and ERP systems. His practical experience includes numerous operations research and process optimization projects for operations design, logistics, scheduling, and supply chain management. Before joining the Fez National School of Business and Management, he was a professor of economics and management preparatory classes since 2009. Dr. Bentalha's influential research is cited in supply chain management and logistics. His publication record includes over 60 papers in prestigious academic journals such as the International Journal of Logistics Systems and Management, International Journal of Business Performance Management, International Journal of Business and Technology Studies, Journal of Environmental Issues and Climate Change, and Economical Managerial Alternatives. He is also the author of three books: Netnography: Principles, Method, and Ethics, Internal Audit of Treasury and Cash Management: Report on the Audit of Treasury and Integrating Intelligence and Sustainability in Supply Chains. Professor Bentalha specializes in supply chain and operations management, operations research, and service management. He is passionate about integrating knowledge across disciplines to solve real-world problems. He has delivered numerous invited plenary talks, keynotes, and panel discussions at conferences and global webinars. He serves as Associate Editor of the Economical Managerial Alternatives journal and is on the editorial boards and advisory committees of several other publications. He regularly presents his research and has chaired organized program committees, and served on advisory boards for over 60 international conferences in supply chain management, operations, management studies, control theory, and information science. He is a member of numerous scientific committees of international journals and conferences. Since 2009, Mr. Bentalha has helped sundry companies implement an audit system, resolve issues, and improve performance.

Lhoussaine Alla is a professor in management sciences at the National School of Applied Sciences, researcher at the LAREMEF laboratory, Sidi Mohamed Ben Abdellah University, Fez, Morocco. He is a permanent professor of various marketing management modules at Sidi Mohamed Ben Abdellah University in Fez (Morocco) and in several public and private business graduate schools. He is an accredited professional expert in the fields of Training Engineering, Professional Coaching and Mentoring young entrepreneurs. After a PhD thesis on creating value for the customer and its impact on the overall performance of companies, Prof. Lhoussaine Alla invested more in scientific research in marketing, through various scientific contributions in the form of participation in international conferences and symposia and scientific publications, in various themes inherent to Marketing (customer value creation, customer behaviour, e-marketing, customer experience, sales performance, Marketing Data Analytics, territorial marketing, territorial attractiveness, territorial economic intelligence, ...), finance (stock market performance, financial analysts, financing package,...), entrepreneurship (entrepreneur profile, startups, entrepreneurial resilience,...), tourism (tourist attractiveness, visitor behaviour, tourism, performance of tourist destinations,...), logistics (SC, SCM, SSCM, Green SCM, logistics performance,...). Prof. Lhoussaine ALLA is also (co)editor of the e-book "Integrating intelligence and sustainability into supply chains. IGI Global. DOI: 10.4018/979-8-3693-0225-5" and Publication Director of the scientific journal "Managerial and Economic Alternatives - AME", indexed on the IMIST portal, . Prof. Lhoussaine ALLA is also (co)coordinator of many scientific events devoted to marketing research. As much scientific and educational accumulation that the publisher aims to mobilize as potential to bring together researchers and imminent experts to share the fruits of their relevant and innovative research in implementation, supervision, evaluation, Mastery and reinvention of innovative solutions developed through Data Engineering and AI in Marketing. He is now co-editor of two collective works that will be published via IGI Global: 1. Applying Qualitative Research Methods to Science and Management. 2. Utilizing Technology to Manage Territories. Prof Lhoussaine is the coordinator of a team of young researchers, very dynamic and talented, with complementary profiles, creating a friendly space for high-level scientific research.

* * *

Adil Cherkaoui is Professor HDR, Deputy Director of the Research Laboratory in Development Economics and Organizational Governance (LAREDGO), Faculty of Legal, Economic and Social Sciences - Ain Chock, Hassan II University, Casablanca, Morocco. His research focuses on CSR, diversity management, corporate governance and sustainable finance. He received the 2018 Prize for the best scientific article published in French-speaking emerging countries, awarded by the United Nations PRI Quebec Network, as well as the 2018 Prize for the best pedagogical case study in management and business ethics, awarded by Al Akhawayn University. He is a member of the scientific committee of several indexed journals. Author of numerous scientific publications.

Laila Cherradi is a professor at the Department of Mathematics and Business, University Ibn Zohr, Morocco. She also a member of the Performance Management of Public, Private and Social Economy Organizations research laboratory, at National School of Business and Management, Université Ibnou Zohr, Morocco. Her research interests include the promotion of carbon finance in developing countries, social and solidarity marketing and entrepreneurial motivation.

Abdelhay El Mouatassem is a PhD student in sociology at Sidi Mohamed Ben Abdellah University in Morocco.

Saadeddine Igamane is a professor-researcher in sociology at Sidi Mohammed Ben Abdellah University in Fès, Morocco. He received interdisciplinary training in political science, economics, and sociology, earning his Ph.D. in sociology with a focus on Moroccan entrepreneurs. Currently, he is a member of the sociology and psychology laboratory and serves as the coordinator for the master's program in "Sociology of Social and Solidarity Economy" at the University of Fès. Additionally, he is a member of the Arab Council for Social Sciences, member of the CODESRIA (Council for the Development of Social Science Research in Africa), member of the ISA (International Sociological Association) and member of the AISLF (International Association of French-speaking Sociologists). Saadeddine Igamane holds the position of president at the Interdisciplinary Institute for Social Sciences. His research interests include economic sociology, labor sociology, precarity, migration, and policy evaluation in the fields of women's and youth employment, as well as in social and solidarity economy. He is an expert in research methods and the analysis of quantitative and qualitative data for national and international organizations. Igamane has coordinated several publications and published indexed scientific articles on cooperative work, public policy evaluation, precarious middle class, and professional precarity among youth and women in the fields of education and entrepreneurship, particularly in the cooperative sector.

Essalmani Mourad, aged 37, resides in the historic city of Fès. His academic journey began with his high school diploma in Experimental Sciences, which he earned in 2007. Driven by a passion for economics and management, Mourad pursued a Bachelor's degree in Economic Sciences and Management at the Faculty of Legal, Economic, and Social Sciences in Fès, graduating in 2010. Eager to deepen his understanding of finance and econometrics, Mourad enrolled in a Master's program in Finance and Applied Econometrics at the same faculty. During this time, he not only honed his skills in econometrics but also nurtured a growing passion for research. His commitment to the field led him to consider an academic career, prompting him to pursue doctoral studies. In 2015, Mourad joined the Interdisciplinary Research Laboratory in Economics, Finance, and Management of Organizations at the Faculty of Legal, Economic, and Social Sciences in Fès. His doctoral thesis, titled "The Sustainability of Public Debt: An Econometric Study of the Case of Morocco," reflects his dedication to addressing complex economic issues through rigorous research. Mourad ESSALMANI's academic pursuits and research contributions highlight his expertise and commitment to advancing knowledge in the fields of economics and finance.

Youness Oudrhough is a PhD Student at the Laboratory for Research in Development Economics and Organisational Governance (LAREDGO) at the Faculty of Legal, Economic and Social Sciences - Ain Chock, Hassan II University, Casablanca, Morocco. His research focuses on the impact of cognitive biases on the strategic decisions of corporate executives in Morocco, from a behavioural perspective.

Index

A

B

C

D

E

F

G

I

L

M

N

O

P

Q

R

S

T

W

www.ingramcontent.com/pod-product-compliance
Lightning Source LLC
LaVergne TN
LVHW080601300926
849162LV00005B/1020